AF389438

Insulin-Like Growth Factors in Development, Cancers and Aging

Insulin-Like Growth Factors in Development, Cancers and Aging

Editor

Haim Werner

MDPI • Basel • Beijing • Wuhan • Barcelona • Belgrade • Manchester • Tokyo • Cluj • Tianjin

Editor
Haim Werner
Tel Aviv University
Israel

Editorial Office
MDPI
St. Alban-Anlage 66
4052 Basel, Switzerland

This is a reprint of articles from the Special Issue published online in the open access journal *Cells* (ISSN 2073-4409) (available at: https://www.mdpi.com/journal/cells/special_issues/Insulin_Like).

For citation purposes, cite each article independently as indicated on the article page online and as indicated below:

LastName, A.A.; LastName, B.B.; LastName, C.C. Article Title. *Journal Name* **Year**, *Volume Number*, Page Range.

ISBN 978-3-0365-0770-5 (Hbk)
ISBN 978-3-0365-0771-2 (PDF)

Contents

About the Editor

Haim Werner was born in Buenos Aires, Argentina, in 1952. He moved to Israel in 1970, where he began his academic studies. He obtained his B.Sc. (1974) and M.Sc. (1977) degrees in Biology and Genetics from the Hebrew University in Jerusalem and his Ph.D. (1987) degree in Hormone Research from the Weizmann Institute of Science, Rehovot. Between 1987 and 1996 he was a Visiting Scientist at the Diabetes Branch, National Institutes of Health, Bethesda, Maryland, USA. Professor Werner returned to Israel in 1996 and is currently the incumbent of the Lady Davis Chair in Biochemistry and Head of the Shalom and Varda Yoran Institute for Human Genome Research, Sackler School of Medicine, Tel Aviv University. Professor Werner's research focus on the molecular biology of the insulin-like growth factor system. He is the author of over 200 papers.

Editorial

Insulin-Like Growth Factors in Development, Cancers and Aging

Haim Werner

Department of Human Molecular Genetics and Biochemistry, and the Shalom and Varda Yoran
Institute for Human Genome Research, Sackler School of Medicine, Tel Aviv University, Tel Aviv 69978, Israel;
hwerner@tauex.tau.ac.il

Received: 14 October 2020; Accepted: 16 October 2020; Published: 17 October 2020

Since their discovery in the late 1950s, insulin-like growth factors (IGFs) have attracted significant interest in multiple areas of biology and medicine, including endocrinology, pediatrics, growth, metabolism, nutrition, aging, and oncology. IGF1, which was initially identified as the mediator of growth hormone (GH) action, is regarded as a key player in numerous cellular and organismal processes. The signaling pathways elicited by IGF1 have been extensively characterized in biochemical and molecular terms over the past 40 years. However, fundamental questions regarding basic differences between the mechanisms of action of IGF1 and the closely related insulin molecule are yet to be resolved. This Special Issue of *Cells* provides a collection of modern articles dealing with the role of IGF1 in cancer biology, aging, and development. The articles explore basic and clinical aspects of the IGF1 system, including post genomic analyses as well as novel approaches to target the IGF1R in oncology.

The role of IGF binding proteins (IGFBPs) in the regulation of IGF1-stimulated growth has been the focus of intensive research for many years. Forced expression of IGFBPs in transgenic mice, under most circumstances, leads to inhibitory effects on somatic growth. To evaluate the impact of IGFBPs on normal growth, Walz et al. [1] measured IGF1 and IGFBP-2, -3, and -4 levels in the serum of growth-selected mouse models (obese and lean) and expressed these values as a function of longitudinal growth. The authors provide evidence that part of the elevated growth activity during prepubertal growth could be related to the elevated bioactivity of IGF1. Specifically, elevated ratios of IGF1/IGFBPs were established by a delayed increase in IGFBPs compared to a strong increase in IGF1 levels between two and four weeks of age.

As mentioned above, the IGF1 axis plays a key role in aging and longevity. However, the biochemical and molecular mechanisms responsible for the linkage between IGF1 and aging processes are poorly defined. Zhang et al. [2] evaluated age- and sex-adjusted hazards for all-cause mortality and incident age-related diseases in a prospective cohort of older adults (mean age = 76.1 ± 6.8 year) as predicted by baseline total serum IGF1, IGFBP-1, IGFBP-3, and IGF1/IGFBP-3 molar ratio. The authors report that higher IGF1 levels and bioavailability predicted mortality and morbidity risk, supporting the hypothesis that diminished GH-IGF1 signaling may contribute to human longevity and health-span.

Yoshida and Delafontaine [3] provide a comprehensive review of the role of IGF1 and its downstream signaling paths in skeletal muscle atrophy associated with chronic diseases and aging. The authors describe the involvement of autophagy in IGF1-stimulated muscle atrophy and protein degradation. In addition, they emphasize the fact that given the multiple (sometimes opposed) interactions of IGF1 in skeletal muscle, it is often difficult to clearly define the specific role of IGF1 in this tissue. The authors conclude that further studies are required to develop effective strategies to apply IGF1 in order to treat muscle atrophy in humans.

Glucose regulated protein 94 (GRP94) is a ubiquitously expressed chaperone in the endoplasmic reticulum that is required for the proper folding and secretion of IGF1. Argon et al. [4] review the implications of IGF1–GRP94 interaction in the context of idiopathic short stature and suggest that the chaperone machinery can be modulated with small molecules. The net result of this molecular intervention might constitute a novel way to manipulate both IGF1 deficiency as well as conditions of excessive growth factor production. Similarly, IGF1–GRP94 interaction might be relevant in cancer. Thus, differences in the association of IGF1/IGF2 with GRP94 can be exploited for selective tissue targeting of compounds.

Ahmad et al. [5] describe the role of IGF1 in the maintenance of skeletal muscle mass. Specifically, their review focuses on the mechanisms involved in the proliferation of muscle satellite cells as well as the key role of IGF1 in myoblast differentiation during normal growth or regeneration after skeletal muscle injury. The authors state that the development of protocols for the use of IGF1 in muscle-wasting conditions remains an important research challenge.

The mitochondria are key organelles that regulate vital processes in eukaryotic cells. A decline in mitochondrial function is regarded as an important hallmark of aging. Poudel et al. [6] review the evidence that GH and IGF1 regulate mitochondrial mass and function and contribute to specific processes of cellular aging. The authors highlight the involvement of these hormones in mitochondrial biogenesis, ATP production, oxidative stress, and senescence, with a special focus on mitochondrial pathologies during aging.

IGFBP-3 is the best characterized IGF binding protein and its disruption has been linked to a number of pathologies. IGFBP-3 exhibits a number of IGF1-independent activities, ranging from tumor suppressing to tumor promoting effects. Cai et al. [7] describe the identification of TMEM219, an unknown transmembrane protein, as a potential IGFBP-3 interacting protein. Furthermore, they delineate the underlying mechanisms and biological implications of IGFBP-3–TMEM219 interplay. Finally, the authors portray the therapeutic potential of TMEM219 agonists for cancer therapy.

As alluded to above, the IGF1R constitutes a promising target in oncology. Chen et al. [8] review the current state of IGF-targeting approaches and outline the stepwise bioengineering and validation of *IGF-Trap*, a novel anticancer modality that could bypass the limitations of current techniques, including interference with insulin receptor signaling. In vivo, IGF-Trap displays favorable kinetic properties and could reduce metastatic outgrowth of colon and lung cancers in the liver. In addition, Chen and colleagues developed a sensitive IGF kinase receptor-activation (KIRA) assay that serves as a surrogate biomarker for drug efficacy.

The inherent complexity of the IGF1 system is elegantly discussed by Janssen [9]. While the classical view postulated that phosphorylation of tyrosine residues plays a major role in IGF1R activation, there is increasing evidence showing that this dogma was too simplistic and grossly underestimated the downstream complexity of the IGF1R pathways. Janssen discusses the novel concept that IGF1R can be also considered as a functional tyrosine kinase/G-protein coupled receptor (GPCR) hybrid. According to this view, this hybrid is able to integrate kinase signaling with some IGF1R-mediated GPCR features. In summary, the IGF1R is far more complex than previously thought and a big challenge for the future will be to integrate and translate this new knowledge into clinical practice.

Finally, recent developments in the area of IGF-II research are discussed by Blyth et al. [10]. IGF-II is the least investigated ligand of the IGF system and it is unique in that it acts through both the IGF1R and the insulin receptor isoform A (IR-A). The solved structure of IGF-II bound to IGF1R using cryo-electron microscopy is clearly depicted. In addition, comparisons are made with the structures of insulin and IGF1 bound to their cognate receptors. Lastly, the authors discuss future investigations required to develop antagonists of IGF action for cancer treatment.

Funding: This research was funded by a grant from the Israel Science Foundation (Grant 1403/14).

Acknowledgments: HW is the incumbent of the Lady Davis Chair in Biochemistry.

Conflicts of Interest: The authors declare no conflict of interest.

References

1. Walz, M.; Chau, L.; Walz, C.; Sawitzky, M.; Ohde, D.; Brenmoehl, J.; Tuchscherer, A.; Langhammer, M.; Metzger, F.; Höflich, C.; et al. Overlap of Peak Growth Activity and Peak IGF-1 to IGFBP Ratio: Delayed Increase of IGFBPs versus IGF-1 in Serum as a Mechanism to Speed up and down Postnatal Weight Gain in Mice. *Cells* **2020**, *9*, 1516. [CrossRef] [PubMed]
2. Zhang, W.B.; Aleksic, S.; Gao, T.; Weiss, E.F.; Demetriou, E.; Verghese, J.; Holtzer, R.; Barzilai, N.; Milman, S. Insulin-like Growth Factor-1 and IGF Binding Proteins Predict All-Cause Mortality and Morbidity in Older Adults. *Cells* **2020**, *9*, 1368. [CrossRef] [PubMed]
3. Yoshida, T.; Delafontaine, P. Mechanisms of IGF-1-Mediated Regulation of Skeletal Muscle Hypertrophy and Atrophy. *Cells* **2020**, *9*, 1970. [CrossRef] [PubMed]
4. Argon, Y.; Bresson, S.E.; Marzec, M.T.; Grimberg, A. Glucose-Regulated Protein 94 (GRP94): A Novel Regulator of Insulin-Like Growth Factor Production. *Cells* **2020**, *9*, 1844. [CrossRef] [PubMed]
5. Ahmad, S.S.; Ahmad, K.; Lee, E.J.; Lee, Y.H.; Choi, I. Implications of Insulin-Like Growth Factor-1 in Skeletal Muscle and Various Diseases. *Cells* **2020**, *9*, 1773. [CrossRef] [PubMed]
6. Poudel, S.B.; Dixit, M.; Neginskaya, M.; Nagaraj, K.; Pavlov, E.; Werner, H.; Yakar, S. Effects of GH/IGF on the Aging Mitochondria. *Cells* **2020**, *9*, 1384. [CrossRef] [PubMed]
7. Cai, Q.; Dozmorov, M.; Oh, Y. IGFBP-3/IGFBP-3 Receptor System as an Anti-Tumor and Anti-Metastatic Signaling in Cancer. *Cells* **2020**, *9*, 1261. [CrossRef] [PubMed]
8. Chen, Y.M.; Qi, S.; Perrino, S.; Hashimoto, M.; Brodt, P. Targeting the IGF-Axis for Cancer Therapy: Development and Validation of an IGF-Trap as a Potential Drug. *Cells* **2020**, *9*, 1098. [CrossRef] [PubMed]
9. Janssen, J.A.M.J.L. New Insights from IGF-IR Stimulating Activity Analyses: Pathological Considerations. *Cells* **2020**, *9*, 862. [CrossRef] [PubMed]
10. Blyth, A.J.; Kirk, N.S.; Forbes, B.E. Understanding IGF-II Action through Insights into Receptor Binding and Activation. *Cells* **2020**, *9*, 2276. [CrossRef]

Publisher's Note: MDPI stays neutral with regard to jurisdictional claims in published maps and institutional affiliations.

Article

Overlap of Peak Growth Activity and Peak IGF-1 to IGFBP Ratio: Delayed Increase of IGFBPs Versus IGF-1 in Serum as a Mechanism to Speed up and down Postnatal Weight Gain in Mice

Michael Walz [1], Luong Chau [1], Christina Walz [1], Mandy Sawitzky [1], Daniela Ohde [1], Julia Brenmoehl [1], Armin Tuchscherer [2], Martina Langhammer [2], Friedrich Metzger [3], Christine Höflich [4] and Andreas Hoeflich [1,*]

1   Institute of Genome Biology, Leibniz-Institute for Farm Animal Biology (FBN), 18196 Dummerstorf, Germany; walz.michael@fbn-dummerstorf.de (M.W.); chau@fbn-dummerstorf.de (L.C.); walz@fbn-dummerstorf.de (C.W.); mandysawitzky@gmx.de (M.S.); ohde@fbn-dummerstorf.de (D.O.); brenmoehl@fbn-dummerstorf.de (J.B.)
2   Institute of Genetics and Biometry, Leibniz-Institute for Farm Animal Biology (FBN), 18197 Dummerstorf, Germany; atuchsch@fbn-dummerstorf.de (A.T.); martina.langhammer@fbn-dummerstorf.de (M.L.)
3   Versameb AG, 4057 Basel, Switzerland; friedrich.metzger@versameb.com
4   Ligandis UG, 18276 Gülzow-Prüzen, Germany; christine.hoeflich@ligandis.de
*   Correspondence: hoeflich@fbn-dummerstorf.de; Tel.: +49-(0)38208-68744; Fax: +49-(0)38208-68-702

Received: 14 April 2020; Accepted: 17 June 2020; Published: 22 June 2020

Abstract: Forced expression of insulin-like growth factor binding proteins (IGFBPs) in transgenic mice has clearly revealed inhibitory effects on somatic growth. However, by this approach, it cannot be solved if or how IGFBPs rule insulin-like growth factor (IGF)-dependent growth under normal conditions. In order to address this question, we have used growth-selected mouse models (obese and lean) and studied IGF-1 and IGFBPs in serum with respect to longitudinal growth activity in males and females compared with unselected controls. In mice of both genders, body weights were recorded and daily weight gains were calculated. Between 2 and 54 weeks of age, serum IGF-1 was determined by ELISA and intact IGFBP-2, -3 and -4 were quantified by Western ligand blotting. The molar ratio of IGF-1 to the sum of IGFBP-2 to -4 was calculated for all groups and plotted against the daily weight gain curve. Growth-selected mice are characterized by higher daily weight gains and extended periods of elevated growth activity if compared to matched unselected controls. Therefore, adult mice from the obese and lean groups can achieve more than twofold increased body weight in both genders ($p < 0.001$). Between 2 and 11 weeks of age, in obese and lean mice of both genders, serum IGF-1 concentrations are increased more prominently if compared to unselected controls ($p < 0.001$). Instead, substantial decreases of IGFBPs, particularly of IGFBP-2, are observed in males and females of all groups at the age of 2 to 4 weeks ($p < 0.001$). Due to the strong increase of IGF-1 but not of IGFBPs between two and four weeks of age, the ratio of IGF-1 to IGFBP-2 to -4 in serum significantly increased in all groups and genders ($p < 0.05$). Notably, the IGF-1 to IGFBP ratio was higher in male and female obese mice if compared to unselected controls ($p < 0.05$).

Keywords: longitudinal study; IGFBP; mouse models

---

1. Introduction

Long-term selection for high body weight goes back to 1930, when Goodale initiated an experiment to explore the boundaries of growth in mice [1]. After 35 generations of selection, the mice had increased their body weight from ≈25 g to ≈43 g (+72%). Most probably due to inbreeding effects, additional selection for 49 generations did not further increase body weight to a significant extent [2]. Starting from an outbred background and under avoidance of inbreeding in the present selection experiment, substantial increases (+144%) of male body weight at the age of six weeks were achieved after 146 generations of selection in the obese mouse line (DU6) [3]. This finding not only underlines the potential of non-inbred backgrounds for functional genome analysis but even more importantly proves the idea that growth is a complex trait regulated by a multitude of effectors [4]. Here we have used two separate growth selected mouse models for the study of longitudinal regulation of the IGF-system. Accordingly, mice long-term selected for high body mass [3,5,6] (obese model; DU6), and a second mouse model selected for high protein mass [7,8] (lean model; DU6P) were compared to unselected controls [9,10].

Clearly, the GH–IGF system is highly responsive to growth selection; specific effects have been described on the level of DNA, mRNA, and protein with respect to the GH–IGF system in model animals or farm animals [11–15], and many of these studies have particularly addressed the biomarker potential of IGF-1 or assessed single time points. In human subjects, longitudinal concentrations have been provided both for IGF-1 [16] and for IGFBP-3 [17]. In order to estimate the bioactivity of IGF-1, reference levels for the ratio of IGF-1 to IGFBP-3 were also calculated for male and female subjects from a larger population longitudinally [17]. In biological matrices and in the circulation, IGF-1 bioactivity is not only regulated by IGFBP-3, and therefore the inclusion of additional IGFBPs enables a more comprehensive view e.g., on the control of IGF-1 dependent growth. For the hypothesis-free assessment of IGFBPs in a given matrix, Western ligand blotting (WLB) technique can be applied [18]. By this method, it is possible to include all IGFBPs present and detectable in a given sample. Perhaps even more important is the fact that WLB delivers structural information of a given IGFBP [19]. Thereby the information provided by WLB is related to a specific molecular weight (e.g., intact IGFBP-3), whereas other methods do not have this power. This fundamental feature of WLB is getting more and more important, as we understand that IGFBP-proteolysis represents a fundamental process of physiological growth control related to IGFs [20,21] or in cancer [22]. Just recently, an IGFBP-3 protease has been described as an effector of free IGF-1 in children and adolescents [23]. Accordingly, the inclusion of structural information could tremendously improve the biomarker value of IGFBPs [19]. Here we compared intact IGFBPs quantified by WLB with longitudinal concentrations of IGF-1 and, for the first time, discuss IGF-1 to IGFBP ratio based on structurally validated biomarker information of IGFB-2, -3 and -4 in serum.

2. Materials and Methods

2.1. Animals, Husbandry, and Study Design

In the present study, long-term selected non-inbred mouse lines established at the Leibniz Institute for Farm Animal Biology (FBN) were used. Two lines were selected for high male body mass at the age of 42 days (DU6: obese model) or high protein amount (DU6P: lean model) at the same age of 42 days after birth. These long-term selected mouse lines were originally based on the genetic background of the unselected control line Fzt:DU [9,10]. The control mouse line (Fzt:DU) was developed by random mating procedures during the experiment. Husbandry, mode of selection, and phenotypical features of the three mouse lines have been described in detail before [3,8]. In brief, all mice were maintained under semi-barrier conditions with free access to chow (breeding diet 1314, Altromin, Lage, Germany) and water. In order to assess longitudinal levels of IGFBPs in serum from male and female mice, we used serum produced by Sawitzky et al. [8]. In the course of this study, male and female mice from all three lines were dissected at the age of 2, 4, 7, 11, 16, 29, 42, and 54 weeks after birth, and serum was frozen until further use. The experiment was designed with 8 animals per group. Due to elevated mortality, only 4 male obese mice reached an age of 54 weeks, resulting in a total sample number of N = 380. In addition, body weights were recorded from all mice included in this study. Daily weight gain was calculated from intrapolated daily weights extracted from the Gompertz growth curve (Y = YM*(Y0/YM)^(exp(−K*X))). The experiments were performed in adherence to national and international laws and were further approved by the National Animal Protection Board Mecklenburg-Vorpommern (file number: LALLF M-V/TSD/7221.3-1.2-037/06).

2.2. Longitudinal Analysis of IGFs and IGFBPs in Mouse Serum

In serum from male (N = 188) and female (N = 192) mice between 2 and 54 weeks of age, IGF-1 was quantified by ELISA as described before [24]. In all samples, IGFBPs were quantified by Western ligand blotting as already described [25] with exceptions as described here. Serum was denatured for 5 min at 95 °C in sample buffer containing 10% sucrose, 2% sodium dodecyl sulfate (SDS), and 62.5 mM Tris (pH 6.8) and loaded on 12%-SDS/polyacrylamide gels. For quantification of IGFBP-2, -3, and -4, dilution series of human recombinant IGFBP-2, -3, and -4 were included with each run. After electrophoresis, proteins were blotted from the gel to solid carrier membranes (polyvinyl fluoride, Millipore, Schwalbach, Germany). The membranes were incubated using human recombinant IGF-2 radiolabeled with iodine-125 overnight at 4 °C. After five consecutive repetitions of washing in phosphate-buffered saline (pH 7.4) for 15 min, membranes were exposed to Storage Phosphorimager screens mounted on plates for 8 h. The signals were quantified using the Phosphor-Imager Storm (Molecular Dynamics, Sunnyvale, CA, USA). Quantification was achieved using ImageQuant software (GE Healthcare, Marlborough, MA, USA). Regression coefficients from standard dilutions were higher than 0.99 (http://www.ligandis.de/index.php?id=20&L=1). Intraassay and interassay variations were <15% and <20% for all IGFBPs, as published before [25]. Lower limits of quantification also as published before [25] were 0.25 ng for IGFBP-2 and 1 ng for IGFBP-3 and IGFBP-4. Using the software GraphPad Prism version 8.4.2, three samples were identified as outliers (GraphPad Prism) and therefore excluded from further analysis. Accordingly, in male controls at an age of 2 weeks, in obese males at an age of 42 weeks, and in female controls at an age of 42, only 7 samples per group were included (N = 377).

2.3. Statistical Analyses

Statistical analyses were performed using the SAS software for Windows, version 9.4 (Copyright, SAS Institute Inc., Cary, NC, USA). IGF-1 and IGFBP and growth data were analyzed by analyses of variance (ANOVA) using the MIXED procedure in SAS/STAT software. The ANOVA model contained the fixed factors group (levels: obese, lean, control), gender (levels: female, male), age (levels: weeks 2, 4, 7, 11, 16, 29, 42, 54), and their interactions.

Least square means (LS means) and their standard errors (SE) were computed for each fixed effect in the models described above, and all pairwise differences between LS means were tested using the Tukey–Kramer procedure. The SLICE Statement of the MIXED procedure was used to perform partitioned analyses of the LSM for all interactions. Effects and differences with p-values < 0.05 were considered significant.

3. Results

3.1. Longitudinal Growth in Non-Inbred Mouse Models

Body weight was recorded in mice selected for high body mass at the age of 42 days (obese mouse model), in mice selected for high protein amount (lean mouse model), and in unselected controls (Figure 1). At an age of 11 weeks in females and 16 weeks in males, body weights were significantly different between obese, lean, and control mice. Within groups, daily weight gains were highest in male or female controls at the age of 24.9 days or 20.9 days after birth. Lean mice elevated daily weight increases until an age of 26.3 days in males and 24.5 days in females. In obese mice, the daily weight gains peaked at the age of 26.7 days in male and 24.2 days in female mice. The absolute amount of daily weight increase amounted to 0.635 g/d and 0.59 g/d in male and female unselected controls, 1.99 g/d and 1.52 g/d in lean male and female mice, but 2.2 g/d and 1.7 g/d in obese male and female mice.

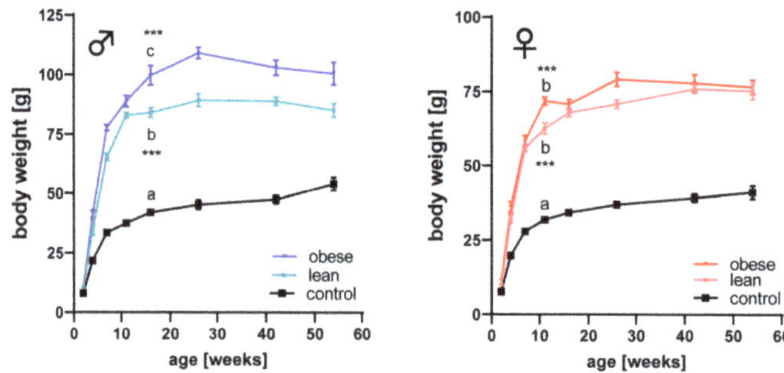

Figure 1. Body weight in male (left panel) and female (right panel) mice selected for high body weight (obese), for high protein amount (lean), and in unselected controls at an age of 2, 4, 7, 11, 16, 29, 42, and 54 weeks. (mean ± SEM; $n = 8$; due to high mortality, sample number was reduced to $n = 4$ at an age of 54 weeks in male obese mice; different letters (a, b, and c) indicate significant differences also in different genetic groups per gender, ***: $p < 0.001$; identical letters indicate no statistically significant difference).

3.2. Effects of Age and Growth Selection on the Concentrations of IGF-1

As a main effect of growth selection independent of age, IGF-1 was significantly increased in lean mice ($p < 0.001$) of both genders but only in obese male mice ($p < 0.01$) if compared to sex-matched unselected controls (Figure 2). As an effect of age and genetic group, in male obese mice, a significant ($p < 0.001$) increase of IGF-1 concentrations in serum was present between 2 and 4 weeks of age. In male lean mice, a similar increase was found between 2 and 7 weeks of age ($p < 0.001$). In males from both growth-selected mouse lines, IGF-1 concentrations decreased between 4 or 7 and 29 weeks of age ($p < 0.01$). Moreover, as an interaction of age and genetic group, in female obese and lean mice, increases of IGF-1 concentrations were found between 2 and 7 weeks of age ($p < 0.001$). However, a significant decrease of IGF-1 concentrations over time was only found in obese female mice between 7 and 54 weeks of age ($p < 0.001$).

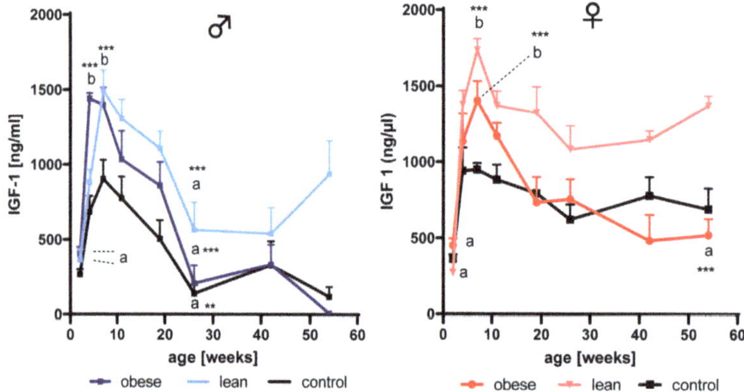

Figure 2. Concentrations of insulin-like growth factor (IGF)-1 in serum from mice selected for high body weight (obese), for high protein amount (lean), and in unselected controls of both genders at an age of 2, 4, 7, 11, 16, 29, 42, and 54 weeks. Different letters (a and b) indicate significant differences also with respect to different genetic groups in each gender; identical letters indicate no statistically significant difference. Sample information is provided by Figure 1 (mean ± SEM; **: $p < 0.01$; ***: $p < 0.001$).

3.3. Effects of Age and Gender on Levels of IGFBP2- to 4

By direct comparison of longitudinal IGFBP profiles in serum from male and female unselected non-inbred mice (data not shown), IGFBP-2, -3, and -4 exhibited gender-related features: if compared to age-matched females, male mice had higher concentrations of IGFBP-2 ($p < 0.01$) and -3 ($p < 0.001$) at an age of 16 and 42 weeks, respectively, but lower concentrations of IGFBP-4 in serum ($p < 0.001$) at the age of 54 weeks.

3.3.1. IGFBP-2

As a main effect of age in all female mice, IGFBP-2 was reduced between weeks 2 and 4 ($p < 0.01$), increased between weeks 4 and 11 ($p < 0.001$), and reduced between weeks 11 and 16 ($p < 0.001$; Figure 3). As an effect of age in all male mice, IGFBP-2 also was reduced between weeks 2 and 4 ($p < 0.001$), increased from week 4 until week 16 ($p < 0.001$), and then decreased from week 16 to week 26 ($p < 0.001$). The effects of age in selected mouse lines are depicted in Figures 3 and 4 (interactions of age and genetic group).

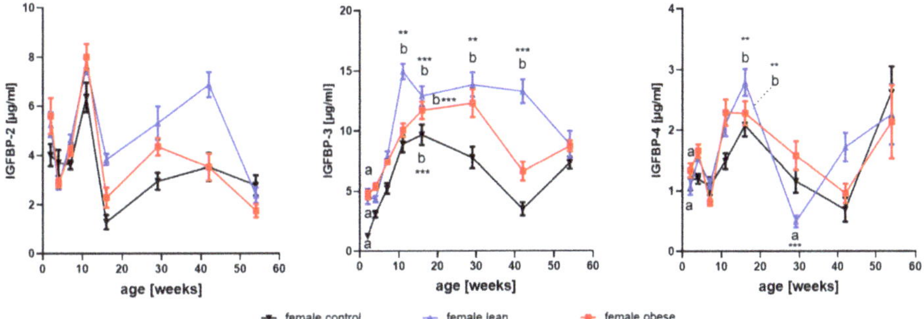

Figure 3. Concentrations of insulin-like growth factor binding protein (IGFBP) -2, -3, and -4 in serum from female mice selected for high body weight (obese), for high protein amount (lean), and in unselected controls at an age of 2, 4, 7, 11, 16, 29, 42, and 54 weeks (mean ± SEM; n ≥ 7; different letters (a and b) indicate significant effects also if different genetic groups were compared; identical letters indicate no statistically significant difference; **: $p < 0.01$; ***: $p < 0.001$).

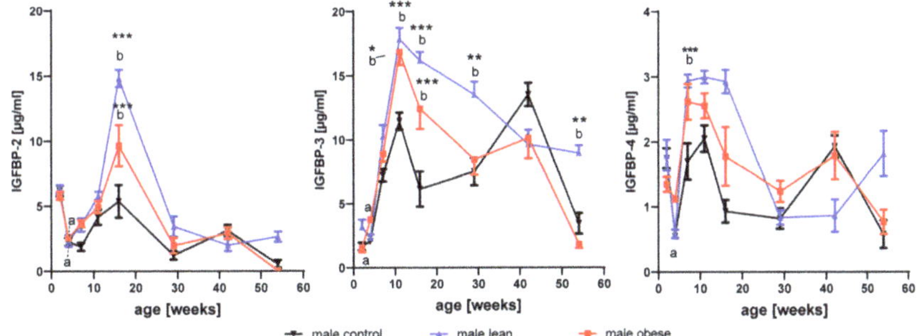

Figure 4. Concentrations of IGFBP-2, -3, and -4 in serum from male mice selected for high body weight (obese), for high protein amount (lean), and in unselected controls at an age of 2, 4, 7, 11, 16, 29, 42, and 54 weeks (mean ± SEM; $n = 8$ with the exception of obese male at an of 42 weeks and 54 weeks with $n = 7$ and $n = 4$, respectively; different letters (a and b) indicate significant effects of age; identical letters indicate no statistically significant difference also if different genetic groups were compared; *: $p < 0.05$; **: $p < 0.01$; ***: $p < 0.001$).

3.3.2. IGFBP-3

As an effect of age in all female mice, a substantial increase of IGFBP-3 (Figure 4) in serum was observed between 2 and 16 weeks of age ($p < 0.001$). In addition, a reduction was observed in all female mice between week 16 and week 54 ($p < 0.001$). Similarly in all male mice, an increase of IGFBP-3 in serum between weeks 2 and 11 ($p < 0.001$) and a decrease between weeks 11 and 54 ($p < 0.001$) was observed. Significant effects of age are presented for separate mouse lines in Figures 3 and 4 (interactions of age, gender, and mouse line).

3.3.3. IGFBP-4

As an effect of age, IGFBP-4 was significantly decreased between weeks 2 and 7 ($p < 0.05$), increased between week 7 and 11 ($p < 0.05$) and decreased between week 11 and 42 ($p < 0.001$). In all male mice, IGFBP-4 was increased between weeks 4 and 11 ($p < 0.01$) and decreased between weeks 11 and 29 ($p < 0.001$). Again, significant differences present in isolated mouse lines are depicted in Figures 3 and 4.

3.4. Effects of Growth Selection on the Concentrations of IGFBP-2 to -4

As a main effect of growth selection and irrespective of age, obese female mice had significantly higher levels of IGFBP-3 in serum ($p < 0.001$). Independent of age, selection for high protein accretion increased IGFBP-3 and IGFBP-2 ($p < 0.001$). In males, selection for high body mass and selection for high protein mass had an effect on the concentrations of IGFBP-2 and IGFBP-3 in serum ($p < 0.05$) over all age groups.

As an interaction of genetic group and age, in growth-selected obese and lean male mice (Figure 4), levels of IGFBP-3 at the age of 11 and 16 weeks were increased if compared to age-matched unselected controls ($p < 0.05$). In lean male mice, IGFBP-3 remained on a higher level also at later time points, with significant differences if compared to obese male mice and unselected controls at the age of 29 and 54 weeks ($p < 0.05$). Growth selection further stimulated the increase of IGFBP-2 from younger ages to week 16, observed in unselected controls, resulting in about 3-fold increased levels of IGFBP-2 in serum from lean male mice ($p < 0.001$). Between 4 and 11 weeks of age, a significant increase of IGFBP-4 was observed ($p < 0.01$) in all males independent of line. In all genetic groups and in both genders, the postnatal increase of IGFBP-3 in serum is lagging behind the increases of IGF-1.

3.5. Longitudinal Molar Ratio of IGF-1 and IGFBP Concentrations in Serum

In order to estimate the longitudinal molar ratio of IGF-1 with respect to the IGFBPs detected in serum by Western ligand blotting, the concentrations of IGF-1 and IGFBP-2 to -4 were corrected for their respective molecular weights (IGF-1: 7.5 kDa, IGFBP-2: 32 kDa, IGFBP-3: 41 kDa, IGFBP-4: 24 kDa). The longitudinal molar ratio of IGF-1 versus the sum of IGFBP-2, IGFBP-3, and IGFBP-4 is presented in Figure 5. Neither in female nor in male mice of all genetic groups, IGF-1 was in molar excess over the sum of IGFBP-2 to -4. As an effect of age and genetic group, between weeks 2 and 4, there was a significant increase in all groups and genders ($p < 0.001$) with the exception of female controls ($p < 0.05$). At the age of 4 weeks, obese male and female mice have higher ratios of IGF-1/IGFBPs if compared to unselected controls ($p < 0.05$).

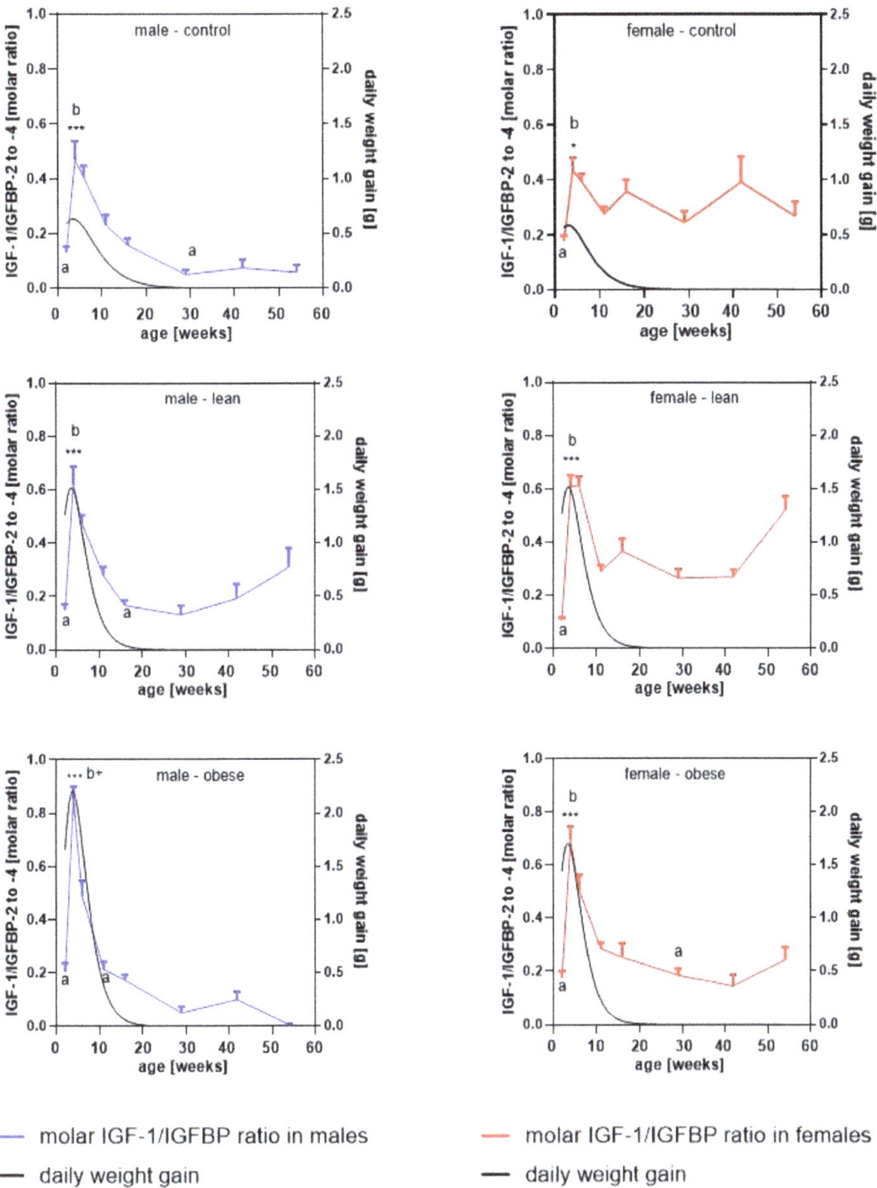

Figure 5. Molar ratio of IGF-1 to the sum of IGFBP-2, -3, and -4 (left Y-axis) present in serum from mice selected for high body weight (obese), for high protein amount (lean), and in unselected controls at an age of 2, 4, 7, 11, 16, 29, 42, and 54 weeks ($n \geq 7$ with the exception of obese male at 54 weeks with $n = 4$; different letters indicate significant effects of age; identical letters indicate no statistically significant difference; *: $p < 0.05$; ***: $p < 0.001$; b+: significantly different if compared to unselected controls, with $p < 0.001$ for males and $p < 0.05$ for females). On the right Y-axis, daily weight increases, intrapolated from the body weight data in Figure 1 by the Gompertz function, were included.

4. Discussion

Functional genome analysis in genotype-based (transgenic or knockout) mouse models has identified multiple functions of the IGF system for somatic growth [26–28]. By contrast, descriptive studies using phenotype-derived mouse models have been used to a much lesser extent. Nevertheless, phenotype-derived models can provide important information on growth regulation under physiological conditions and may also be useful for the identification and validation of biomarkers.

For the establishment of those models, phenotype selection in mice was initiated in 1976 in Dummerstorf (Germany) based on a mixed genetic background comprising four different inbred and four different outbred mouse models [10]. Growth selection has been performed with respect to male body weight at the age of 42 days, resulting in an obese model characterized by extreme body weight and marked obesity resulting in the DU6 mouse model [5]. In addition, a lean mouse model (DU6P) has been developed from the identical genetic background by selection for high protein amount in the whole body [29]. Line-specific accumulation of body fat and the accretion of muscle mass over time is described elsewhere [3,8]. Here we describe endocrine parameters of IGF-related growth in male and female mice from obese and lean mice, compared to unselected controls, in a longitudinal setting. For the analysis, we included exclusively intact IGFBPs (IGFBP-2 to -4) detectable in serum by Western ligand blotting.

Line-, gender- and age-specific growth characteristics were identified by the statistical model with significantly higher body mass in obese versus lean mice. Higher body mass in growth-selected mice is reflected by higher daily weight gains and prolonged pubertal growth if compared to unselected controls.

As published before [29] and confirmed here, after birth, obese and lean mice are characterized by substantial increases of serum IGF-1 concentrations if compared to controls, which might nicely explain higher growth activity in both mouse models. However, IGF-1 and GH have common and independent effects during early postnatal growth [30]. Therefore, we have to consider the IGF-independent effects of growth hormone during the earlier postnatal growth period but also the effects of IGF-2. The potential effects of the embryonal growth factor IGF-2 [31], which also can have positive effects on body weight after birth [32], need to be studied in a separate study. Although daily weight gain declined between 17 and 27.5 days of age in all mice included in that study, the concentrations of IGF-1 in the circulation remained elevated until an age of at least 7 weeks. The elevated levels of IGF-1 therefore cannot explain the massive reductions of growth activity in mice between 4 and 7 weeks of age. In female but not male human subjects, IGF-1 concentrations also lag behind the peak weight gain in males and females [16,33]. Accordingly, highest IGF-1 serum concentrations were found at the age of about 14.6 years in male and female human subjects [17], whereas peak weight gain was referred to an age of 12 years in females and 14 years in males [34]. The clear decrease of serum IGF-1 concentrations, at least in male mice between week four and week 26 of age, identifies peak-like kinetic for serum IGF-1 concentrations as also found in human subjects [16]. In female mice, a decrease of serum IGF-1 concentrations is less clear and was only found in obese mice. In unselected controls, a peak-like pattern is virtually absent. This is a clear difference if compared to human subjects, where serum IGF-1 concentrations were clearly lower in adults compared to younger subjects [16].

In order to understand why daily weight gain was reversed in the presence of high or elevated IGF-1 serum concentrations, we studied serum IGFBP concentrations in all samples. In all male genetic groups, IGFBP-2 concentrations were reduced directly after birth between two and four weeks of age (effect of age by gender for males: $p < 0.001$). The reductions in male mice between two and four weeks were characterized by outmost uniformity as the curves were overlaying each other, and the standard deviations were comparably low. Another clear feature of longitudinal IGFBP-expression was seen in strong increases of IGFBP-3 in male and female mice between four and 11 weeks of age. In human subjects of both genders, IGFBP-3 increased during the growth period until the age of 20 years [17,35].

In lean and obese male mice, serum IGFBP-2 concentrations were elevated at week 16 after birth compared to earlier time points. In female mice, a postnatal increase occurred one month earlier at week 11 and was significant compared to two, four, and seven weeks of age, independent of the genetic group (data not shown). Due to gender-specific patterns of longitudinal IGFBP-2 concentrations in serum, females from all genetic groups had significantly lower levels of IGFBP-2 compared with their male littermates at the age of 16 weeks. In human serum, concentrations of IGFBP-2 are one order of magnitude lower compared to IGFBP-3 and decrease from childhood to adolescence [36], similar to mice. With advanced age in humans, concentrations of IGFBP-2 in serum increase, with the highest levels found during senescence [37,38]. Accordingly, we may have similarities between mice and humans only during the initial postnatal growth period.

Similar to IGFBP-2, serum levels of IGFBP-4 were reduced between weeks 2 and 4 in male mice but significantly increased in both genders between week 4 and 16 independent of group ($p < 0.001$). In healthy human subjects, serum IGFBP-4 concentrations did not change with age [39].

The molar ratio of IGF-1 to the molar sum of all IGFBPs identified in serum by Western ligand blotting was characterized by significant increases between week 2 and week 4 in all groups. Thereby, the IGF/IGFBP ratios reached their lifetime maxima when growth activity was also high in mice. In obese mice, at the age of four weeks, the IGF-1/IGFBP ratio was also significantly increased compared to unselected controls, which may explain at least part of the higher growth activity under conditions of growth selection. The overlay of IGF-1/IGFBP ratios with the daily weight increases are only partial in unselected controls of both genders. This may indicate that the extended periods of growth activity in growth-selected mice may be related to elevated IGF-1 and/or elevated IGF-1 bioactivity. As provided by data from larger groups of mice (data not shown), growth-selected mice are heavier already at the time of birth, where we could not identify higher IGF-1 or IGF-1/IGFBP ratios. From work in genotype-derived mouse models, we know that in particular, IGF-II and GH or other hormones like insulin have an effect on early growth and development [30]. As also mentioned earlier, the potential roles of these hormones have to be addressed in future studies. In all groups and genders, the kinetics of serum IGFBP-3 concentrations are lagging behind those of serum IGF-1 concentrations. Thereby, a mechanism may be generated for the establishment of acutely high IGF-1/IGFBP ratios. By contrast, a delayed increase of IGFBP-3 versus IGF-1 was definitely not described in humans. According to published reference levels [17], male and female children at the age of about five years already had an increase of 75% of their maximal IGFBP-3 during adulthood. Thus, a delayed increase of IGFBP-3 versus IGF-1 was definitely not described in humans [16,17], which could be due to species differences or the different analytical systems used.

To date, it is unclear how such a shift is established in mice. In general, the altered expression of IGFBPs on the level of RNA and/or protein or altered stability of IGFBPs in biological matrices may be causative of altered concentrations of IGFBPs and thus altered bioactivity of IGF-1. An involvement of IGFBP-proteolysis for the control of height attainment has been suggested by Marouli et al. [4]. In this study, an allele of stanniocalcin 2 was characterized, which was less efficient in blocking proteolytic activity of PAPP-A. Just recently, it was demonstrated that PAPPA-2, which represents a candidate gene for growth regulation in mice as well [40], is significantly decreased during childhood and negatively correlated with intact IGFBP-3 in humans [23].

However, this study has distinct limitations. First of all, due to the longitudinal and initially descriptive approach, the number in every single age group was comparably small and further reduced by higher mortality with age, particularly in male obese mice. Accordingly, the highest age group with 4 replicates in obese males can be considered less reliable. In the future, higher sample numbers should be chosen in selected groups also for confirmatory studies. Furthermore, IGFBP-3 is present in a ternary complex, whereas IGFBP-2 and IGFBP-4 are present in binary complexes only. Therefore, differential pharmacokinetic properties can be expected for the different types of complexes. Therefore, the present study followed a simplistic approach by combining IGFBP-2 to -4 for the estimation of IGF-1 to IGFBP ratios. Future studies also would have to consider the concentrations of IGFBP-1, -5, and -6, which were not detected by Western ligand blotting due to lower sensitivity compared e.g., to ELISA. Furthermore, the different compounds from the GH/IGF-system are inter-related, as GH and IGF-1 are particular determinants of IGFBP concentrations. Finally, the physiological relevance of IGF to IGFBP ratios less than 1 needs to be addressed in future studies. This could be achieved by the analysis of free IGF-1 or IGF-related bioactivity in animals characterized by different IGF to IGFBP ratios.

To summarize, we have characterized longitudinal concentrations of IGF-1 and intact IGFBP-2 to -4 in serum from two different mouse lines selected for high growth and unselected controls in both genders. We compared the IGF-1/IGFBP ratios with daily weight gain and were able to provide evidence that part of the elevated growth activity during prepubertal growth in normal and growth-selected mice could be related to elevated bioactivity of IGF-1. Elevated ratios of IGF-1/IGFBPs are established by a delayed increase of IGFBPs compared to strong increases of IGF-1 between 2 and 4 weeks of age.

We therefore may be in a position to distinguish two phases of IGF-1 related growth during early postnatal development: acceleration of postnatal growth by elevated serum IGF-1 concentrations followed by a phase of deceleration due to the delayed increase of IGFBPs in serum.

Author Contributions: Conceptualization, A.H.; methodology, M.W., L.C., and C.H.; software, A.T. and A.H.; validation, A.T. and A.H.; formal analysis, M.W., L.C., A.T., and C.H.; investigation, M.W., L.C., C.H.; resources, M.S. and M.L.; data curation, A.H. and M.L.; writing—original draft preparation, A.H.; writing—review and editing, M.W, C.W., D.O., J.B., A.T., M.L., F.M., C.H., and A.H.; visualization, A.H.; supervision, C.W. and A.H.; project administration, A.H.; funding acquisition, A.H. All authors have read and agreed to the published version of the manuscript.

Funding: M.W. is financially supported by the H. Wilhelm Schaumann Stiftung.

Acknowledgments: The publication of this article was funded by the Open Access Fund of the Leibniz Association and the Open Access Fund of the Leibniz Institute for Farm Animal Biology (FBN). The authors thank the staff of the mouse facility for excellent animal care and excellent technical assistance, respectively. We further thank S.Sänger (F. Hoffmann-La Roche, Basel, Switzerland) for excellent support in the IGF-I analysis.

Conflicts of Interest: C.H. and A.H. are related to Ligandis UG. The other authors do not have any potential conflicts of interest to declare. Friedrich Metzger declares relations to Versameb AG.

References

1. Goodale, H.D. A Study of the Inheritance of Body Weight in the Albino Mouse by Selection. *J. Hered.* **1938**, *29*, 101–112. [CrossRef]

2. Wilson, S.P.; Goodale, H.D.; Kyle, W.H.; Godfrey, E.F. Long term selection for bodyweight in mice. *J. Hered.* **1971**, *62*, 228–234. [CrossRef] [PubMed]

3. Renne, U.; Langhammer, M.; Brenmoehl, J.; Walz, C.; Zeissler, A.; Tuchscherer, A.; Piechotta, M.; Wiesner, R.J.; Bielohuby, M.; Hoeflich, A. Lifelong Obesity in a Polygenic Mouse Model Prevents Age- and Diet-Induced Glucose Intolerance– Obesity Is No Road to Late-Onset Diabetes in Mice. *PLoS ONE* **2013**, *8*, e79788. [CrossRef] [PubMed]

4. Marouli, E.; The EPIC-InterAct Consortium; Graff, M.; Medina-Gomez, C.; Lo, K.S.; Wood, A.R.; Kjaer, T.R.; Fine, R.; Lu, Y.; Schurmann, C. Rare and low-frequency coding variants alter human adult height. *Nature* **2017**, *542*, 186–190. [CrossRef]

5. Brockmann, A.G.; Haley, C.S.; Renne, U.; Knott, A.S.; Schwerin, M. Quantitative trait loci affecting body weight and fatness from a mouse line selected for extreme high growth. *Genetics* **1998**, *150*, 369–381.

6. Bunger, L.; Laidlaw, A.; Bulfield, G.; Eisen, E.J.; Medrano, J.F.; Bradford, G.E.; Pirchner, F.; Renne, U.; Schlote, W.; Hill, W.G. Inbred lines of mice derived from long-term growth selected lines: Unique resources for mapping growth genes. *Mamm. Genome Off. J. Int. Mamm. Genome Soc.* **2001**, *12*, 678–686. [CrossRef]

7. Bunger, L.; Renne, U.; Dietl, G.; Kuhla, S. Long-term selection for protein amount over 70 generations in mice. *Genet. Res.* **1998**, *72*, 93–109. [CrossRef]

8. Sawitzky, M.; Zeissler, A.; Langhammer, M.; Bielohuby, M.; Stock, P.; Hammon, H.M.; Görs, S.; Metges, C.C.; Stoehr, B.J.M.; Bidlingmaier, M.; et al. Phenotype Selection Reveals Coevolution of Muscle Glycogen and Protein and PTEN as a Gate Keeper for the Accretion of Muscle Mass in Adult Female Mice. *PLoS ONE* **2012**, *7*, e39711. [CrossRef]

9. Schüler, L. Der Mäuseauszuchtstamm Fzt:DU und seine Anwendung als Modell in der Tierzuchtforschung. *Arch. Tierz.* **1985**, *28*, 357–363.

10. Dietl, G.; Langhammer, M.; Renne, U. Model simulations for genetic random drift in the outbred strain Fzt: DU. *Arch. Tierz.* **2004**, *47*, 595–604. [CrossRef]

11. Hoeflich, A.; Bunger, L.; Nedbal, S.; Renne, U.; Elmlinger, M.W.; Blum, W.F.; Bruley, C.; Kolb, H.J.; Wolf, E. Growth selection in mice reveals conserved and redundant expression patterns of the insulin-like growth factor system. *Gen. Comp. Endocrinol.* **2004**, *136*, 248–259. [CrossRef] [PubMed]

12. Jia, J.; Ahmed, I.; Liu, L.; Liu, Y.; Xu, Z.; Duan, X.; Li, Q.; Dou, T.; Gu, D.; Rong, H.; et al. Selection for growth rate and body size have altered the expression profiles of somatotropic axis genes in chickens. *PLoS ONE* **2018**, *13*, e0195378. [CrossRef] [PubMed]

13. Pas, M.T.; Freriksen, J.; Van Bijnen, A.; Gerritsen, C.; Bosch, T.V.D.; Harders, F.; Verburg, F.; Visscher, A.; De Greef, K. Selection for growth rate or against back fat thickness in pigs is associated with changes in growth hormone axis plasma protein concentration and mRNA level. *Domest. Anim. Endocrinol.* **2001**, *20*, 165–184. [CrossRef]

14. Anh, N.T.L.; Kunhareang, S.; Duangjinda, M. Association of Chicken Growth Hormones and Insulin-like Growth Factor Gene Polymorphisms with Growth Performance and Carcass Traits in Thai Broilers. *Asian Australas. J. Anim. Sci.* **2015**, *28*, 1686–1695. [CrossRef]

15. El-Magd, M.A.; Saleh, A.A.; Nafeaa, A.A.; El-Komy, S.M.; Afifi, M. Polymorphisms of the IGF1 gene and their association with growth traits, serum concentration and expression rate of IGF1 and IGF1R in buffalo. *J. Zhejiang Univ. Sci. B* **2017**, *18*, 1064–1074. [CrossRef]

16. Bidlingmaier, M.; Friedrich, N.; Emeny, R.T.; Spranger, J.; Wolthers, O.D.; Roswall, J.; Körner, A.; Obermayer-Pietsch, B.; Hübener, C.; Dahlgren, J.; et al. Reference Intervals for Insulin-like Growth Factor-1 (IGF-I) From Birth to Senescence: Results From a Multicenter Study Using a New Automated Chemiluminescence IGF-I Immunoassay Conforming to Recent International Recommendations. *J. Clin. Endocrinol. Metab.* **2014**, *99*, 1712–1721. [CrossRef]

17. Friedrich, N.; Wolthers, O.D.; Arafat, A.; Emeny, R.T.; Spranger, J.; Roswall, J.; Kratzsch, J.; Grabe, H.J.; Hübener, C.; Pfeiffer, A.F.; et al. Age- and Sex-Specific Reference Intervals Across Life Span for Insulin-Like Growth Factor Binding Protein 3 (IGFBP-3) and the IGF-I to IGFBP-3 Ratio Measured by New Automated Chemiluminescence Assays. *J. Clin. Endocrinol. Metab.* **2014**, *99*, 1675–1686. [CrossRef]

18. Hossenlopp, P.; Segovia, B.; Lassarre, C.; Roghani, M.; Bredon, M.; Binoux, M. Evidence of Enzymatic Degradation of Insulin-Like Growth Factor-Binding Proteins in the 150K Complex during Pregnancy. *J. Clin. Endocrinol. Metab.* **1990**, *71*, 797–805. [CrossRef]

19. Hoeflich, A.; David, R.; Hjortebjerg, R. Current IGFBP-Related Biomarker Research in Cardiovascular Disease—We Need More Structural and Functional Information in Clinical Studies. *Front. Endocrinol.* **2018**, *9*, 388. [CrossRef]

20. Lawrence, J.B.; Oxvig, C.; Overgaard, M.T.; Sottrup-Jensen, L.; Gleich, G.J.; Hays, L.G.; Yates, J.R.; Conover, C.A. The insulin-like growth factor (IGF)-dependent IGF binding protein-4 protease secreted by human fibroblasts is pregnancy-associated plasma protein-a. *Proc. Natl. Acad. Sci. USA* **1999**, *96*, 3149–3153. [CrossRef]

21. Botkjaer, J.A.; Noer, P.R.; Oxvig, C.; Yding Andersen, C. A common variant of the pregnancy-associated plasma protein-a (pappa) gene encodes a protein with reduced proteolytic activity towards igf-binding proteins. *Sci. Rep.* **2019**, *9*, 13231. [CrossRef]

22. Conover, C.A.; Oxvig, C. 40 YEARS OF IGF1: PAPP-A and cancer. *J. Mol. Endocrinol.* **2018**, *61*, T1–T10. [CrossRef] [PubMed]

23. Fujimoto, M.; Khoury, J.C.; Khoury, P.R.; Kalra, B.; Kumar, A.; Sluss, P.; Oxvig, C.; Hwa, V.; Dauber, A. Anthropometric and biochemical correlates of PAPP-A2, free IGF-I, and IGFBP-3 in childhood. *Eur. J. Endocrinol.* **2020**, *182*, 363–374. [CrossRef] [PubMed]

24. Saenger, S.; Goeldner, C.; Frey, J.; Ozmen, L.; Ostrowitzki, S.; Spooren, W.; Ballard, T.; Prinssen, E.; Borroni, E.; Metzger, F. PEGylation enhances the therapeutic potential for insulin-like growth factor I in central nervous system disorders. *Growth Horm. IGF Res.* **2011**, *21*, 292–303. [CrossRef]

25. Wirthgen, E.; Höflich, C.; Spitschak, M.; Helmer, C.; Brand, B.; Langbein, J.; Metzger, F.; Hoeflich, A. Quantitative Western ligand blotting reveals common patterns and differential features of IGFBP-fingerprints in domestic ruminant breeds and species. *Growth Horm. IGF Res.* **2016**, *26*, 42–49. [CrossRef]

26. Efstratiadis, A. Genetics of mouse growth. *Int. J. Dev. Biol.* **1998**, *42*, 955–976.

27. Pintar, J.E.; Cerro, J.A.; Wood, T.L. Genetic Approaches to the Function of Insulin-Like Growth Factor-Binding Proteins during Rodent Development. *Horm. Res.* **1996**, *45*, 172–177. [CrossRef] [PubMed]

28. Leroith, D.; Roberts, J.C.T. Insulin-like Growth Factors. *Ann. N. Y. Acad. Sci.* **1993**, *692*, 1–9. [CrossRef]
29. Timtchenko, D.; Kratzsch, J.; Sauerwein, H.; Wegner, J.; Souffrant, W.; Schwerin, M.; Brockmann, G. Fat storage capacity in growth-selected and control mouse lines is associated with line-specific gene expression and plasma hormone levels. *Int. J. Obes.* **1999**, *23*, 586–594. [CrossRef] [PubMed]
30. Lupu, F.; Terwilliger, J.D.; Lee, K.; Segre, G.V.; Efstratiadis, A. Roles of Growth Hormone and Insulin-like Growth Factor 1 in Mouse Postnatal Growth. *Dev. Biol.* **2001**, *229*, 141–162. [CrossRef]
31. Baker, J.; Liu, J.P.; Robertson, E.J.; Efstratiadis, A. Role of insulin-like growth factors in embryonic and postnatal growth. *Cell* **1993**, *75*, 73–82. [CrossRef]
32. Moerth, C.; Schneider, M.R.; Renner-Mueller, I.; Blutke, A.; Elmlinger, M.W.; Erben, R.G.; Camacho-Hübner, C.; Hoeflich, A.; Wolf, E. Postnatally Elevated Levels of Insulin-Like Growth Factor (IGF)-II Fail to Rescue the Dwarfism of IGF-I-Deficient Mice except Kidney Weight. *Endocrinology* **2007**, *148*, 441–451. [CrossRef] [PubMed]
33. Rogol, A.D.; A Clark, P.; Roemmich, J.N. Growth and pubertal development in children and adolescents: Effects of diet and physical activity. *Am. J. Clin. Nutr.* **2000**, *72*, S521–S528. [CrossRef] [PubMed]
34. Tanner, J.M. *Fetus into Man: Physical Growth from Conception to Maturity*; Harward University Press: Cambridge, MA, USA, 1989.
35. Bereket, A.; Turan, S.; Omar, A.; Berber, M.; Ozen, A.; Akbenlioglu, C.; Haklar, G. Serum IGF-I and IGFBP-3 Levels of Turkish Children during Childhood and Adolescence: Establishment of Reference Ranges with Emphasis on Puberty. *Horm. Res. Paediatr.* **2006**, *65*, 96–105. [CrossRef]
36. Smith, W.J.; Nam, T.J.; Underwood, L.E.; Busby, W.H.; Celnicker, A.; Clemmons, D.R. Use of insulin-like growth factor-binding protein-2 (igfbp-2), igfbp-3, and igf-i for assessing growth hormone status in short children. *J. Clin. Endocrinol. Metab.* **1993**, *77*, 1294–1299.
37. Mattsson, A.; Svensson, D.; Schuett, B.; Osterziel, K.J.; Ranke, M.B. Multidimensional reference regions for IGF-I, IGFBP-2 and IGFBP-3 concentrations in serum of healthy adults. *Growth Horm. IGF Res.* **2008**, *18*, 506–516. [CrossRef]
38. Beld, A.W.V.D.; Carlson, O.D.; Doyle, M.E.; Rizopoulos, D.; Ferrucci, L.; Van Der Lely, A.J.; Egan, J.M. IGFBP-2 and aging: A 20-year longitudinal study on IGFBP-2, IGF-I, BMI, insulin sensitivity and mortality in an aging population. *Eur. J. Endocrinol.* **2019**, *180*, 109–116. [CrossRef]
39. Van Doorn, J.; Cornelissen, A.J.F.H.; Van Buul-Offers, S.C. Plasma levels of insulin-like growth factor binding protein-4 (IGFBP-4) under normal and pathological conditions. *Clin. Endocrinol.* **2001**, *54*, 655–664. [CrossRef]
40. Christians, J.K.; Hoeflich, A.; Keightley, P.D. PAPPA2, an Enzyme That Cleaves an Insulin-Like Growth-Factor-Binding Protein, Is a Candidate Gene for a Quantitative Trait Locus Affecting Body Size in Mice. *Genetics* **2006**, *173*, 1547–1553. [CrossRef]

Article

Insulin-like Growth Factor-1 and IGF Binding Proteins Predict All-Cause Mortality and Morbidity in Older Adults

William B. Zhang [1,†], **Sandra Aleksic** [1,†], **Tina Gao** [1], **Erica F. Weiss** [2], **Eleni Demetriou** [3], **Joe Verghese** [2,4], **Roee Holtzer** [2,3], **Nir Barzilai** [1,5] and **Sofiya Milman** [1,5,*]

[1] Department of Medicine, Division of Endocrinology, Institute for Aging Research, Albert Einstein College of Medicine, Bronx, NY 10461, USA; william.zhang@einsteinmed.org (W.B.Z.); saleksic@montefiore.org (S.A.); tina.gao@einsteinmed.org (T.G.); nir.barzilai@einsteinmed.org (N.B.)

[2] Department of Neurology, Albert Einstein College of Medicine, Bronx, NY 10461, USA; EWEISS@montefiore.org (E.F.W.); joe.verghese@einsteinmed.org (J.V.); roee.holtzer@yu.edu (R.H.)

[3] Ferkauf Graduate School of Psychology, Yeshiva University, New York, NY 10033, USA; eleni.demetriou@mail.yu.edu

[4] Department of Medicine, Division of Geriatrics, Institute for Aging Research, Albert Einstein College of Medicine, Bronx, NY 10461, USA

[5] Department of Genetics, Albert Einstein College of Medicine, Bronx, NY 10461, USA

* Correspondence: sofiya.milman@einsteinmed.org

† These authors equally contributed to this work.

Received: 10 May 2020; Accepted: 28 May 2020; Published: 1 June 2020

Abstract: While the growth hormone/insulin-like growth factor-1 (GH/IGF-1) pathway plays essential roles in growth and development, diminished signaling via this pathway in model organisms extends lifespan and health-span. In humans, circulating IGF-1 and IGF-binding proteins 3 and 1 (IGFBP-3 and 1), surrogate measures of GH/IGF-1 system activity, have not been consistently associated with morbidity and mortality. In a prospective cohort of independently-living older adults ($n = 840$, mean age 76.1 ± 6.8 years, 54.5% female, median follow-up 6.9 years), we evaluated the age- and sex-adjusted hazards for all-cause mortality and incident age-related diseases, including cardiovascular disease, diabetes, cancer, and multiple-domain cognitive impairment (MDCI), as predicted by baseline total serum IGF-1, IGF-1/IGFBP-3 molar ratio, IGFBP-3, and IGFBP-1 levels. All-cause mortality was positively associated with IGF-1/IGFBP-3 molar ratio (HR 1.28, 95% CI 1.05–1.57) and negatively with IGFBP-3 (HR 0.82, 95% CI 0.680–0.998). High serum IGF-1 predicted greater risk for MDCI (HR 1.56, 95% CI 1.08–2.26) and composite incident morbidity (HR 1.242, 95% CI 1.004–1.538), whereas high IGFBP-1 predicted lower risk for diabetes (HR 0.50, 95% CI 0.29–0.88). In conclusion, higher IGF-1 levels and bioavailability predicted mortality and morbidity risk, supporting the hypothesis that diminished GH/IGF-1 signaling may contribute to human longevity and health-span.

Keywords: IGF-1; IGFBP-3; IGFBP-1; older adults; longevity; health-span; age-related disease; cognitive impairment; diabetes

---

1. Introduction

The rise in age-related diseases and disability that accompany advanced age presents a burden to economies, health care systems, and individuals worldwide [1]. Evidence from model organisms demonstrates that aging is a biologically regulated process that can be modulated to extend lifespan and health-span [2]. The evolutionarily conserved growth hormone/insulin-like growth factor-1 (GH/IGF-1) pathway, which plays essential roles in growth, development, and metabolism, has been recognized as one key regulator of aging [3]. In invertebrates, attenuated signaling in pathways homologous to mammalian IGF-1 dramatically extends lifespan [4,5]. Similarly, mutant rodents with diminished GH and/or IGF-1 secretion or signaling exhibit lifespan extension of 25–60% [6–10]. They also display extended health-span, including delayed age-related impairments in cognition, musculoskeletal function, glucose homeostasis, immunosenescence and cancer [6–8,11–14]. On the other hand, overexpression of GH/IGF-1 in transgenic rodents accelerates age-related pathologies and dramatically reduces lifespan [15,16].

However, the role of the GH/IGF-1 system in human aging and longevity is still uncertain. In humans, profound reduction or enhancement in GH/IGF-1 signaling has consequences for survival and health that replicate some of those noted in experimental models. For instance, individuals with GH receptor deficiency are protected from lethal malignancies and type 2 diabetes, although their lifespan is not prolonged [17]. On the other hand, patients with acromegaly, characterized by hypersecretion of GH, have increased risk for premature cardiovascular disease, diabetes, malignancy, and mortality [18–20]. Nonetheless, epidemiologic studies investigating the relationship between circulating levels of IGF-1, which are used as a proxy for the activity of GH/IGF-1 axis in humans [21], and clinical outcomes have yielded inconsistent findings. While our group and others found inverse relationship between IGF-1 levels and survival [22,23], a number of studies reported positive [24,25] or null [26] associations. Furthermore, IGF-1 levels have been reported to have opposite effects on risk for different age-related diseases. For instance, lower levels of IGF-1 were associated with increased risk for cardiovascular disease [27,28], while high IGF-1 levels were related to increased risk for cancer [29]; although the findings were not consistent in all studies [30].

Several reasons may contribute to the inconsistent findings above. First, the activity of the GH/IGF-1 axis and levels of circulating IGF-1 are affected by acute [31] and chronic illness [32]; this introduces the possibility of "reverse causation" in some studies conducted in high-risk populations [25]. Second, numerous epidemiologic studies investigated associations between IGF-1 and morbidity and mortality in cohorts with wide age-ranges under the assumption that the effect of IGF-1 would be similar in younger and older adults [33,34]. Furthermore, total IGF-1 does not represent bioavailable IGF-1 [21]: Almost all circulating IGF-1 is bound to six IGF-binding proteins (IGFBP-1-6), leaving <1% of IGF-1 in a free form, bioavailable to bind to its receptors [35]. In addition to providing a long-lasting pool of circulating IGF-1, IGFBPs closely regulate biological functions of IGF-1 through controlled inhibition and promotion of IGF-1 interactions with its receptor [36]. Since measuring free IGF-1 remains challenging [37], the molar ratio of total IGF-1 to IGFBP-3, the most abundant IGFBP in circulation, is commonly used as a proxy for bioavailable IGF-1 [21]. Finally, other elements of the GH/IGF-1 system, including IGFBP-3 and IGFBP-1, have been implicated in human disease and survival, independent of IGF-1 [38,39]. Therefore, our aim was to prospectively investigate the associations between several components of the GH/IGF-1 pathway, including total IGF-1, IGF-1/IGFBP-3 molar ratio, IGFBP-3, and IGFBP-1, with mortality and incidence of major age-associated diseases in a cohort of independently-living older adults with majority in general good health at enrollment.

2. Materials and Methods

2.1. Human Cohort Data Acquisition

LonGenity is an ongoing longitudinal study initiated in 2008, that seeks to identify genotypes and phenotypes that protect from age-related diseases and promote exceptional longevity in humans [40]. The LonGenity cohort is composed of Ashkenazi Jewish older adults and about half of the cohort has a parental history of exceptional longevity, defined as having at least one parent survive to 95 years of age. Other inclusion criteria include baseline age ≥ 65 years or older and being free of significant cognitive impairment at baseline. Study participants are extensively characterized at annual visits, which include medical history and neurocognitive testing. Baseline IGF-1 and related protein measurements were available for 877 (54.5% female) study participants. Among this group, 37 individuals only completed the baseline visit and thus, were excluded from this analysis. Among the 840 participants included in this study, 20 did not have complete physical examinations as they either declined or were unable to do so due to mobility issues and were therefore missing body mass index (BMI) measurements. Fasting blood samples were also collected biannually at follow up visits. The LonGenity study was approved by the institutional review board (IRB) at the Albert Einstein College of Medicine. Informed consent was obtained from all study participants.

2.2. Biochemical Measurements

Biochemical measurements were performed as previously described [41]. Total IGF-1 levels were measured by liquid chromatography/mass spectrometry at Quest Diagnostics Nichols Institute laboratories (Quest, San Juan Capistrano, CA, USA) in serum collected at baseline and subsequently stored at -80 °C. For IGF-1, the limit of quantification (LOQ) was 15.6 ng/mL and the coefficient of variance (CV) was 3.3%, 3.1%, 2.8%, and 5% for the low (mean 57.2 ng/mL), medium (mean 248 ng/mL), high (mean 447.1 ng/mL) Bio-Rad quality controls and pooled human serum in-house control (mean 104 ng/mL), respectively. IGFBP-3 levels were measured at Quest with a chemiluminescent immuno-metric assay (Siemens Immulite 2000; Siemens Healthineers AG, Erlangen, Bavaria, Germany). IGFBP-1 levels were measured at Quest with a radioimmunoassay. For IGFBP-1, the LOQ was 5 ng/mL, and the CV was 9.3%, 10.1%, and 8.5% for the low (mean 19.2 ng/mL), medium (mean 53.5 ng/mL) and high (mean 111.3 ng/mL) controls, respectively. For IGFBP-3 the LOQ was 0.5 mg/L, and the CV was 5.1%, 6.1%, and 6.5% for the low (mean 0.90 mg/L), high (mean 3.56 mg/L), and in-house controls, respectively. The IGF-1/IGFBP-3 molar ratio was calculated by dividing measured serum total IGF-1 and IGFBP-3 levels by their molecular weights (7649 Daltons and 31,673 Daltons, respectively), and then calculating the ratio between the two quantities [42,43]. Insulin was measured by radioimmunoassay at the Albert Einstein College of Medicine Biomarker Analytic Research Core (BARC). Due to limitations in sample volume, IGFBP-3 was measured in 828 subjects, IGF-1 in 761, IGFBP-1 in 728 subjects, and insulin in 801 subjects.

2.3. Disease Definitions

Three of the age-associated morbidities in this study, cardiovascular disease, diabetes, and cancer, were selected because they represent major causes of morbidity and mortality in the aging population [44]. These morbidities were defined using a combination of self-reported questionnaire data, medical records, and laboratory results. Cardiovascular disease was defined as having a history of myocardial infarction, stroke, or cardiac procedure such as percutaneous coronary intervention or coronary artery bypass grafting surgery. Diabetes was defined as a self-reported history of diabetes, a fasting blood glucose of 126 mg/dL or greater, or a hemoglobin A1C level of 6.5% or greater. Cancer was defined as a self-reported history of any malignancy, excluding non-melanoma skin cancers.

We also investigated multiple-domain cognitive impairment (MDCI), as it was previously shown that individuals with MDCI with memory involvement have high rates of progression to Alzheimer's disease [45], which is a major cause of morbidity and mortality in older individuals [44]. MDCI was assigned by the neuropsychology team under the direction of the study neuropsychologist. Annual neurocognitive batteries evaluating memory, language, visuospatial, attention, and executive cognitive domains were double scored and age normed. As controversy remains about optimal cut-off scores [46], performances 1.5 SD below the age-appropriate mean were defined as impaired [45]. Multiple-domain cognitive impairment assignment was made when participants had impaired performance on at least one measure of memory and impaired performance on at least one measure in another cognitive domain.

Composite incident morbidity was a composite outcome defined as onset of either cardiovascular disease, diabetes, cancer, or MDCI during study follow-up. To maximize power for the incident morbidity analysis, individuals with pre-existing disease were included in the analysis and were monitored for the onset of an additional age-associated morbidity. For the single disease analyses, individuals with that particular disease at baseline were excluded from the analysis (e.g., individuals with baseline diabetes were excluded from all analyses of incident diabetes but were included in the incident morbidity analysis).

2.4. Statistical Analysis

Statistical analysis was performed using custom scripts in Python (version 3.6), a general-purpose programming language. For all analyses tracking a particular biochemical measurement—IGF-1, IGFBP-1, IGFBP-3, or IGF-1/IGFBP-3 molar ratio—the participants were dichotomized into high and low groups using sex-specific medians of their baseline measurements. The median IGF-1 for men was 123 ng/mL and for women it was 105.5 ng/mL. Thus, the "low IGF-1" group was formed by combining men and women with IGF-1 levels below the sex-specific medians and the "high IGF-1" group was formed by combining men and women with IGF-1 levels at or above the sex-specific medians. "High" and "low" groups for analyses of IGFBP-1 (median 14 ng/mL in males, 18 ng/mL in females), IGFBP-3 (median 3.5 mg/L in males, 4.1 mg/L in females), and IGF-1/IGFBP-3 ratio (median 0.15 in males, 0.11 in females) were all formed by this procedure.

For comparison of baseline characteristics, normally distributed continuous variables (age and biochemical measurements) were compared using a two-tailed student's *t*-test. Normality was confirmed by visual inspection of the histograms. Categorical variables were compared using either a chi-squared test of homogeneity (deaths) or a binomial test (number of participants). Results were considered statistically significant at p-value < 0.05.

Unadjusted survival curves for "high" and "low" groups of participants were generated using the Kaplan–Meier method for censored data and the survival curves were compared using log-rank tests. In addition, Cox proportional hazards models adjusted for sex and age at study enrollment were fit to the dichotomized biochemical measures and a clinical outcome of interest. The analyses were further stratified by sex, in order to identify any sex-specific differences in the association between dichotomized biochemical measures and clinical outcomes of interest. Interaction between age and dichotomized biochemical measures were investigated in mortality models stratified by median age at enrollment. Additionally, models were adjusted for dichotomized IGF-1 levels to investigate the independent associations between IGFBPs and clinical outcomes. BMI and insulin levels were included as covariates in models that predicted incidence of diabetes.

3. Results

3.1. Baseline Characteristics of Study Cohort

The study included 840 subjects (54.5% female), whose baseline characteristics are shown in Table 1 and in Tables S1 and S2. The median follow-up time for mortality was 6.9 years (interquartile range 4.6–8.5 years). The average age of the cohort was 76.1 ± 6.8 years, with no significant difference between men and women ($p = 0.39$). Men, on average, had higher serum IGF-1 levels ($p < 0.001$) and IGF-1/IGFBP-3 molar ratios ($p < 0.001$), but lower IGFBP-1 ($p < 0.001$) and IGFBP-3 ($p < 0.001$) compared to women. The average age was significantly older in subjects with low IGF-1 compared to those with high IGF-1 in the combined cohort ($p = 0.04$) and among men in the sex-stratified analysis ($p = 0.02$), (Table S2). Over the course of follow-up, 13.9% of study participants died. At baseline, prevalence of morbidities was as follows: cardiovascular disease 12.7%, diabetes 10.5%, cancer 22.1%, and MDCI 3.0% (Table S1).

Table 1. Baseline characteristics of study cohort. All *p*-values are for comparisons between males and females.

	All	Male	Female	*p*-Value
Number of Individuals, *n* (%)	840	382 (45.5)	458 (54.5)	0.01
Deaths, *n* (%)	117 (13.9)	65 (17.0)	52 (11.4)	0.02
Age (years), mean ± SD	76.1 ± 6.8	76.4 ± 7.0	76.0 ± 6.7	0.39
BMI (kg/m^2), mean ± SD, *n* = 820	27.6 ± 4.7	27.9 ± 3.9	27.3 ± 5.3	0.053
Insulin (mIU/L), mean ± SD, *n* = 801	15.4 ± 12.3	16.6 ± 15.7	14.5 ± 8.3	0.02
IGF-1 (ng/mL), mean ± SD, *n* = 761	117 ± 38	127 ± 39	108 ± 36	<0.001
IGFBP-1 (ng/mL), mean ± SD, *n* = 728	19 ± 15	17 ± 14	21 ± 15	<0.001
IGFBP-3 (mg/L), mean ± SD, *n* = 828	3.9 ± 1.0	3.6 ± 0.9	4.2 ± 1.0	<0.001
IGF-1/IGFBP-3 Molar Ratio, *n* = 749	0.13 ± 0.04	0.15 ± 0.04	0.11 ± 0.03	<0.001

3.2. IGF-Associated Proteins and Mortality: Low IGFBP-3 and High IGF-1/IGFBP-3 Molar Ratio Predict Mortality Risk

In unadjusted analysis, baseline IGF-1 levels were not predictive of mortality risk (Figure 1a–c). Upon adjustment for baseline age, we observed a non-significant trend towards higher mortality hazard with high IGF-1 levels in women (HR = 1.28, 95% CI 0.96–1.71, *p* = 0.09; Figure 2c). High levels of IGFBP-1, compared to low levels, were associated with significantly higher mortality risk in the overall cohort (*p* < 0.001) and among men (*p* < 0.001), (Figure 1d–f), but the associations became non-significant upon adjustment for age and sex (Figure 2a–c). On the other hand, high levels of IGFBP-3 predicted a lower mortality risk in an unadjusted analysis of the overall cohort (*p* < 0.001), as well as in men (*p* = 0.005) and women (*p* = 0.003) (Figure 1g–i). The difference in mortality risk between subjects with high vs. low IGFBP-3 remained significant upon adjusting for age and sex in the overall cohort (HR 0.82, 95% CI 0.680–0.998, *p* = 0.048), while in sex-stratified analysis associations retained the same direction but lost statistical significance (Figure 2a–c). Further adjustment for IGF-1 did not significantly alter the association between IGFBP-3 and mortality in the overall cohort (HR 0.71, 95% CI 0.56–0.89, *p* = 0.003) and strengthened it in women (HR 0.60, 95% CI 0.43–0.84, *p* = 0.003), but the association in men remained non-significant (HR 0.85, 95% CI 0.62–1.17 *p* = 0.32). High IGF-1/IGFBP-3 molar ratio, which is an estimate of circulating free IGF-1, was associated with higher mortality risk in the overall cohort (*p* = 0.002) and in women (*p* = 0.003) in unadjusted analysis (Figure 1j–l), and these associations persisted upon adjustment for age and sex (HR 1.28, 95% CI 1.05–1.57, *p* = 0.02 and HR 1.53, 95% CI 1.12–2.09, *p* = 0.007, respectively) (Figure 2a–c). After exclusion of 5 participants (4 males, 1 female) who died during the first year of follow-up, associations between IGF-1 and related proteins with mortality remained largely unchanged (data not shown).

Stratification by median participant baseline age showed consistent associations of IGF-related proteins and mortality between the two age groups (Figure S1), with the exception of IGFBP-1, which was significantly positively associated with mortality in younger (HR 1.60, 95% CI 1.04–2.46, *p* = 0.03) but not in older participants (HR 0.98, 95% CI 0.78–1.22, *p* = 0.85). High IGF-1, on the other hand, was more strongly associated with mortality among older women (HR 1.47, 95% CI 1.03–2.08, *p* = 0.03) compared with younger women (HR 1.10, 95% CI 0.64–1.87, *p* = 0.73).

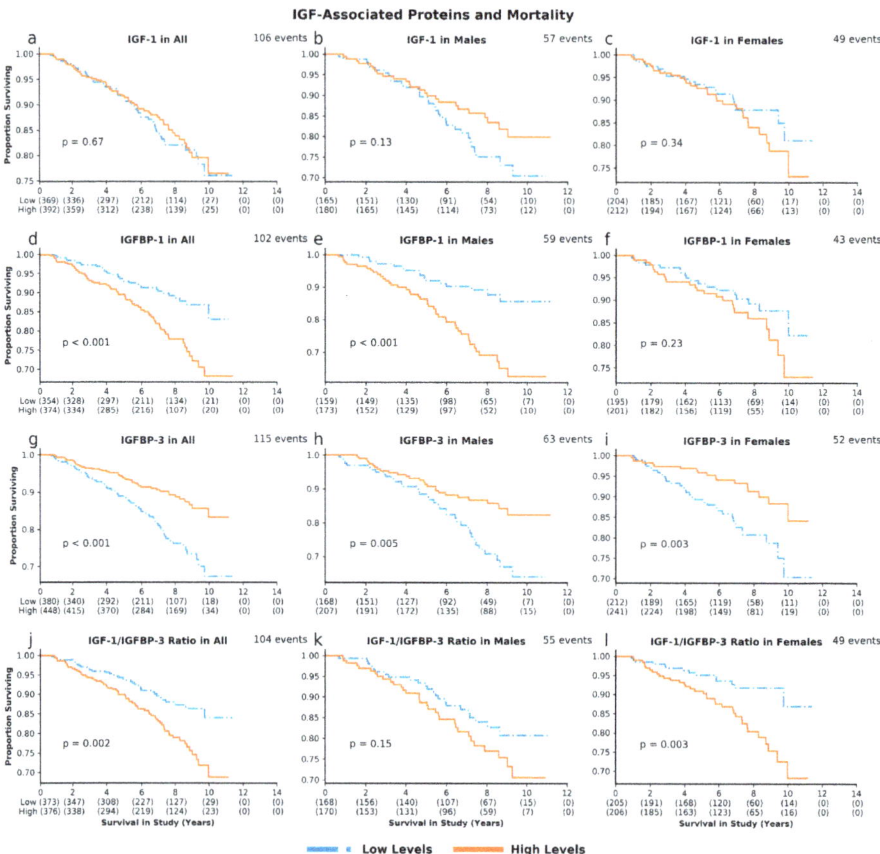

Figure 1. Insulin-like growth factor (IGF)-associated proteins and mortality. Unadjusted survival curves for individuals with high and low levels of IGF-1 (**a**. combined cohort; **b**. males; **c**. females), IGFBP-1 (**d–f**), IGFBP-3 (**g–i**), and IGF-1/IGFBP-3 molar ratio (**j–l**).

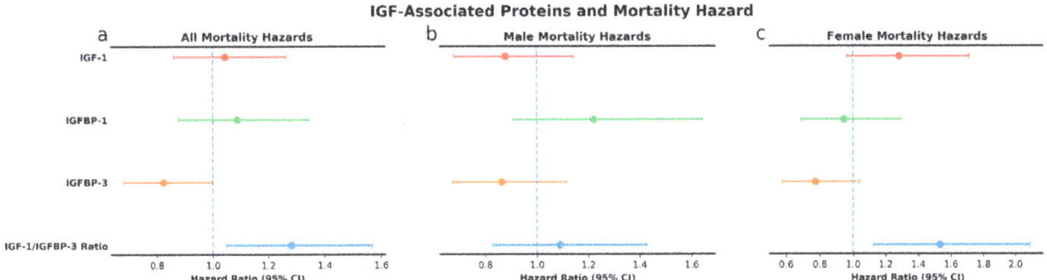

Figure 2. IGF-associated proteins and mortality hazard. Sex and age-adjusted survival hazards for combined cohort (**a**) and age-adjusted survival hazard for males (**b**) and females (**c**) with high levels of IGF-associated proteins as compared to individuals with low levels.

3.3. IGF-Associated Proteins and Morbidity: High IGF-1 Predicts Risk for MDCI and Age-Related Composite Morbidity while Low IGFBP-1 Predicts Risk for Diabetes

High IGF-1 levels, compared to low IGF-1 levels, were associated with greater risk for incident MDCI in the overall cohort ($p = 0.04$) and in men ($p = 0.045$). These associations remained significant after adjusting for baseline age and sex, with HR 1.56, 95% CI 1.08–2.26, $p = 0.02$ and HR 1.81, 95% CI 1.04–3.16, $p = 0.04$ for MDCI in the overall cohort and in men, respectively. Similarly, we observed a greater risk with high IGF-1 for composite incident morbidity in the overall cohort ($p = 0.04$) and in men ($p = 0.03$) (Figure 3), which remained significant after adjustments (HR 1.242, 95% CI 1.004–1.538, $p = 0.046$ and HR 1.44, 95% CI 1.04–2.01, $p = 0.03$, respectively; Figure 4). High IGFBP-1, compared to low IGFBP-1 level, was also associated with higher risk for incident MDCI in men ($p = 0.004$), but not in the overall cohort ($p = 0.11$) or in women ($p = 0.43$), (Figure 5). After adjusting for age and sex, the association between IGFBP-1 level and MDCI hazard became non-significant (Figure 6). On the other hand, high IGFBP-1 was associated with reduced diabetes risk in the overall cohort in unadjusted analysis ($p = 0.01$), (Figure 5). In age- and sex-adjusted analysis, high IGFBP-1 remained significantly associated with protection from incident diabetes in the overall cohort (HR 0.50, 95% CI 0.29–0.88, $p = 0.02$) and in men (HR 0.31, 95% CI 0.10–0.92, $p = 0.03$; Figure 6). These associations persisted upon inclusion of IGF-1 as a covariate in the age- and sex-adjusted model (HR 0.50, 95% CI 0.29–0.89, $p = 0.02$ in the overall cohort; HR 0.30, 95% CI 0.10–0.89, $p = 0.03$ in men). However, when body mass index (BMI) and insulin levels were added to the model, the association between IGFBP-1 and diabetes was attenuated and no longer significant (HR 0.47, 95% CI 0.21–1.03, $p = 0.06$ in the overall cohort; HR 0.66, 95% CI 0.26–1.67, $p = 0.38$ in men). Levels of IGFBP-3 were not significantly associated with risk of any of the investigated age-related diseases (Figures S2 and S3). Associations between high IGF-1/IGFBP-3 ratio and MDCI risk were in the same directions as those between IGF-1 and MDCI, but they reached statistical significance only among women in both unadjusted ($p = 0.04$) and age-adjusted analyses (HR 1.81, 95% CI 1.03–3.21, $p = 0.04$), (Figures S4 and S5).

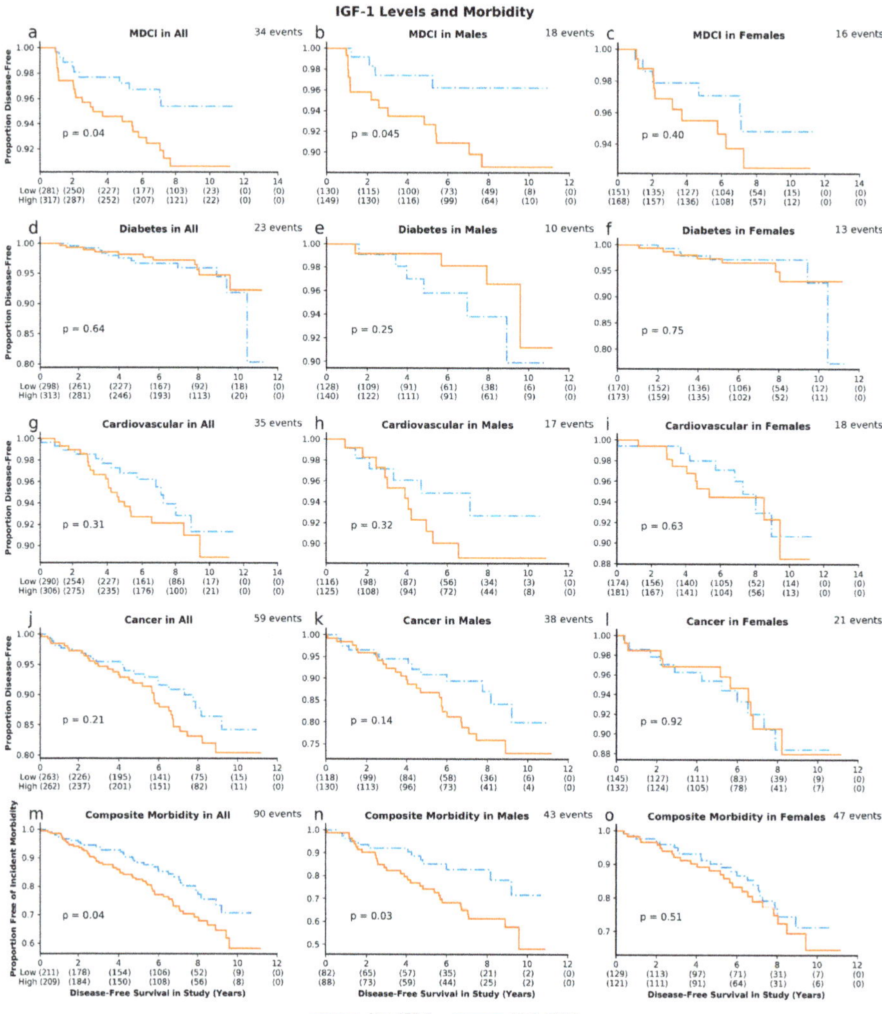

Figure 3. IGF-1 levels and morbidity. Unadjusted survival curves for multiple-domain cognitive impairment (MDCI) (**a**. combined cohort; **b**. males; **c**. females), diabetes (**d–f**), cardiovascular disease (**g–i**), cancer (**j–l**), and composite incident morbidity (**m–o**) in individuals with high and low levels of IGF-1.

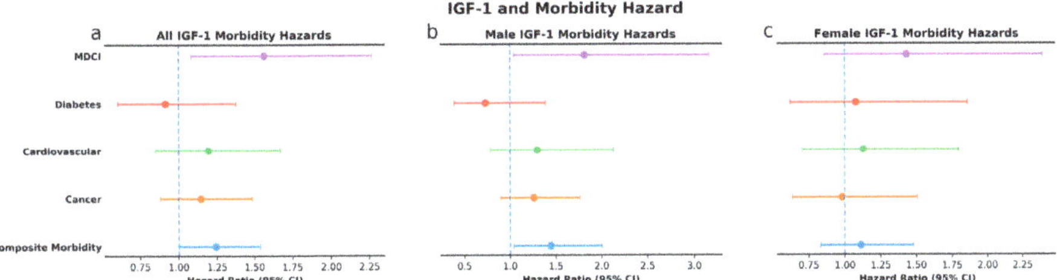

Figure 4. IGF-1 and morbidity hazard. Sex and age-adjusted morbidity hazards for all individuals in cohort (**a**), and age-adjusted morbidity hazards for males (**b**) and females (**c**) with high levels of IGF-1 as compared to individuals with low levels.

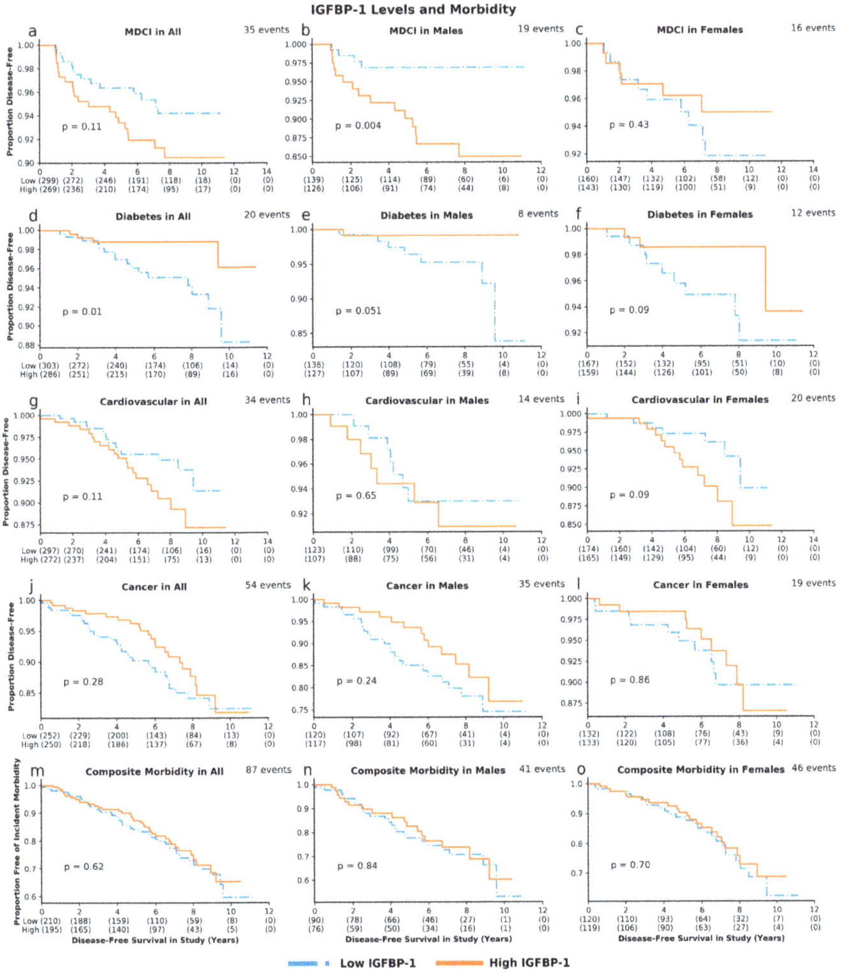

Figure 5. IGFBP-1 levels and morbidity. Unadjusted survival curves for MDCI (**a**. combined cohort; **b**. males; **c**. females), diabetes (**d–f**), cardiovascular disease (**g–i**), cancer (**j–l**), and composite incident morbidity (**m–o**) in individuals with high and low levels of IGFBP-1.

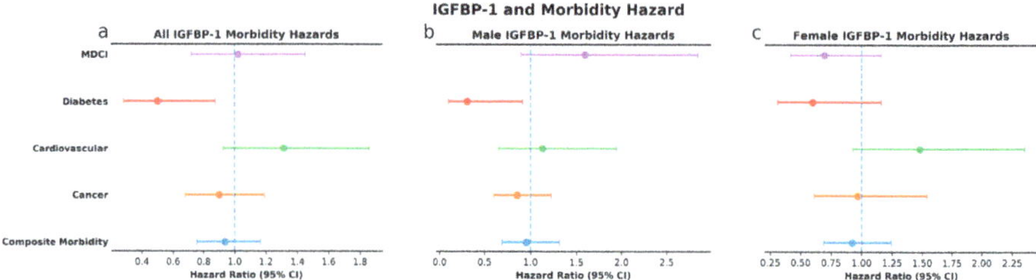

Figure 6. IGFBP-1 and morbidity hazard. Sex and age-adjusted morbidity hazards for all individuals (**a**), and age-adjusted morbidity hazard for males (**b**) and females (**c**) with high levels of IGFBP-1 as compared to individuals with low levels.

4. Discussion

In a longitudinal cohort of independently living older adults in generally good health, we found that IGF-1 and associated proteins predicted all-cause mortality and incidence of age-related diseases, including MDCI, diabetes, and composite incident morbidity. High IGF-1/IGFBP-3 molar ratio, which is considered an estimate of bioavailable IGF-1 [21], predicted a 28% greater risk of mortality, while high baseline IGF-1 level predicted a 56% greater risk for MDCI and a 24% greater risk for composite incident morbidity. The rationale for studying a morbidity composite that includes several major diseases is that aging is a risk factor for all age-related diseases. Therefore, a biological process that accelerates aging is expected to increase the risk for multiple age-associated diseases [47]. Our results confirm findings from model organisms [6–8] and cohorts with exceptional longevity [22,23,41], which demonstrated that attenuated IGF-1 levels or bioavailability were predictive of extended lifespan and health-span. While these findings are consistent with several other epidemiologic studies [48], we provide additional evidence for the role of GH/IGF-1 axis in mortality and morbidity, specifically among older adults. We have also shown that IGFBP-3 and IGFBP-1 predict mortality and diabetes, respectively. This contributes to the growing body of evidence that IGFBPs, in addition to their classical roles in regulating IGF-1 bioavailability, may also exert independent effects on lifespan and health-span.

The results from our study support the theory that diminished IGF-1 levels and bioavailability promote longevity and prolonged health-span in humans. The longevity-promoting mechanisms of diminished GH/IGF-1 signaling are well-studied in animal models and include improved stress defense, autophagy and cell survival via reduced PI3K/Akt and mTOR signaling [49,50]. Data from human studies have also shown that GH receptor deficiency improves defense from oxidative stress in healthy tissues and promotes apoptosis in neoplastic cells [17]. In genetic studies, mutations in the IGF-1 receptor that result in partial IGF-1 resistance [51] and polymorphisms in genes in insulin/IGF-1 signaling pathway [52,53], were associated with exceptional longevity. In exceptionally long-lived human cohorts, our group and others have shown that lower levels of IGF-1 and IGF-1/IGFBP-3 molar ratio predict longer survival [22,23], better cognitive function [41], and better functional status [23]. On the other hand, studies in older individuals who were not of exceptionally old age have shown inconsistent results. While one study in community-dwelling older adults found lower IGF-1 levels to be associated with decreased mortality [48], other studies in individuals with high cardiovascular risk found associations with increased mortality [24,25], or null results [26,38]. Our results offer additional evidence in support of lower IGF-1 levels being associated with reduced mortality and may bring us closer to resolving these inconsistencies.

It is important to note that the same IGF-1 level can represent different physiological states depending on the context and population studied. For instance, IGF-1 levels may be low due to an acute illness [31] or chronic disease [32], which could lead to findings of an inverse association between IGF-1 levels and mortality as a result of "reverse causation". On the other hand, low IGF-1 level may reflect a lifelong diminished IGF-1 signaling due to genetic variants that may confer longevity. In fact, the relationship between IGF-1 levels and mortality may be bimodal in a heterogeneous population, as shown in a meta-analysis that included 12 studies and more than 14,000 subjects [54]. The association of low IGF-1 levels with mortality may reflect the presence of chronic disease, while the association of high IGF-1 levels with mortality might reflect life-long higher IGF-1 exposure. Our cohort was in good overall health, with relatively low prevalence of chronic diseases [55,56] and our findings were confirmed upon exclusion of those who died within the first year of follow-up. Therefore, the associations between IGF-1 levels and mortality were unlikely to be affected by pre-existing comorbid conditions and suggest protection by reduced IGF-1 signaling per se. The age of the cohort should also be taken into consideration. Since IGF-1 levels naturally decline with age [57], a low IGF-1 level in a younger individual may reflect an underlying disease or accelerated aging, whereas a low IGF-1 level in an older individual may reflect healthy physiology. Thus, age-interaction is important to consider in any analysis. Since our cohort was composed only of older individuals, it would not have been surprising not to find interactions between age and IGF-1-associated measures in prediction of mortality risk. However, even in this older cohort (mean age 76.0 for females) we detected a signal for greater hazard of mortality with higher IGF-1 in a subgroup of females above median age (mean age 81.8 years), which further supports the theory that high IGF-1 may be particularly detrimental in older individuals. In our cohort, low baseline IGF-1 bioavailability and levels predicted both delayed occurrence of age-associated morbidities and longer survival, supporting the notion that diminished IGF-1 signaling is associated with delayed aging. Furthermore, by conducting our analysis in a relatively healthy cohort of older age, we minimized many potential confounders.

Higher IGF-1 levels in our cohort were associated with incident MDCI. While the role of IGF-1 system in cognitive aging has been extensively studied, prior findings have not been conclusive [58]. Cross-sectional studies in middle-aged and older individuals have reported both positive [59,60] and negative [41,61] correlations between circulating IGF-1 levels and cognitive performance. Prospective studies have similarly shown conflicting findings. A study in older women, using phone-based neurocognitive assessments, found positive association between baseline IGF-1 and future cognitive performance [62]. The opposite was found in men of similar age in a study that used more comprehensive, in-person neurocognitive evaluations [61]. Some of these conflicting findings may be attributed to heterogeneity between the study populations, methods of cognitive assessment, and definitions of cognitive outcomes [58]. The comprehensive in-person neurocognitive assessments and diagnosis of MDCI, established by a neuropsychologist, increase confidence in the validity of our findings. Furthermore, the biological effects of IGF-1 on the brain may vary depending on age and type of insult [58]. IGF-1 promotes neurogenesis, synaptogenesis, myelination, and cell survival, which are important for brain development and repair after an acute injury [63,64]. On the other hand, IGF-1 increases oxidative stress and inhibits both autophagy and stress responses, leading to diminished cell resilience and accumulation of aberrant proteins and other cellular debris [49,65]. Consistent with these experimental data, interventional trials did not confirm protective cognitive effects of IGF-1 in older adults [66,67]. Presence of high levels of IGF-1 is therefore beneficial for the brain during youth and after an acute insult, but may be detrimental during aging and in evolving neurodegenerative diseases [58], which is supported by our results. As prior studies have shown that older individuals with MDCI with memory involvement have a high rate of conversion to Alzheimer's disease [45], it will be important to further explore the role of IGF-1 in progression from cognitive impairment to Alzheimer's disease in cohorts with larger number of participants and/or longer follow-up.

Our findings reaffirmed the negative association between IGFBP-3 and all-cause mortality previously noted by other studies [33,34,38]. Furthermore, we confirmed that the association between IGFBP-3 and all-cause mortality is independent from IGF-1 levels, as previously suggested [33,34,38]. These epidemiologic observations of IGFBP-3's independent effects are supported by experimental evidence and possibly involve two different mechanisms. First, the functional nuclear localization sequence of IGFBP-3 allows it to enter the nucleus [68], where it has been shown to alter gene expression [69]. Second, IGFBP-3 may bind to a cell-surface receptor lipoprotein receptor-related protein-1 (LRP-1), which was shown to mediate inhibitory effects of IGFBP-3 on cellular growth [70]. Experimental [71,72] and epidemiologic studies [34] suggest that IGFBP-3 may exert its protective effects by reducing cancer-related mortality. However, not all studies have been consistent [73,74] and power limitations in our study precluded cause-specific mortality analysis.

While low levels of IGFBP-1 have been associated with increased risk for diabetes in middle-aged individuals [75–77], our study is among the first to show that low IGFBP-1 levels may predict diabetes in older adults. Circulating IGFBP-1, which is produced mainly by the liver, is normally suppressed in postprandial state by hyperinsulinemia [78] and increased glycolysis [79]. As the levels of IGFBP-1 fluctuate throughout the day in response to feeding and fasting, it acutely regulates the availability of free IGF-1, which has insulin-sensitizing effects [80]. In adipose tissue, IGFBP-1 inhibits IGF-1 stimulated proliferation of preadipocytes [81], resulting in reduced fat mass [82]. At the same time, IGFBP-1 may promote insulin secretion and glucose uptake independently of IGF-1, via binding to a cell-surface integrin receptor [83]. In line with these experimental findings, several cross-sectional and prospective epidemiologic studies have associated low IGFBP-1 levels with obesity [84], high fasting insulin [75], impaired glucose tolerance, and diabetes [75,76,85] in middle-aged individuals. However, prospective data on the association between IGFBP-1 and diabetes risk in older adults are scarce. In our cohort of older adults, we found in an age-adjusted analysis that low IGFBP-1 predicted risk for diabetes. This association persisted after adjusting for IGF-1 but was attenuated with inclusion of BMI and insulin in the model. These findings indicate that the protective effects of high IGFBP-1 against diabetes may be partly mediated by lower BMI and related enhanced insulin sensitivity in individuals with high IGFBP-1.

The sex-stratified analysis highlighted that some of the studied associations may be sex-specific or preferential. For instance, we found that IGF-1/IGFBP-3 molar ratio may be a better predictor of mortality in women than in men, similar to the findings from many rodent [6,10,11] and human [22] studies. It is well known that levels of various elements of the GH/IGF-1 system vary between men and women. Men have higher levels of IGF-1 and IGF-1/IGFBP-3 molar ratio, whereas women have higher levels of IGFBP-3 and IGFBP-1 [21,86]. Women also have diminished physiologic response to GH, which translates into lower IGF-1 levels and alterations in body water and fat content [87]. However, it is not established whether these differences contribute to observed divergence in mortality and morbidity between men and women. Future studies are needed to clarify if hormonal or other sex-specific factors interact with signaling in the GH/IGF-1 pathway.

Although our study possesses many unique strengths, it also has some limitations. Assessment of GH/IGF-1 pathway activity in humans remains challenging due to the inherent complexity of this biological system [36]. Additionally, it has been noted that total IGF-1 level may be an imperfect proxy for bioavailable IGF-1 [88]. However, the high-affinity with which IGF-1 binds to IGFBPs has limited the development of a reliable laboratory assay for measuring free IGF-1 [37]. We therefore used IGF-1/IGFBP-3 molar ratio as an estimate of free IGF-1, similar to a number of previous studies [23,89–91]. IGF-1/IGFBP-3 ratio has been shown to positively correlate with free IGF-1 [92] and has been associated with a number of clinical outcomes, including functional status in nonagenarians [23], metabolic disease [89,90], and neoplastic diseases [91,93]. Regardless of the selected measure, a single measure of IGF-1 level and its associated proteins does not capture adequately the life-long exposure to IGF-1; thus, a longitudinal study with repeated measures would be needed to investigate the role of IGF-1 trajectories in human longevity and health-span. Another important member of the GH/IGF-1 system and the most abundant IGF in circulation is IGF-2. IGF-2, which has been implicated in disease [91,94], binds to the same IGFBPs and receptors as IGF-1, although at lower affinity, and signals via shared pathways with IGF-1 [95]; however, assessment of IGF-2 levels was out of scope of this study. Additionally, due to general good health of our cohort [55,56], there were relatively few incident disease events, which limited our power to study some of the age-related disease outcomes, in particular in sex-stratified analyses. However, the fact that our cohort was in good health allowed us to interpret our findings more reliably in the context of healthy aging and to minimize confounding that may arise from alterations in the GH/IGF-1 axis as a result of disease. Furthermore, all study samples were collected in the morning and under fasting conditions. This was particularly relevant for the interpretation of IGFBP-1 levels, which normally fluctuate in relation to prandial status; yet, not all prior studies have been able to establish these conditions [39].

In conclusion, our findings indicate that higher IGF-1 levels and/or bioavailability are predictive of mortality and morbidity risk. These results support the hypothesis that diminished signaling via GH/IGF-1 pathway may contribute to longevity and health-span in humans. If the detrimental effects of high IGF-1 signaling in older adults are confirmed by larger studies with longer follow-up time, then the GH/IGF-1 pathway may represent a promising target for therapies that delay aging. A monoclonal antibody that targets IGF-1 receptor (IGF-1R) and decreases IGF-1 signaling has already been shown to increase health-span and lifespan in middle-age female mice [96]. In fact, several FDA approved drugs that inhibit GH/IGF-1 signaling are currently in clinical use for other indications. For example, pegvisomant, a growth hormone receptor antagonist, is used for normalizing IGF-1 levels in acromegaly [97] and teprotumumab, which antagonizes IGF-1R, is used to treat thyroid eye disease [98]. These drugs could be readily repurposed for slowing aging in clinical trials. The findings in our study highlight the relevance of this evolutionarily conserved longevity pathway in human aging, and underscore the importance of future studies. In particular, investigating the longitudinal trajectories of circulating IGF-1 and associated proteins and genetically quantifying GH/IGF-1 signaling could serve to strengthen the causal connection between the GH/IGF-1 pathway and human aging.

Supplementary Materials: The following are available online at http://www.mdpi.com/2073-4409/9/6/1368/s1, Table S1: Baseline prevalence of age-related morbidities; Table S2: Baseline Characteristics of High and Low IGF-1 Groups; Figure S1: IGF-Associated Proteins and Mortality Hazard, Age-Stratified; Figure S2: IGFBP-3 Levels and Morbidity; Figure S3: IGFBP-3 and Morbidity Hazard; Figure S4: IGF-1/IGFBP-3 Molar Ratios and Morbidity; Figure S5: IGF-1/IGFBP-3 Molar Ratios and Morbidity Hazard.

Author Contributions: Conceptualization, W.B.Z., S.A., S.M., and N.B.; methodology, W.B.Z., S.A. and S.M.; software, W.B.Z.; validation, W.B.Z. and S.A.; formal analysis, W.B.Z.; investigation, W.B.Z. and S.A.; resources, T.G. and S.M.; data curation, W.B.Z., S.A., T.G., E.F.W., and E.D.; writing—original draft preparation, S.A. and W.B.Z.; writing—review and editing, S.A., W.B.Z., S.M., E.F.W., R.H., N.B., and J.V.; visualization, W.B.Z.; supervision, S.M. and N.B.; project administration, S.M. and T.G.; funding acquisition, S.M., J.V., and N.B. All authors have read and agreed to the published version of the manuscript.

Funding: This research was funded by NIH/NIA K23AG051148 (S.M.), R01AG061155 (S.M.), P30AG038072 (N.B.), R01AG057909 (N.B.), R01AG044829-01A1 (N.B., J.V.), R01AG050448 (R.H.), T32AG023475-18 (W.B.Z.); NIH/NINDS 2R37NS43209 (E.F.W.), 1UG3NS105565–01 (E.F.W.), R01NS109023 (R.H.); NIH/NCATS Einstein/Montefiore CTSA UL1TR002556 (S.A.); American Federation for Aging Research (S.M.); Paul F. Glenn Center for Biology of Aging Research (N.B.) and NY State Department of Health Center of Excellence for Alzheimer's Disease (E.F.W.).

Conflicts of Interest: The authors declare no conflict of interest.

References

1. Nikolich-Zugich, J.; Goldman, D.P.; Cohen, P.R.; Cortese, D.; Fontana, L.; Kennedy, B.K.; Mohler, M.J.; Olshansky, S.J.; Perls, T.; Perry, D.; et al. Preparing for an Aging World: Engaging Biogerontologists, Geriatricians, and the Society. *J. Gerontol. A Biol. Sci. Med. Sci.* **2016**, *71*, 435–444. [CrossRef] [PubMed]

2. Guarente, L.; Kenyon, C. Genetic pathways that regulate ageing in model organisms. *Nature* **2000**, *408*, 255–262. [CrossRef] [PubMed]

3. Sonntag, W.E.; Csiszar, A.; deCabo, R.; Ferrucci, L.; Ungvari, Z. Diverse roles of growth hormone and insulin-like growth factor-1 in mammalian aging: Progress and controversies. *J. Gerontol. A Biol. Sci. Med. Sci.* **2012**, *67*, 587–598. [CrossRef] [PubMed]

4. Kenyon, C.; Chang, J.; Gensch, E.; Rudner, A.; Tabtiang, R. A C. elegans mutant that lives twice as long as wild type. *Nature* **1993**, *366*, 461–464. [CrossRef] [PubMed]

5. Broughton, S.J.; Piper, M.D.; Ikeya, T.; Bass, T.M.; Jacobson, J.; Driege, Y.; Martinez, P.; Hafen, E.; Withers, D.J.; Leevers, S.J.; et al. Longer lifespan, altered metabolism, and stress resistance in Drosophila from ablation of cells making insulin-like ligands. *Proc. Natl. Acad. Sci. USA* **2005**, *102*, 3105–3110. [CrossRef] [PubMed]

6. Brown-Borg, H.M.; Borg, K.E.; Meliska, C.J.; Bartke, A. Dwarf mice and the ageing process. *Nature* **1996**, *384*, 33. [CrossRef]

7.   Ikeno, Y.; Bronson, R.T.; Hubbard, G.B.; Lee, S.; Bartke, A. Delayed occurrence of fatal neoplastic diseases in ames dwarf mice: Correlation to extended longevity. *J. Gerontol. A Biol. Sci. Med. Sci.* **2003**, *58*, 291–296. [CrossRef]

8.   Flurkey, K.; Papaconstantinou, J.; Miller, R.A.; Harrison, D.E. Lifespan extension and delayed immune and collagen aging in mutant mice with defects in growth hormone production. *Proc. Natl. Acad. Sci. USA* **2001**, *98*, 6736–6741. [CrossRef]

9.   Coschigano, K.T.; Clemmons, D.; Bellush, L.L.; Kopchick, J.J. Assessment of growth parameters and life span of GHR/BP gene-disrupted mice. *Endocrinology* **2000**, *141*, 2608–2613. [CrossRef]

10. Holzenberger, M.; Dupont, J.; Ducos, B.; Leneuve, P.; Geloen, A.; Even, P.C.; Cervera, P.; Le Bouc, Y. IGF-1 receptor regulates lifespan and resistance to oxidative stress in mice. *Nature* **2003**, *421*, 182–187. [CrossRef]

11. Selman, C.; Lingard, S.; Choudhury, A.I.; Batterham, R.L.; Claret, M.; Clements, M.; Ramadani, F.; Okkenhaug, K.; Schuster, E.; Blanc, E.; et al. Evidence for lifespan extension and delayed age-related biomarkers in insulin receptor substrate 1 null mice. *FASEB J.* **2008**, *22*, 807–818. [CrossRef]

12. Kinney, B.A.; Meliska, C.J.; Steger, R.W.; Bartke, A. Evidence that Ames dwarf mice age differently from their normal siblings in behavioral and learning and memory parameters. *Horm. Behav.* **2001**, *39*, 277–284. [CrossRef]

13. Ikeno, Y.; Hubbard, G.B.; Lee, S.; Cortez, L.A.; Lew, C.M.; Webb, C.R.; Berryman, D.E.; List, E.O.; Kopchick, J.J.; Bartke, A. Reduced incidence and delayed occurrence of fatal neoplastic diseases in growth hormone receptor/binding protein knockout mice. *J. Gerontol. A Biol. Sci. Med. Sci.* **2009**, *64*, 522–529. [CrossRef]

14. Spadaro, O.; Goldberg, E.L.; Camell, C.D.; Youm, Y.H.; Kopchick, J.J.; Nguyen, K.Y.; Bartke, A.; Sun, L.Y.; Dixit, V.D. Growth Hormone Receptor Deficiency Protects against Age-Related NLRP3 Inflammasome Activation and Immune Senescence. *Cell Rep.* **2016**, *14*, 1571–1580. [CrossRef]

15. Wolf, E.; Kahnt, E.; Ehrlein, J.; Hermanns, W.; Brem, G.; Wanke, R. Effects of long-term elevated serum levels of growth hormone on life expectancy of mice: Lessons from transgenic animal models. *Mech. Ageing Dev.* **1993**, *68*, 71–87. [CrossRef]

16. Bartke, A.; Chandrashekar, V.; Bailey, B.; Zaczek, D.; Turyn, D. Consequences of growth hormone (GH) overexpression and GH resistance. *Neuropeptides* **2002**, *36*, 201–208. [CrossRef]

17. Guevara-Aguirre, J.; Balasubramanian, P.; Guevara-Aguirre, M.; Wei, M.; Madia, F.; Cheng, C.W.; Hwang, D.; Martin-Montalvo, A.; Saavedra, J.; Ingles, S.; et al. Growth hormone receptor deficiency is associated with a major reduction in pro-aging signaling, cancer, and diabetes in humans. *Sci. Transl. Med.* **2011**, *3*, 70ra13. [CrossRef]

18. Colao, A.; Ferone, D.; Marzullo, P.; Lombardi, G. Systemic complications of acromegaly: Epidemiology, pathogenesis, and management. *Endocr. Rev.* **2004**, *25*, 102–152. [CrossRef]

19. Dal, J.; Leisner, M.Z.; Hermansen, K.; Farkas, D.K.; Bengtsen, M.; Kistorp, C.; Nielsen, E.H.; Andersen, M.; Feldt-Rasmussen, U.; Dekkers, O.M.; et al. Cancer Incidence in Patients With Acromegaly: A Cohort Study and Meta-Analysis of the Literature. *J. Clin. Endocrinol. Metab.* **2018**, *103*, 2182–2188. [CrossRef]

20. Dekkers, O.M.; Biermasz, N.R.; Pereira, A.M.; Romijn, J.A.; Vandenbroucke, J.P. Mortality in acromegaly: A metaanalysis. *J. Clin. Endocrinol. Metab.* **2008**, *93*, 61–67. [CrossRef]

21. Friedrich, N.; Wolthers, O.D.; Arafat, A.M.; Emeny, R.T.; Spranger, J.; Roswall, J.; Kratzsch, J.; Grabe, H.J.; Hübener, C.; Pfeiffer, A.F.H.; et al. Age- and Sex-Specific Reference Intervals Across Life Span for Insulin-Like Growth Factor Binding Protein 3 (IGFBP-3) and the IGF-I to IGFBP-3 Ratio Measured by New Automated Chemiluminescence Assays. *J. Clin. Endocrinol. Metab.* **2014**, *99*, 1675–1686. [CrossRef] [PubMed]

22. Milman, S.; Atzmon, G.; Huffman, D.M.; Wan, J.; Crandall, J.P.; Cohen, P.; Barzilai, N. Low insulin-like growth factor-1 level predicts survival in humans with exceptional longevity. *Aging Cell* **2014**, *13*, 769–771. [CrossRef] [PubMed]

23. Van der Spoel, E.; Rozing, M.P.; Houwing-Duistermaat, J.J.; Slagboom, P.E.; Beekman, M.; de Craen, A.J.; Westendorp, R.G.; van Heemst, D. Association analysis of insulin-like growth factor-1 axis parameters with survival and functional status in nonagenarians of the Leiden Longevity Study. *Aging (Albany Ny)* **2015**, *7*, 956–963. [CrossRef] [PubMed]

24. Schutte, A.E.; Conti, E.; Mels, C.M.; Smith, W.; Kruger, R.; Botha, S.; Gnessi, L.; Volpe, M.; Huisman, H.W. Attenuated IGF-1 predicts all-cause and cardiovascular mortality in a Black population: A five-year prospective study. *Eur. J. Prev. Cardiol.* **2016**, *23*, 1690–1699. [CrossRef]

25. Bourron, O.; Le Bouc, Y.; Berard, L.; Kotti, S.; Brunel, N.; Ritz, B.; Leclercq, F.; Tabone, X.; Drouet, E.; Mulak, G.; et al. Impact of age-adjusted insulin-like growth factor 1 on major cardiovascular events after acute myocardial infarction: Results from the fast-MI registry. *J. Clin. Endocrinol. Metab.* **2015**, *100*, 1879–1886. [CrossRef]

26. Kaplan, R.C.; McGinn, A.P.; Pollak, M.N.; Kuller, L.; Strickler, H.D.; Rohan, T.E.; Xue, X.; Kritchevsky, S.B.; Newman, A.B.; Psaty, B.M. Total insulinlike growth factor 1 and insulinlike growth factor binding protein levels, functional status, and mortality in older adults. *J. Am. Geriatr Soc.* **2008**, *56*, 652–660. [CrossRef]

27. Juul, A.; Scheike, T.; Davidsen, M.; Gyllenborg, J.; Jorgensen, T. Low serum insulin-like growth factor I is associated with increased risk of ischemic heart disease: A population-based case-control study. *Circulation* **2002**, *106*, 939–944. [CrossRef]

28. Saber, H.; Himali, J.J.; Beiser, A.S.; Shoamanesh, A.; Pikula, A.; Roubenoff, R.; Romero, J.R.; Kase, C.S.; Vasan, R.S.; Seshadri, S. Serum Insulin-Like Growth Factor 1 and the Risk of Ischemic Stroke: The Framingham Study. *Stroke A J. Cereb. Circ.* **2017**, *48*, 1760–1765. [CrossRef]

29. Renehan, A.G.; Zwahlen, M.; Minder, C.; O'Dwyer, S.T.; Shalet, S.M.; Egger, M. Insulin-like growth factor (IGF)-I, IGF binding protein-3, and cancer risk: Systematic review and meta-regression analysis. *Lancet* **2004**, *363*, 1346–1353. [CrossRef]

30. Kaplan, R.C.; McGinn, A.P.; Pollak, M.N.; Kuller, L.H.; Strickler, H.D.; Rohan, T.E.; Cappola, A.R.; Xue, X.; Psaty, B.M. Association of total insulin-like growth factor-I, insulin-like growth factor binding protein-1 (IGFBP-1), and IGFBP-3 levels with incident coronary events and ischemic stroke. *J. Clin. Endocrinol. Metab.* **2007**, *92*, 1319–1325. [CrossRef]

31. Ross, R.; Miell, J.; Freeman, E.; Jones, J.; Matthews, D.; Preece, M.; Buchanan, C. Critically ill patients have high basal growth hormone levels with attenuated oscillatory activity associated with low levels of insulin-like growth factor-I. *Clin. Endocrinol.* **1991**, *35*, 47–54. [CrossRef] [PubMed]

32. Juul, A. Serum levels of insulin-like growth factor I and its binding proteins in health and disease. *Growth Horm. Igf. Res.* **2003**, *13*, 113–170. [CrossRef]

33. Saydah, S.; Graubard, B.; Ballard-Barbash, R.; Berrigan, D. Insulin-like growth factors and subsequent risk of mortality in the United States. *Am. J. Epidemiol.* **2007**, *166*, 518–526. [CrossRef] [PubMed]

34. Friedrich, N.; Haring, R.; Nauck, M.; Ludemann, J.; Rosskopf, D.; Spilcke-Liss, E.; Felix, S.B.; Dorr, M.; Brabant, G.; Volzke, H.; et al. Mortality and serum insulin-like growth factor (IGF)-I and IGF binding protein 3 concentrations. *J. Clin. Endocrinol. Metab.* **2009**, *94*, 1732–1739. [CrossRef]

35. Jogie-Brahim, S.; Feldman, D.; Oh, Y. Unraveling insulin-like growth factor binding protein-3 actions in human disease. *Endocr. Rev.* **2009**, *30*, 417–437. [CrossRef]

36. Allard, J.B.; Duan, C. IGF-Binding Proteins: Why Do They Exist and Why Are There So Many? *Front. Endocrinol. (Lausanne)* **2018**, *9*, 117. [CrossRef]

37. Faix, J.D. Principles and pitfalls of free hormone measurements. *Best Pract. Res. Clin. Endocrinol. Metab.* **2013**, *27*, 631–645. [CrossRef]

38. Yeap, B.B.; Chubb, S.A.; McCaul, K.A.; Ho, K.K.; Hankey, G.J.; Norman, P.E.; Flicker, L. Associations of IGF1 and IGFBPs 1 and 3 with all-cause and cardiovascular mortality in older men: The Health In Men Study. *Eur. J. Endocrinol. Eur. Fed. Endocr. Soc.* **2011**, *164*, 715–723. [CrossRef]

39. Laughlin, G.A.; Barrett-Connor, E.; Criqui, M.H.; Kritz-Silverstein, D. The prospective association of serum insulin-like growth factor I (IGF-I) and IGF-binding protein-1 levels with all cause and cardiovascular disease mortality in older adults: The Rancho Bernardo Study. *J. Clin. Endocrinol. Metab.* **2004**, *89*, 114–120. [CrossRef]

40. Gubbi, S.; Schwartz, E.; Crandall, J.; Verghese, J.; Holtzer, R.; Atzmon, G.; Braunstein, R.; Barzilai, N.; Milman, S. Effect of Exceptional Parental Longevity and Lifestyle Factors on Prevalence of Cardiovascular Disease in Offspring. *Am. J. Cardiol.* **2017**, *120*, 2170–2175. [CrossRef]

41. Perice, L.; Barzilai, N.; Verghese, J.; Weiss, E.F.; Holtzer, R.; Cohen, P.; Milman, S. Lower circulating insulin-like growth factor-I is associated with better cognition in females with exceptional longevity without compromise to muscle mass and function. *Aging (Albany Ny)* **2016**, *8*, 2414–2424. [CrossRef]

42. The UniProt Consortium. UniProt: A worldwide hub of protein knowledge. *Nucleic Acids Res.* **2019**, *47*, D506–D515. [CrossRef]

43. Rinderknecht, E.; Humbel, R.E. The amino acid sequence of human insulin-like growth factor I and its structural homology with proinsulin. *J. Biol. Chem.* **1978**, *253*, 2769–2776.

44. Johnson, N.B.; Hayes, L.D.; Brown, K.; Hoo, E.C.; Ethier, K.A. CDC National Health Report: Leading causes of morbidity and mortality and associated behavioral risk and protective factors–United States, 2005–2013. *MMWR Suppl.* **2014**, *63*, 3–27.

45. Tabert, M.H.; Manly, J.J.; Liu, X.; Pelton, G.H.; Rosenblum, S.; Jacobs, M.; Zamora, D.; Goodkind, M.; Bell, K.; Stern, Y.; et al. Neuropsychological prediction of conversion to Alzheimer disease in patients with mild cognitive impairment. *Arch. Gen. Psychiatry* **2006**, *63*, 916–924. [CrossRef]

46. Jak, A.J.; Bondi, M.W.; Delano-Wood, L.; Wierenga, C.; Corey-Bloom, J.; Salmon, D.P.; Delis, D.C. Quantification of five neuropsychological approaches to defining mild cognitive impairment. *Am. J. Geriatr. Psychiatry* **2009**, *17*, 368–375. [CrossRef]

47. MacNee, W.; Rabinovich, R.A.; Choudhury, G. Ageing and the border between health and disease. *Eur. Respir. J.* **2014**, *44*, 1332–1352. [CrossRef]

48. Andreassen, M.; Raymond, I.; Kistorp, C.; Hildebrandt, P.; Faber, J.; Kristensen, L.O. IGF1 as predictor of all cause mortality and cardiovascular disease in an elderly population. *Eur. J. Endocrinol. Eur. Fed. Endocr. Soc.* **2009**, *160*, 25–31. [CrossRef]

49. Barzilai, N.; Huffman, D.M.; Muzumdar, R.H.; Bartke, A. The critical role of metabolic pathways in aging. *Diabetes* **2012**, *61*, 1315–1322. [CrossRef]

50. Bartke, A.; Darcy, J. GH and ageing: Pitfalls and new insights. *Best Pract. Res. Clin. Endocrinol. Metab.* **2017**, *31*, 113–125. [CrossRef]

51. Suh, Y.; Atzmon, G.; Cho, M.O.; Hwang, D.; Liu, B.; Leahy, D.J.; Barzilai, N.; Cohen, P. Functionally significant insulin-like growth factor I receptor mutations in centenarians. *Proc. Natl. Acad. Sci. USA* **2008**, *105*, 3438–3442. [CrossRef]

52. Van Heemst, D.; Beekman, M.; Mooijaart, S.P.; Heijmans, B.T.; Brandt, B.W.; Zwaan, B.J.; Slagboom, P.E.; Westendorp, R.G. Reduced insulin/IGF-1 signalling and human longevity. *Aging Cell* **2005**, *4*, 79–85. [CrossRef]

53. Pawlikowska, L.; Hu, D.; Huntsman, S.; Sung, A.; Chu, C.; Chen, J.; Joyner, A.H.; Schork, N.J.; Hsueh, W.C.; Reiner, A.P.; et al. Association of common genetic variation in the insulin/IGF1 signaling pathway with human longevity. *Aging Cell* **2009**, *8*, 460–472. [CrossRef]

54. Burgers, A.M.; Biermasz, N.R.; Schoones, J.W.; Pereira, A.M.; Renehan, A.G.; Zwahlen, M.; Egger, M.; Dekkers, O.M. Meta-analysis and dose-response metaregression: Circulating insulin-like growth factor I (IGF-I) and mortality. *J. Clin. Endocrinol. Metab.* **2011**, *96*, 2912–2920. [CrossRef]

55. Centers for Disease Control and Prevention. *National Diabetes Statistics Report*; Centers for Disease Control and Prevention, U.S. Dept of Health and Human Services: Atlanta, GA, USA, 2020.

56. Villarroel, M.A.B.D.; Jen, A. Tables of Summary Health Statistics for U.S. Adults: 2018 National Health Interview Survey. National Center for Health Statistics, 2019. Available online: http://www.cdc.gov/nchs/nhis/SHS/tables.htm (accessed on 27 April 2020).

57. Yamamoto, H.; Sohmiya, M.; Oka, N.; Kato, Y. Effects of aging and sex on plasma insulin-like growth factor I (IGF-I) levels in normal adults. *Acta Endocrinol.* **1991**, *124*, 497–500. [CrossRef]

58. Gubbi, S.; Quipildor, G.F.; Barzilai, N.; Huffman, D.M.; Milman, S. 40 YEARS of IGF1: IGF1: The Jekyll and Hyde of the aging brain. *J. Mol. Endocrinol.* **2018**, *61*, T171–T185. [CrossRef]

59. Al-Delaimy, W.K.; von Muhlen, D.; Barrett-Connor, E. Insulinlike growth factor-1, insulinlike growth factor binding protein-1, and cognitive function in older men and women. *J. Am. Geriatr. Soc.* **2009**, *57*, 1441–1446. [CrossRef]

60. Wennberg, A.M.V.; Hagen, C.E.; Machulda, M.M.; Hollman, J.H.; Roberts, R.O.; Knopman, D.S.; Petersen, R.C.; Mielke, M.M. The association between peripheral total IGF-1, IGFBP-3, and IGF-1/IGFBP-3 and functional and cognitive outcomes in the Mayo Clinic Study of Aging. *Neurobiol. Aging.* **2018**, *66*, 68–74. [CrossRef]

61. Tumati, S.; Burger, H.; Martens, S.; van der Schouw, Y.T.; Aleman, A. Association between Cognition and Serum Insulin-Like Growth Factor-1 in Middle-Aged & Older Men: An 8 Year Follow-Up Study. *PLoS ONE* **2016**, *11*, e0154450. [CrossRef]

62. Okereke, O.; Kang, J.H.; Ma, J.; Hankinson, S.E.; Pollak, M.N.; Grodstein, F. Plasma IGF-I levels and cognitive performance in older women. *Neurobiol. Aging* **2007**, *28*, 135–142. [CrossRef]

63. Liang, G.; Cline, G.W.; Macica, C.M. IGF-1 stimulates de novo fatty acid biosynthesis by Schwann cells during myelination. *Glia* **2007**, *55*, 632–641. [CrossRef] [PubMed]

64. Nieto-Estevez, V.; Defterali, C.; Vicario-Abejon, C. IGF-I: A Key Growth Factor that Regulates Neurogenesis and Synaptogenesis from Embryonic to Adult Stages of the Brain. *Front. Neurosci.* **2016**, *10*, 52. [CrossRef]

65. Gontier, G.; George, C.; Chaker, Z.; Holzenberger, M.; Aid, S. Blocking IGF Signaling in Adult Neurons Alleviates Alzheimer's Disease Pathology through Amyloid-beta Clearance. *J. Neurosci.* **2015**, *35*, 11500–11513. [CrossRef] [PubMed]

66. Sevigny, J.J.; Ryan, J.M.; van Dyck, C.H.; Peng, Y.; Lines, C.R.; Nessly, M.L. Growth hormone secretagogue MK-677: No clinical effect on AD progression in a randomized trial. *Neurology* **2008**, *71*, 1702–1708. [CrossRef] [PubMed]

67. Friedlander, A.L.; Butterfield, G.E.; Moynihan, S.; Grillo, J.; Pollack, M.; Holloway, L.; Friedman, L.; Yesavage, J.; Matthias, D.; Lee, S.; et al. One year of insulin-like growth factor I treatment does not affect bone density, body composition, or psychological measures in postmenopausal women. *J. Clin. Endocrinol. Metab.* **2001**, *86*, 1496–1503. [CrossRef] [PubMed]

68. Schedlich, L.J.; Le Page, S.L.; Firth, S.M.; Briggs, L.J.; Jans, D.A.; Baxter, R.C. Nuclear import of insulin-like growth factor-binding protein-3 and -5 is mediated by the importin beta subunit. *J. Biol. Chem.* **2000**, *275*, 23462–23470. [CrossRef]

69. Schedlich, L.J.; O'Han, M.K.; Leong, G.M.; Baxter, R.C. Insulin-like growth factor binding protein-3 prevents retinoid receptor heterodimerization: Implications for retinoic acid-sensitivity in human breast cancer cells. *Biochem. Biophys. Res. Commun.* **2004**, *314*, 83–88. [CrossRef]

70. Huang, S.S.; Ling, T.Y.; Tseng, W.F.; Huang, Y.H.; Tang, F.M.; Leal, S.M.; Huang, J.S. Cellular growth inhibition by IGFBP-3 and TGF-beta1 requires LRP-1. *FASEB J.* **2003**, *17*, 2068–2081. [CrossRef]

71. Cohen, P.; Lamson, G.; Okajima, T.; Rosenfeld, R.G. Transfection of the human insulin-like growth factor binding protein-3 gene into Balb/c fibroblasts inhibits cellular growth. *Mol. Endocrinol.* **1993**, *7*, 380–386. [CrossRef]

72. Valentinis, B.; Bhala, A.; DeAngelis, T.; Baserga, R.; Cohen, P. The human insulin-like growth factor (IGF) binding protein-3 inhibits the growth of fibroblasts with a targeted disruption of the IGF-I receptor gene. *Mol. Endocrinol.* **1995**, *9*, 361–367. [CrossRef]

73. Martin, J.L.; Weenink, S.M.; Baxter, R.C. Insulin-like growth factor-binding protein-3 potentiates epidermal growth factor action in MCF-10A mammary epithelial cells. Involvement of p44/42 and p38 mitogen-activated protein kinases. *J. Biol. Chem.* **2003**, *278*, 2969–2976. [CrossRef] [PubMed]

74. Lofqvist, C.; Chen, J.; Connor, K.M.; Smith, A.C.; Aderman, C.M.; Liu, N.; Pintar, J.E.; Ludwig, T.; Hellstrom, A.; Smith, L.E. IGFBP3 suppresses retinopathy through suppression of oxygen-induced vessel loss and promotion of vascular regrowth. *Proc. Natl. Acad. Sci. USA* **2007**, *104*, 10589–10594. [CrossRef] [PubMed]

75. Rajpathak, S.N.; He, M.; Sun, Q.; Kaplan, R.C.; Muzumdar, R.; Rohan, T.E.; Gunter, M.J.; Pollak, M.; Kim, M.; Pessin, J.E.; et al. Insulin-like growth factor axis and risk of type 2 diabetes in women. *Diabetes* **2012**, *61*, 2248–2254. [CrossRef] [PubMed]

76. Petersson, U.; Ostgren, C.J.; Brudin, L.; Brismar, K.; Nilsson, P.M. Low levels of insulin-like growth-factor-binding protein-1 (IGFBP-1) are prospectively associated with the incidence of type 2 diabetes and impaired glucose tolerance (IGT): The Soderakra Cardiovascular Risk Factor Study. *Diabetes Metab.* **2009**, *35*, 198–205. [CrossRef]

77. Lewitt, M.S.; Hilding, A.; Ostenson, C.G.; Efendic, S.; Brismar, K.; Hall, K. Insulin-like growth factor-binding protein-1 in the prediction and development of type 2 diabetes in middle-aged Swedish men. *Diabetologia* **2008**, *51*, 1135–1145. [CrossRef]

78. Suikkari, A.M.; Koivisto, V.A.; Rutanen, E.M.; Yki-Jarvinen, H.; Karonen, S.L.; Seppala, M. Insulin regulates the serum levels of low molecular weight insulin-like growth factor-binding protein. *J. Clin. Endocrinol. Metab.* **1988**, *66*, 266–272. [CrossRef]

79. Snyder, D.K.; Clemmons, D.R. Insulin-dependent regulation of insulin-like growth factor-binding protein-1. *J. Clin. Endocrinol. Metab.* **1990**, *71*, 1632–1636. [CrossRef]

80. Clemmons, D.R. Role of IGF-binding proteins in regulating IGF responses to changes in metabolism. *J. Mol. Endocrinol.* **2018**, *61*, T139–T169. [CrossRef]

81. Siddals, K.W.; Westwood, M.; Gibson, J.M.; White, A. IGF-binding protein-1 inhibits IGF effects on adipocyte function: Implications for insulin-like actions at the adipocyte. *J. Endocrinol.* **2002**, *174*, 289–297. [CrossRef]

82. Rajkumar, K.; Modric, T.; Murphy, L.J. Impaired adipogenesis in insulin-like growth factor binding protein-1 transgenic mice. *J. Endocrinol.* **1999**, *162*, 457–465. [CrossRef]

83. Haywood, N.J.; Cordell, P.A.; Tang, K.Y.; Makova, N.; Yuldasheva, N.Y.; Imrie, H.; Viswambharan, H.; Bruns, A.F.; Cubbon, R.M.; Kearney, M.T.; et al. Insulin-Like Growth Factor Binding Protein 1 Could Improve Glucose Regulation and Insulin Sensitivity Through Its RGD Domain. *Diabetes* **2017**, *66*, 287–299. [CrossRef] [PubMed]

84. Wheatcroft, S.B.; Kearney, M.T. IGF-dependent and IGF-independent actions of IGF-binding protein-1 and -2: Implications for metabolic homeostasis. *Trends Endocrinol. Metab.* **2009**, *20*, 153–162. [CrossRef] [PubMed]

85. Heald, A.H.; Cruickshank, J.K.; Riste, L.K.; Cade, J.E.; Anderson, S.; Greenhalgh, A.; Sampayo, J.; Taylor, W.; Fraser, W.; White, A.; et al. Close relation of fasting insulin-like growth factor binding protein-1 (IGFBP-1) with glucose tolerance and cardiovascular risk in two populations. *Diabetologia* **2001**, *44*, 333–339. [CrossRef] [PubMed]

86. Unden, A.L.; Elofsson, S.; Brismar, K. Gender differences in the relation of insulin-like growth factor binding protein-1 to cardiovascular risk factors: A population-based study. *Clin. Endocrinol.* **2005**, *63*, 94–102. [CrossRef]

87. Span, J.P.; Pieters, G.F.; Sweep, F.G.; Hermus, A.R.; Smals, A.G. Gender differences in rhGH-induced changes in body composition in GH-deficient adults. *J. Clin. Endocrinol. Metab.* **2001**, *86*, 4161–4165. [CrossRef] [PubMed]

88. Frystyk, J. Free insulin-like growth factors – measurements and relationships to growth hormone secretion and glucose homeostasis. *Growth Horm. IGF Res.* **2004**, *14*, 337–375. [CrossRef]

89. Drogan, D.; Schulze, M.B.; Boeing, H.; Pischon, T. Insulin-Like Growth Factor 1 and Insulin-Like Growth Factor-Binding Protein 3 in Relation to the Risk of Type 2 Diabetes Mellitus: Results From the EPIC-Potsdam Study. *Am. J. Epidemiol.* **2016**, *183*, 553–560. [CrossRef]

90. Sierra-Johnson, J.; Romero-Corral, A.; Somers, V.K.; Lopez-Jimenez, F.; Malarstig, A.; Brismar, K.; Hamsten, A.; Fisher, R.M.; Hellenius, M.L. IGF-I/IGFBP-3 ratio: A mechanistic insight into the metabolic syndrome. *Clin. Sci.* **2009**, *116*, 507–512. [CrossRef]

91. Gao, Y.; Katki, H.; Graubard, B.; Pollak, M.; Martin, M.; Tao, Y.; Schoen, R.E.; Church, T.; Hayes, R.B.; Greene, M.H.; et al. Serum IGF1, IGF2 and IGFBP3 and risk of advanced colorectal adenoma. *Int. J. Cancer* **2012**, *131*, E105–E113. [CrossRef]

92. Marzullo, P.; Di Somma, C.; Pratt, K.L.; Khosravi, J.; Diamandis, A.; Lombardi, G.; Colao, A.; Rosenfeld, R.G. Usefulness of different biochemical markers of the insulin-like growth factor (IGF) family in diagnosing growth hormone excess and deficiency in adults. *J. Clin. Endocrinol. Metab.* **2001**, *86*, 3001–3008. [CrossRef]

93. Soubry, A.; Il'yasova, D.; Sedjo, R.; Wang, F.; Byers, T.; Rosen, C.; Yashin, A.; Ukraintseva, S.; Haffner, S.; D'Agostino, R., Jr. Increase in circulating levels of IGF-1 and IGF-1/IGFBP-3 molar ratio over a decade is associated with colorectal adenomatous polyps. *Int. J. Cancer* **2012**, *131*, 512–517. [CrossRef] [PubMed]

94. Cheng, Y.W.; Idrees, K.; Shattock, R.; Khan, S.A.; Zeng, Z.; Brennan, C.W.; Paty, P.; Barany, F. Loss of imprinting and marked gene elevation are 2 forms of aberrant IGF2 expression in colorectal cancer. *Int. J. Cancer* **2010**, *127*, 568–577. [CrossRef] [PubMed]

95. Bergman, D.; Halje, M.; Nordin, M.; Engstrom, W. Insulin-like growth factor 2 in development and disease: A mini-review. *Gerontology* **2013**, *59*, 240–249. [CrossRef] [PubMed]

96. Mao, K.; Quipildor, G.F.; Tabrizian, T.; Novaj, A.; Guan, F.; Walters, R.O.; Delahaye, F.; Hubbard, G.B.; Ikeno, Y.; Ejima, K.; et al. Late-life targeting of the IGF-1 receptor improves healthspan and lifespan in female mice. *Nat. Commun.* **2018**, *9*, 2394. [CrossRef] [PubMed]

97. Van der Lely, A.J.; Kopchick, J.J. Growth hormone receptor antagonists. *Neuroendocrinology* **2006**, *83*, 264–268. [CrossRef]

98. Douglas, R.S.; Kahaly, G.J.; Patel, A.; Sile, S.; Thompson, E.H.Z.; Perdok, R.; Fleming, J.C.; Fowler, B.T.; Marcocci, C.; Marino, M.; et al. Teprotumumab for the Treatment of Active Thyroid Eye Disease. *N. Engl. J. Med.* **2020**, *382*, 341–352. [CrossRef]

Review

Understanding IGF-II Action through Insights into Receptor Binding and Activation

Andrew J. Blyth [1], Nicholas S. Kirk [2,3] and Briony E. Forbes [1,*]

[1] Department of Medical Biochemistry, Flinders Health and Medical Research Institute, Flinders University, Bedford Park 5042, Australia; andrew.blyth@flinders.edu.au
[2] Walter and Eliza Hall Institute of Medical Research, Parkville 3052, Victoria, Australia; kirk.n@wehi.edu.au
[3] Department of Medical Biology, Faculty of Medicine, Dentistry and Health Sciences, University of Melbourne, Parkville 3050, Victoria, Australia
[*] Correspondence: briony.forbes@flinders.edu.au

Received: 18 September 2020; Accepted: 7 October 2020; Published: 12 October 2020

Abstract: The insulin-like growth factor (IGF) system regulates metabolic and mitogenic signaling through an intricate network of related receptors and hormones. IGF-II is one of several hormones within this system that primarily regulates mitogenic functions and is especially important during fetal growth and development. IGF-II is also found to be overexpressed in several cancer types, promoting growth and survival. It is also unique in the IGF system as it acts through both IGF-1R and insulin receptor isoform A (IR-A). Despite this, IGF-II is the least investigated ligand of the IGF system. This review will explore recent developments in IGF-II research including a structure of IGF-II bound to IGF-1R determined using cryo-electron microscopy (cryoEM). Comparisons are made with the structures of insulin and IGF-I bound to their cognate receptors. Finally discussed are outstanding questions in the mechanism of action of IGF-II with the goal of developing antagonists of IGF action in cancer.

Keywords: IGF-II; insulin-like growth factor; IGF-1R; insulin receptor; IR-A; structural studies; receptor activation

1. Introduction

The insulin-like growth factor (IGF) system controls metabolic and mitogenic responses in mammalian cells and importantly regulates embryonic growth and development as well as adult growth [1]. The IGF system is regulated by three structurally similar ligands, IGF-I, IGF-II and insulin (Figure 1). These ligands act via one or more of the three related receptor tyrosine kinases: the two splice variants of the insulin receptor (IR-A and IR-B) and the type 1 Insulin-like growth factor receptor (IGF-1R). IR-B signaling is responsible for the classic IR metabolic activities. IGF-II is unique in that it can activate both IGF-1R and IR-A to promote cell growth and survival. However, of the three ligands, the molecular mechanisms underlying IGF-II action are the least understood. For this reason, this review will focus on IGF-II. There is some evidence to suggest that IGF-II, IGF-I and insulin can promote shared and unique signaling outcomes through IGF-1R and IR [2,3]. However, IGF-II specific actions are generally attributed to tissue specific expression. This review will highlight new discoveries regarding IGF-II, including a cryo-electron microscopy (cryoEM) structure of IGF-II bound to IGF-1R that has provided vital information on the structure and function of IGF-II.

IGF-II plays important roles in fetal growth and development, when it is most abundant [4,5]. Notably, IGF-II fetal plasma concentrations are several fold higher than that of IGF-I [6]. Knockout of *Igf2* leads to a 60% reduction in weight at birth [7]. IGF-II serum concentrations in many mammalian species decline rapidly after birth [8–10]. Interestingly, in adult mice, IGF-II serum levels are barely detectable, whereas in humans it is the more abundant IGF ligand [11]. In humans, the *IGF2* is maternally imprinted and only expressed from the paternal allele. Gain of methylation at the regulatory H19 locus on the paternal allele causes underexpression of IGF-II and results in undergrowth syndromes (Russell-Silver syndrome), which can include a variety of phenotypes including prenatal growth deficiency, facial dysmorphic features and developmental delay [12,13]. Alternatively, overexpression of IGF-II can produce an overgrowth syndrome (Beckwith–Wiedemann syndrome), which can include macroglossia, macrosomia, and abdominal wall defects [12,13]. For a detailed review of the genetic regulation of IGF-II expression in physiology and disease refer to [14].

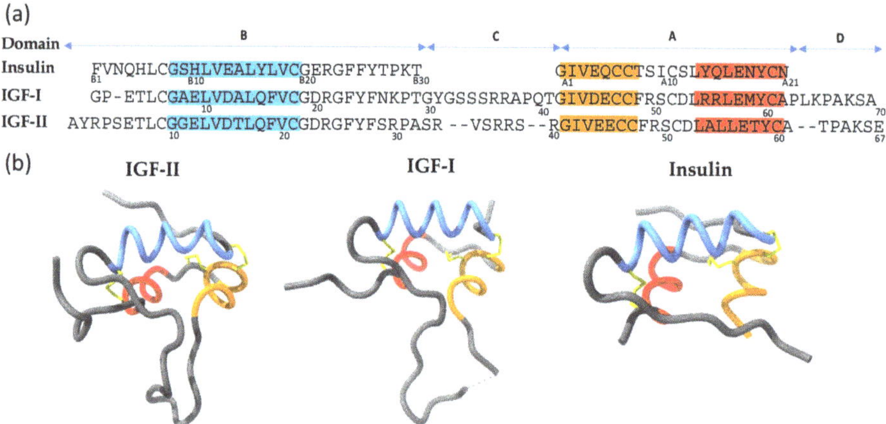

Figure 1. (**a**) Sequence alignment of IGF-II, IGF-I and insulin proteins. Domains are indicated above. Each peptide has three alpha helices; B-chain helix 1 (Blue), A-chain helix 2 (orange) and A-chain helix 3 (red). Residue numbers are indicated below each sequence. (**b**) Ribbon structures of the IGF and insulin proteins (PDB: 1IGL, 1GZR and 1MSO respectively). The three disulfide bonds in each protein are represented in yellow.

At the tissue level, IGF-II promotes cell growth and survival. It regulates bone growth by promoting proper timing of chondrocyte maturation and perichondrial cell differentiation and survival [15]. Overexpression of *Igf2* in smooth muscle and pancreatic beta cells results in the development of cardiovascular defects and type 2 diabetes [16,17]. Conversely, knockout of placental *Igf2* leads to reduced placental growth and fetal growth restriction [18]. IGF-II is most abundant in the fetal and adult brain, primarily produced by the choroid plexus but also the leptomeninges and endothelial cells [19–23]. IGF-II has been identified in cerebral spinal fluid and has been found to promote neurogenesis in the subventricular and subgranular zone of the adult brain [24–26]. Several investigations have also identified that IGF-II promotes stem cell self-renewal through activation of IR-A. For example, IGF-II:IR-A signaling supports neural stem cell maintenance and the expansion of neural progenitor cells [27]. This role in stem cell renewal extends to other tissues, as identified using stem cell specific knockout of *Igf2* in young adults in which growth of intestinal stem cells is also inhibited [28].

IGF-II action is highly regulated by its interaction with soluble IGF binding proteins, including IGF-II specific IGFBP-6. IGFBPs retain IGF-II in circulation and deliver it to target tissues [29]. In addition, the type 2 IGF receptor (IGF-2R, also called cation-independent mannose-6-phosphate receptor) is responsible for the control of circulating IGF-II levels, by binding to IGF-II with high affinity and targeting it for lysosomal degradation [30,31].

2. IGF-II and Cancer

It is well established that abnormal function of the IGF system promotes growth and metastasis of the 3 most commonly diagnosed cancers: breast, prostate and colorectal [32–34]. It also promotes growth and survival of brain, thyroid and ovarian cancers among others [11,14]. Specifically, dysregulation of IGF-II expression has been associated with cancer progression [11]. IGF-II expression is often upregulated in these cancers [33–35] and often results in both autocrine and paracrine effects [36]. For example, in the MDA-MB-157 breast cancer cell line, autocrine production of IGF-II stimulates cell growth though IR-A activation while expression in stromal and epithelial tissue of breast cancer specimens acts in both autocrine and paracrine manners [37]. Loss of imprinted IGF-II expression has been documented in many forms of cancer, leading to increased levels of intratumoural IGF-II, thereby promoting cell growth and tumorigenesis [34,38,39]. Interestingly, the mechanism by which loss of imprinting occurs has recently been investigated and found to involve overexpression of an intronic miRNA (miR-483-5p) found within the *IGF2* gene [40]. miR-483-5p increases IGF-II transcription at the fetal promoter [40].

In cancer, IGF-II can act via IGF-1R and/or IR-A and these autocrine/paracrine signaling loops are regularly observed [41]. IGF-1R, which promotes cell growth and survival, is also commonly upregulated in cancers such as breast, colorectal and prostate cancer [35,41,42]. In contrast to IR-B that signals through metabolic pathways, IR-A has mitogenic signaling capabilities that are important during development when IR-A is most abundantly expressed [33]. IR-A is only expressed at very low levels in most adult cells [43]. However, in malignant cells, including breast, thyroid, colon and prostate cancer, IR is over expressed, and IR-A is the predominant isoform [33,44]. IGF-II:IR-A signaling also supports maintenance of tumour stem and progenitor cells [45,46]. Concomitant upregulation of both IGF-II and IR-A signaling thus provides cancer cells and tumour stem cells with an additional growth and survival mechanism [11].

3. IGF-II Signaling

The biological processes that IGF-II promotes result from activation of signaling pathways through its binding to the extracellular region of IR-A or IGF-1R. The overall mechanisms of binding of IGF-II, IGF-I and insulin to IGF-1R and IR are conserved. Receptor binding results in structural rearrangement of the receptor (further discussed below) causing autophosphorylation of the tyrosine kinase (TK) domains on the intracellular region of the receptor [47,48]. Extensive studies conducted by Cabail et al. [49] have determined that in the unbound state, each monomer is autoinhibited by self-interaction of the activation loop within its TK active site, thereby precluding the binding of ATP. Upon ligand binding, structural rearrangement occurs allowing the juxtamembrane (JM) domain of one monomer to interact with the TK domain of the opposite monomer. This releases the autoinhibitory state and allows for the binding of ATP and subsequent substrate phosphorylation.

The first signaling step upon IGF-II, IGF-I, and insulin binding to their cognate receptors involves phosphorylation of three tyrosine residues within the activation loop of the TK domain (IGF-1R: Y1131, Y1135, and Y1136 and IR: Y1158, Y1162 and Y1163) [47,50]. Subsequently, residue Y950 (IGF-1R) or its equivalent Y960 (IR) is phosphorylated [51]. This creates a docking site for IR substrates (IRS) and Shc (Figure 2), which are then phosphorylated [52]. Subsequent to receptor activation, two main signaling pathways are activated, the phosphoinositide 3-kinase (PI3K)-Akt/protein kinase B (PKB) pathway, responsible for metabolic responses and the Ras-mitogen-activated protein kinase (MAPK) pathway, resulting in mitogenic responses (cell growth, differentiation, and gene expression) [48,53,54].

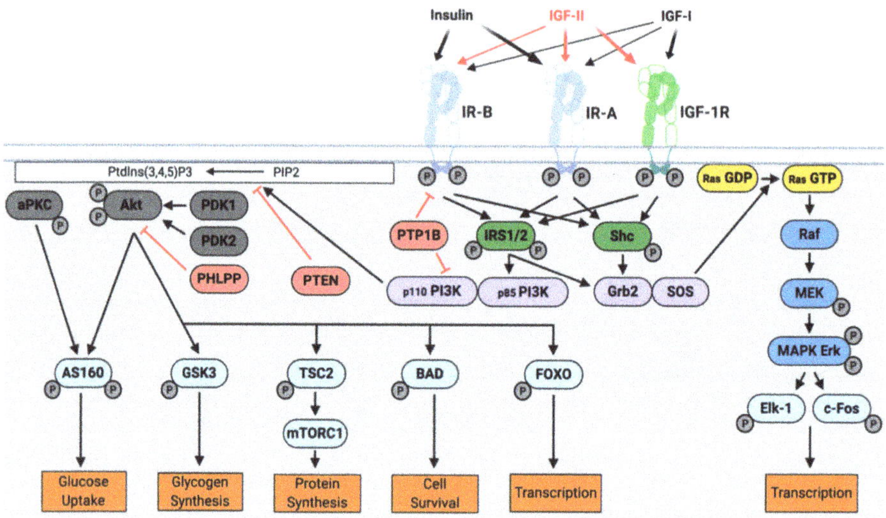

Figure 2. The Insulin and IGF system. Insulin, IGF-I and IGF-II bind with different affinities to IR-B, IR-A and IGF-1R (indicatd by thickness of arrows). IGF-II binds with high affinity to both IGF-1R and IR-A, and with low affinity to IR-B. Upon receptor binding, a structural change leads to activation of the intracellular tyrosine kinase domain and autophosphorylation (indicated by P). IRS1/2 and Shc adapter proteins are recruited and two main signaling pathways are activated: the Akt/PKB and the Ras/MAPK pathways. Metabolic and mitogenic activities are promoted, respectively. (Adapted from: [48]).

4. How does IGF-II Bind and Activate IGF-1R and IR-A?

In order to understand how IGF-II promotes normal cell growth and survival and to develop ways to inhibit its action in cancer, a detailed knowledge of the molecular mechanisms underlying IGF-II receptor binding and activation is required. Our understanding so far has largely been derived through site-directed mutagenesis and comparative structural studies, with a recent cryoEM study revealing the structure of IGF-II bound to IGF-1R. The details of our current understanding will now follow.

4.1. IGF-II Structure

IGF-II is a 67 amino acid single chain polypeptide with sequence and structural similarity to IGF-I (70 amino acids) and insulin (51 amino acid two-chain peptide) (Figure 1). Sequence alignments of the IGFs and insulin (Figure 1a) reveal 50% sequence homology between the B- and A-domains of the IGFs and the equivalent domains of insulin [1]. Three intrachain disulfide bonds hold together the specific three-dimensional structure, which comprises three α-helices (Figure 1b). IGF-I and IGF-II each comprise four domains: B, C, A and D [55]. Insulin, in contrast, is a two-chained mature protein composed of A and B domains joined together by two inter-chain disulfide bonds and having one intra-chain disulfide bond within the A chain (Figure 1b) [56].

IGF-II contacts the receptor through two surfaces originally defined by site-directed mutagenesis that are named site 1 and site 2. Equivalent residues of IGF-I and insulin are involved in binding IGF-1R and IR, respectively (Table 1).

4.2. Receptor Structure, Mechanism of Binding and Activation

IR-A, IR-B and IGF-1R are similar in amino acid sequence and structure (Figure 3a). The two IR isoforms differ by the expression of exon 11, which consists of 12 amino acids that are absent in IR-A splice variant. The receptors are disulfide-linked (αβ)2 homodimers and the extracellular domains of each αβ monomer assemble in an anti-parallel, Λ-shaped conformation, generating two equivalent ligand binding regions. In the apo (unbound) state, the sites of membrane entry are situated far apart, thereby holding the intracellular tyrosine kinase in an inactive monomeric state (Figure 3b left) [57,58]. Site-directed mutagenesis and structural studies have identified two binding surfaces within each binding region (site 1 and site 2) that represent high- and low-affinity binding sites, respectively. Upon ligand binding, the receptors undergo extensive structural change, whereby the FnIII stalks come close together, permitting dimerization of the intracellular region to release the autoinhibition of the TK domains (Figure 3b right). Notably, such a conformation is as predicted by Kavran et al. [59] to be essential for IGF-1R activation.

Molecular detail of receptor binding in the extracellular domain has been derived from a series of crystallographic and cryoEM studies of IGF-1R and IR in the holo and soluble ectodomain forms. The α-chain C-terminal (αCT) helices of each monomer lie on the L1 surface of the opposing monomers to form site 1 [57,58]. The αCT shifts to accommodate the ligand, which makes contact via its site 1 residues [60,61]. As defined by Weis et al. [62], site 1 contacts made between the ligand and the receptor L1 and αCT domains [62–64]. In the case of IGF-II and IGF-I binding IGF-1R as well as insulin binding IR, the residues identified in site-directed mutagenesis studies correspond to those involved in this site 1 interaction (Table 1). Several additional residues were revealed in these structures to contact the L1 and αCT and can now be defined as site 1 residues (Table 1).

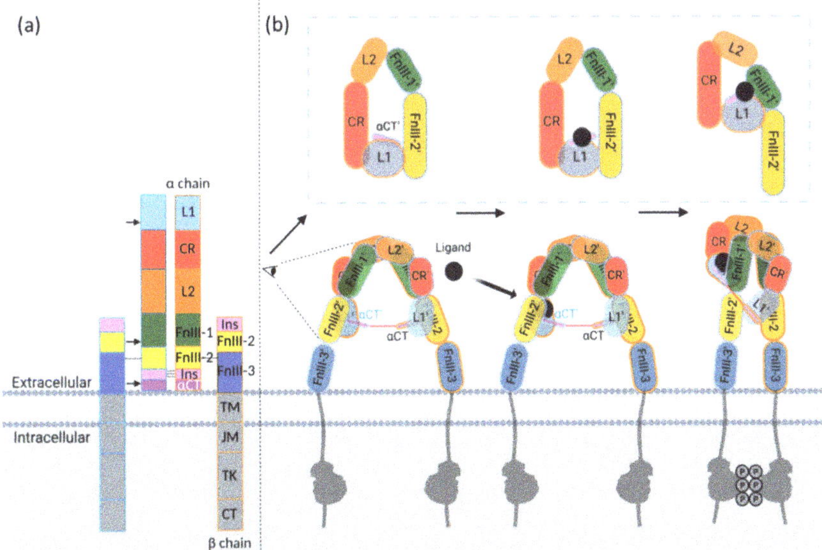

Figure 3. (**a**) Domain structure of IGF-1R and IR receptor tyrosine kinases. Individual αβ monomers are indicated by blue and orange outline. IGF-1R and IR have a high degree of sequence homology and therefore comprise the same domains: first and second leucine-rich repeat domains (L1 and L2), cysteine-rich domain (CR), first, second and third type-III fibronectin -like domains (FnIII-1, 2, and 3), insert domain (ID), α-chain C-terminal domain (αCT), transmembrane domain (TM), juxtamembrane domain, (JM), tyrosine kinase (TK), C-terminal domain (CT). Arrows indicate regions involved in ligand binding. (**b**) Schematic representation of the mechanism of ligand binding. Side view of binding pocket shown in blue dotted box. In the unbound (apo) state (left) the receptor forms an open Λ-shape with FnIII-3 legs situated far apart. Ligand binding is likely to involve a transient interaction (middle) followed by major structural rearrangement forming a J-shape active conformation (right) where the FnIII-3 legs of the receptor are in close proximity. In turn, a structural change occurs in the intracellular domains leading to autophosphorylation by the TK.

Table 1. Binding site 1 and 2 residues of insulin, IGF-I and IGF-II. Residues identified as contacting the receptor through mutagenesis studies (coloured blue), structural studies (red) or both (black). * Residues observed to make transient contact with the FnIII-1 domain of IR [64]. # Residues observed to make transient contact with the FnIII-1 domain of IGF-1R [57]. [a] Asp45Ala IGF-I mutant results in 3-fold decrease in binding affinity [65].

	IGF-II		IGF-I [57,65]		Insulin [64]	
	Cys9		Cys6		CysB7	
	Leu13		Leu10		LeuB11	
	Leu17		Leu14		LeuB15	
	Asp23	[63]	Asp20	[64]	GluB21	[62]
	Ser29		Asn26		ProB28	
	Arg30		Lys27			
	Thr58		Met59		AsnA18	
	Thr62		Lys65			
Site 1	Val14		Val11		ValB12	
	Gln18		Gln15		TyrB16	
	Gly25		Gly22		GlyB23	
	Phe26		Phe23		PheB24	
	Tyr27		Tyr24		PheB25	
	Phe28		Phe25		TyrB26	
			Tyr31			
			Arg36			
		[1,63]	Arg37	[1,64]		[62,65]
	Gly41		Gly42		GlyA1	
	Ile42		Ile43		IleA2	
	Val43		Val44		ValA3	
	-		Asp45 [a]		GluA4	
	Glu45		Glu46		GluA5	
	Phe48		Phe49		ThrA8	
	Tyr59		Tyr60		TyrA19	
	Ala61		Ala62		AsnA21	
Site 2	Glu6	[63]	Glu3	[1]	GlnB4	[62]
	Thr7	[63]	Thr4	[64]	HisB5	[62]
	Cys9	[63]	Cys6	[64]	CysB7	[62]
	Glu12	[63,66]	Glu9 #	[64,67]	HisB10 *	[62,68]
	Asp15	[66]	Asp12	[64,67]	GluB13 *	[62,68]
	Phe19	[66]	Phe16 #	[64,67]	LeuB17 *	[62,68]
	Cys47	[63]	Cys48	[64]	CysA7	[62]
	-		-		SerA12	[62,68]
	Leu53	[66]	Leu54 #	[67]	LeuA13 *	[62,68]
	Glu57	[66]	Glu58 #	[67]	GluA17 *	[62,68]

Recently, a structure of the IGF-II:IGF-1R complex was determined using cryoEM to an average maximum resolution of 3.2 Å (Figure 4b) [63]. The site 1 ligand binding interaction is similar to the previous insulin:IR and IGF-I:IGF-1R structures [64,69,70]. The IGF-II molecule contacts the L1, L2, αCT', and FnIII-1' domains within the head region of the receptor (Figure 4c) [63]. The L1-CR + (αCT') module folds to the top of the receptor, permitting sparse interactions between IGF-II and the membrane- distal loops of FnIII-1', facilitated by an outward rotation of domain L2 from its location in the apo ectodomain. The αCT' helix on the L1 domain surface threads through the IGF-II C-domain loop (residues 33–40). The C-terminal segment of the IGF-II B-domain is displaced from the core of the ligand (in the unbound state) and engages with the receptor to make the site 1 interaction. The B chain of IGF-II is stabilized by an interaction between IGF-II residue Arg30 and the hydroxyl group of IGF-1R residue Tyr28 and possibly a salt bridge between IGF-II residue Arg38 and IGF-1R residue Glu305. The ligand forms a 'clip' on the extended αCT helix in the active conformation, stabilizing a tight interaction between L1-CR-L2 and αCT' with only sparse interactions between the ligand and FnIII-1' (Figure 4c) [63].

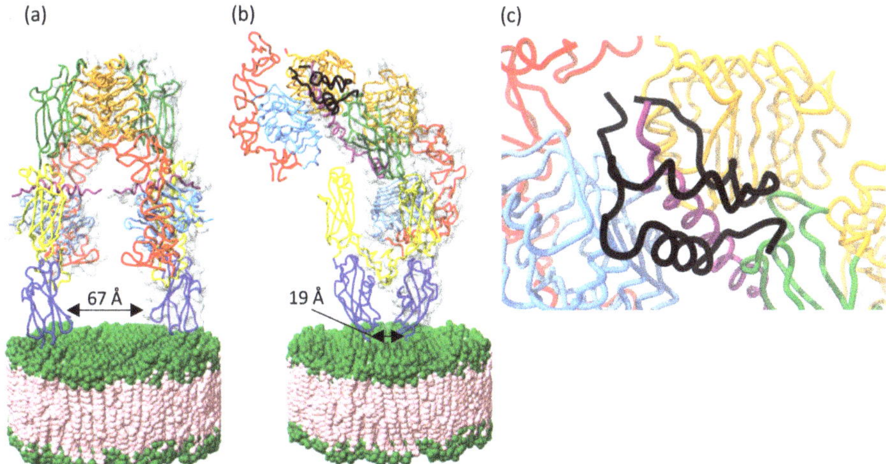

Figure 4. (**a**) Crystal structure of the unbound (apo) IGF-1R ectodomain (PDB: 5U8R). In the apo state the FnIII legs of the receptor are positioned far apart forming an open Λ-shape. Upon IGF-II (black) binding, a major structural rearrangement occurs resulting in a J-shape conformation of the receptor where the FnIII legs are in close proximity. (**b**) The activated conformation (PDB: 6VWI and 6VWJ) is stabilized by the ligand clipping the αCT and L1 domains together, interactions through site 2 on FnIII-1′, and potential salt bridges in the head region facilitated by ligand binding (between Glu687′ (αCT′) and Arg335 (domain L2), between residues Glu693′ (αCT′) and Arg488′ (domain FnIII-1′), and between residues Lys690′ (αCT′) and Asp489′ (domain FnIII-1′)). (**c**) Zoom-in of the site 1 ligand binding region between IGF-II and IGF-1R involving IGF-II B-domain residues; Cys9, Leu13, Val14, Asp15, Leu17, Gln18, Asp23, Phe26, Tyr27, Phe28, Ser29, and Arg30 and the side chains of receptor domain L1 residues Pro5, Ile7, Asp8, Arg10, Asn11, Leu33, Ser35, Ly36, Phe58, and Arg59, and the side chains of receptor αCT′ residues His697′, F701′, Val 702′, and Pro705′. The IGF-II A-domain contacts the receptor αCT′ domain (and not domain L1), with the interaction mediated by the side chains of IGF-II residues Ile42, Val43, Glu44, Phe48, Thr58, Tyr59, and Thr62 and the side chains of receptor αCT′ residues Lys690′, Glu694′, His697′, Asn698′, Phe701′, Val702′, Pro703′, and Arg704′. (**a**–**c**) coloured as in Figure 3.

This overall J-shaped conformation that brings the FnIII stalks together was seen previously in the Weis et al. insulin:IR and Li et al. IGF-I:IGF-1R studies [62,64]. The IGF-II:IGF-1R complex structure [63] was determined using a similar leucine-zippered receptor (IGF-1RZip) to that of the insulin receptor used in the Weis et al. study [62]. The general topology of IGF-1RZip:IGF-II and IRZip:insulin structures also reflects that of the recently reported holoIGF-1R:IGF-I structure [64], providing further evidence that this is the common activated conformation. The asymmetry observed in the activated structure is necessary for negative co-operativity, a hallmark of both IGF-1R and IR ligand binding summarized in a 'harmonic oscillator model' by Kiselyov et al. [71], whereby binding of a second ligand (to the unoccupied receptor binding pocket) accelerates the dissociation of the first bound ligand.

Looking at the molecular detail of the IGF-II:IGF-1R complex structure confirms that the IGF-II site 1 residues identified by site-directed mutagenesis interact with IGF-1R site 1 (summarized in Table 1 and Figure 4). This involves side chains of residues of the B-domain (Cys9, Leu13, Val14, Asp15, Leu17, Gln18, Asp23, Phe26, Tyr27, Phe28, Ser29, and Arg30) contacting the L1 and αCT′ segment and side chains of A-domain residues (Ile42, Val43, Glu44, Phe48, Thr58, Tyr59, and Thr62) contacting the αCT′ segment (but not the L1). These side chain interactions are similar in the IRZip:insulin and holoIGF-1R:IGF-I structures as the ligand sequences are highly conserved in these regions (Table 1).

The major difference in the structures of IGF-1R ectodomain-bound IGF-II and IGF-I occurs in the respective growth factor C-domains. In the receptor complex, the IGF-II C-domain residues 33–36 are disordered, as are the adjacent receptor CR domain residues 258–265, suggesting that the C-domain is too short to form stable interactions with the receptor in this region (Figure 5a) [63]. By contrast, the C-domain of IGF-I in holoIGF-1R:IGF-I is relatively well ordered, with IGF-I residue Tyr31 in its distal loop engaging receptor residues Pro5 and Pro256 (Figure 5a) [64]. Although the resolution of the structure is low at IGF-I residues Arg36 and Arg37, they appear to contact IGF-1R L2 domain. With no equivalent to Tyr31, the IGF-II C-domain instead appears to be stabilized by self-interactions (a salt bridge with IGF-II residue Glu45 near the *N* terminus of the first helix of the IGF-II A-domain, and a polar interaction with the IGF-II residue Ser39).

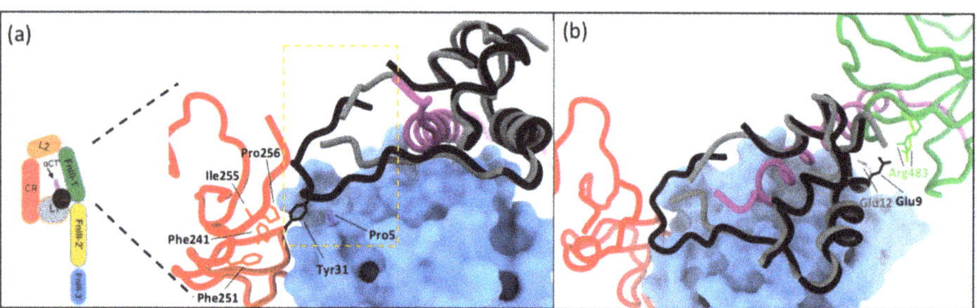

Figure 5. (**a**) Interaction of the C-domain (shown in yellow box) of IGF-II (grey) and IGF-I (black) with IGF-1R (PDB: 6VWI and 6PYH, respectively). Domains of IGF-1R coloured as in Figure 3. Residues of the CR and L1 domain engage with residue Tyr31 of IGF-I. Contact is also made between C-domain residues Arg36 and Arg37 of IGF-I and the L2 domain. There is no equivalent C-domain residue in IGF-II. (**b**) Site 2 contacts involve residues Glu12 of IGF-II and Glu9 IGF-I which contact FnIII-1 domain residue Arg483 (dark green represents the IGF-1:IGF-1R structure and light green the IGF-II:IGF-1R structure). (**a**,**b**) coloured as in Figure 3.

An as yet unexplained observation is the limited correlation of the site-directed mutagenesis data for IGF-II site 2 residues (Table 1) and their involvement in binding in the IGF-II:IGF-1R complex structure (Figure 5b). Of the residues defining site 2 by site-directed mutagenesis, only Glu12 appears to contact the receptor FnIII-1′ in this activated conformation (Figure 5b). In addition, IGF-II B-domain residues Glu6, Thr7, Cys9 and A-domain residues Cys47 and Phe48 are seen to contact the FnIII-1′ in the IGF-II:IGF-1R complex structure, thereby completing the definition of site 2 (Table 1). A similar conundrum was revealed by IGF-I:IGF-1R and insulin:IR complex structures and their corresponding site-directed mutagenesis data (Table 1).

For both IGF-I:IGF-1R and insulin:IR complexes additional structures have been described that have led to a proposed transient interaction of the ligand with a different site on the receptor. This may represent the first site of contact for the ligand or an intermediate site to facilitate conformational change of the ligand and receptor (schematically represented in Figure 3b, middle panel). In the case of insulin, cryoEM structures of insulin-saturated IR constructs [69,70] identified potential transient binding sites on the FnIII-1′ spanning residues Tyr477-488 and 552-554 involving all insulin site 2 residues (Table 1). Such a site has not been reported for IGF:IGF-1R complexes. For IGF-I:IGF-1R, the first ligand-bound ectodomain structure was determined by X-ray crystallography by Xu et al. [58]. This structure was determined by ligand soaking in apo crystals, resulting in an induced fit of IGF-I to the L1- αCT′ binding site. In this structure, the receptor remained in the "apo/legs apart" conformation without the major J-shaped rearrangement. Additional FnIII-2′ contacts (residues 788-792) were observed that involved essentially all IGF-I site 2 residues identified by site-directed mutagenesis except Glu9 and Asp12. It is possible that this interaction represents an IGF-1R transient binding site and suggests a major difference in the activation mechanism between the two receptors. Whether these transient interactions also occur for IGF-II on IGF-1R and IGF-II on IR-A remains to be determined.

In summary, whilst IGF-II binds and activates IGF-1R through a similar mechanism to IGF-I, there are some significant differences that likely explain their different binding affinities. Notably the C-domain interactions are quite different, with IGF-II barely making receptor contact, whereas IGF-I C-domain contributes to binding affinity through several contacts. How this influences ligand specific signaling outcomes is still not understood. Importantly, no structure of IGF-II bound to IR-A has been reported.

5. Conclusions and Implications of Structural Information for Developing Treatments for Disease

IGF-II plays a fundamental role in mammalian growth and fetal development. It is an important regulator of bone growth and promotes cellular growth and survival. While IGF-II is the least investigated ligand of the IGF system, the recently determined structure of IGF-II bound to IGF-1R has certainly advanced our understanding of the mechanism of IGF-II binding and activation. This structural information has confirmed that upon IGF-1R engagement, the receptor undergoes major structural rearrangement, from an open Λ-shape conformation to a J-shaped structure where the legs of the receptor are brought into contact in the active signaling conformation of the receptor. Comparison of IGF-I and IGF-II bound to IGF-1R confirmed that the C-domain of IGF-I contacts the receptor, whereas IGF-II lacks an equivalent contact. While site 1 contacts of IGF-II are in accordance with mutagenesis data, only one site 2 residue is seen to contact the receptor (Glu12) as observed in IGF-I (Glu9) and insulin bound to IR (HisB10). The remaining residues identified by mutagenesis as contacting the receptor may be involved in transient interactions with the receptors. The same transient interaction is expected with IGF-II binding; however, this is yet to be observed.

A detailed understanding of how IGF-II engages with its receptors and confers downstream signaling activation is essential in developing drug therapies that target IGF action in cancer. The relatively minor role of IGF-II in adult cell function means that blocking this pathway as a cancer therapy may have little effect on healthy adult cells whilst slowing cancer cell growth. Currently, most approaches target IGF action by directly blocking binding to IGF-1R:IGF-1R antibodies inhibit ligand binding and stimulate receptor internalisation [34]. Such inhibitors have been shown to reduce growth of IGF-II dependent cancers. However, increases in IGF-II:IR-A signaling can give rise to resistance to treatment [72,73], highlighting the need for inhibitors of IGF-II acting via both IGF-1R and IR-A and the need for structural data of IGF-II bound to IR-A. Such studies will further inform on how IGF-II is uniquely capable of binding and activating both IR-A and IGF-1R with high affinity and will suggest strategies to design inhibitors or allosteric regulators for the treatment of IGF-1R/IR-A regulated disease.

Author Contributions: All authors contributed equally to the writing and review of the manuscript. A.J.B. made all figures. All authors have read and agreed to the published version of the manuscript.

Funding: This research received no external funding.

Acknowledgments: We acknowledge the contribution of Brian Smith (lipid disc structure, Figure 4).

Conflicts of Interest: The authors declare no conflict of interest.

References

1. Denley, A.; Cosgrove, L.J.; Booker, G.W.; Wallace, J.C.; Forbes, B.E. Molecular interactions of the IGF system. *Cytokine Growth Factor Rev.* **2005**, *16*, 421–439. [CrossRef] [PubMed]
2. Versteyhe, S.; Klaproth, B.; Borup, R.; Palsgaard, J.; Jensen, M.; Gray, S.G.; De Meyts, P. IGF-I, IGF-II, and Insulin Stimulate Different Gene Expression Responses through Binding to the IGF-I Receptor. *Front. Endocrinol.* **2013**, *4*, 98. [CrossRef] [PubMed]

3. Morcavallo, A.; Gaspari, M.; Pandini, G.; Palummo, A.; Cuda, G.; Larsen, M.R.; Vigneri, R.; Belfiore, A. Research resource: New and diverse substrates for the insulin receptor isoform A revealed by quantitative proteomics after stimulation with IGF-II or insulin. *Mol. Endocrinol.* **2011**, *25*, 1456–1468. [CrossRef] [PubMed]

4. White, V.; Jawerbaum, A.; Mazzucco, M.B.; Gauster, M.; Desoye, G.; Hiden, U. IGF2 stimulates fetal growth in a sex- and organ-dependent manner. *Pediatric Res.* **2018**, *83*, 183–189. [CrossRef] [PubMed]

5. Randhawa, R.; Cohen, P. The role of the insulin-like growth factor system in prenatal growth. *Mol. Genet. Metab.* **2005**, *86*, 84–90. [CrossRef] [PubMed]

6. Agrogiannis, G.D.; Sifakis, S.; Patsouris, E.S.; Konstantinidou, A.E. Insulin-like growth factors in embryonic and fetal growth and skeletal development (Review). *Mol. Med. Rep.* **2014**, *10*, 579–584. [CrossRef]

7. DeChiara, T.M.; Efstratiadis, A.; Robertson, E.J. A growth-deficiency phenotype in heterozygous mice carrying an insulin-like growth factor II gene disrupted by targeting. *Nature* **1990**, *345*, 78–80. [CrossRef]

8. Stylianopoulou, F.; Efstratiadis, A.; Herbert, J.; Pintar, J. Pattern of the insulin-like growth factor II gene expression during rat embryogenesis. *Development* **1988**, *103*, 497–506.

9. Soares, M.B.; Turken, A.; Ishii, D.; Mills, L.; Episkopou, V.; Cotter, S.; Zeitlin, S.; Efstratiadis, A. Rat insulin-like growth factor II gene: A single gene with two promoters expressing a multitranscript family. *J. Mol. Biol.* **1986**, *192*, 737–752. [CrossRef]

10. Gluckman, P.D.; Butler, J.H. Parturition-related changes in insulin-like growth factors-I and -II in the perinatal lamb. *J. Endocrinol.* **1983**, *99*, 223–232. [CrossRef]

11. Belfiore, A.; Malaguarnera, R.; Vella, V.; Lawrence, M.C.; Sciacca, L.; Frasca, F.; Morrione, A.; Vigneri, R. Insulin Receptor Isoforms in Physiology and Disease: An Updated View. *Endocr. Rev.* **2017**, *38*, 379–431. [CrossRef] [PubMed]

12. Smith, A.C.; Choufani, S.; Ferreira, J.C.; Weksberg, R. Growth regulation, imprinted genes, and chromosome 11p15. 5. *Pediatric Res.* **2007**, *61*, 43–47. [CrossRef] [PubMed]

13. Gicquel, C.; Le Bouc, Y. Hormonal regulation of fetal growth. *Horm. Res.* **2006**, *65* (Suppl. 3), 28–33. [CrossRef]

14. Holly, J.M.P.; Biernacka, K.; Perks, C.M. The Neglected Insulin: IGF-II, a Metabolic Regulator with Implications for Diabetes, Obesity, and Cancer. *Cells* **2019**, *8*, 1207. [CrossRef] [PubMed]

15. Yakar, S.; Werner, H.; Rosen, C.J. Insulin-like growth factors: Actions on the skeleton. *J. Mol. Endocrinol.* **2018**, *61*, T115–T137. [CrossRef]

16. Devedjian, J.C.; George, M.; Casellas, A.; Pujol, A.; Visa, J.; Pelegrín, M.; Gros, L.; Bosch, F. Transgenic mice overexpressing insulin-like growth factor-II in beta cells develop type 2 diabetes. *J. Clin. Investig.* **2000**, *105*, 731–740. [CrossRef]

17. Zaina, S.; Pettersson, L.; Thomsen, A.B.; Chai, C.-M.; Qi, Z.; Thyberg, J.; Nilsson, J. Shortened life span, bradycardia, and hypotension in mice with targeted expression of an Igf2 transgene in smooth muscle cells. *Endocrinology* **2003**, *144*, 2695–2703. [CrossRef]

18. Constância, M.; Hemberger, M.; Hughes, J.; Dean, W.; Ferguson-Smith, A.; Fundele, R.; Stewart, F.; Kelsey, G.; Fowden, A.; Sibley, C.; et al. Placental-specific IGF-II is a major modulator of placental and fetal growth. *Nature* **2002**, *417*, 945–948. [CrossRef]

19. Ferron, S.; Radford, E.; Domingo-Muelas, A.; Kleine, I.; Ramme, A.; Gray, D.; Sandovici, I.; Constancia, M.; Ward, A.; Menheniott, T. Differential genomic imprinting regulates paracrine and autocrine roles of IGF2 in mouse adult neurogenesis. *Nat. Commun.* **2015**, *6*, 1–12. [CrossRef]

20. Charalambous, M.; Menheniott, T.R.; Bennett, W.R.; Kelly, S.M.; Dell, G.; Dandolo, L.; Ward, A. An enhancer element at the Igf2/H19 locus drives gene expression in both imprinted and non-imprinted tissues. *Dev. Biol.* **2004**, *271*, 488–497. [CrossRef]

21. Feil, R.; Walter, J.; Allen, N.D.; Reik, W. Developmental control of allelic methylation in the imprinted mouse Igf2 and H19 genes. *Development* **1994**, *120*, 2933–2943. [PubMed]

22. DeChiara, T.M.; Robertson, E.J.; Efstratiadis, A. Parental imprinting of the mouse insulin-like growth factor II gene. *Cell* **1991**, *64*, 849–859. [CrossRef]

23. Stylianopoulou, F.; Herbert, J.; Soares, M.B.; Efstratiadis, A. Expression of the insulin-like growth factor II gene in the choroid plexus and the leptomeninges of the adult rat central nervous system. *Proc. Natl. Acad. Sci. USA* **1988**, *85*, 141–145. [CrossRef] [PubMed]

24. Ziegler, A.N.; Schneider, J.S.; Qin, M.; Tyler, W.A.; Pintar, J.E.; Fraidenraich, D.; Wood, T.L.; Levison, S.W. IGF-II promotes stemness of neural restricted precursors. *Stem Cells* **2012**, *30*, 1265–1276. [CrossRef] [PubMed]

25. Bracko, O.; Singer, T.; Aigner, S.; Knobloch, M.; Winner, B.; Ray, J.; Clemenson, G.D.; Suh, H.; Couillard-Despres, S.; Aigner, L. Gene expression profiling of neural stem cells and their neuronal progeny reveals IGF2 as a regulator of adult hippocampal neurogenesis. *J. Neurosci.* **2012**, *32*, 3376–3387. [CrossRef]

26. Lehtinen, M.K.; Zappaterra, M.W.; Chen, X.; Yang, Y.J.; Hill, A.D.; Lun, M.; Maynard, T.; Gonzalez, D.; Kim, S.; Ye, P.; et al. The cerebrospinal fluid provides a proliferative niche for neural progenitor cells. *Neuron* **2011**, *69*, 893–905. [CrossRef] [PubMed]

27. Ziegler, A.N.; Levison, S.W.; Wood, T.L. Insulin and IGF receptor signalling in neural-stem-cell homeostasis. *Nat. Rev. Endocrinol.* **2015**, *11*, 161–170. [CrossRef]

28. Ziegler, A.N.; Feng, Q.; Chidambaram, S.; Testai, J.M.; Kumari, E.; Rothbard, D.E.; Constancia, M.; Sandovici, I.; Cominski, T.; Pang, K.; et al. Insulin-like Growth Factor II: An Essential Adult Stem Cell Niche Constituent in Brain and Intestine. *Stem Cell Rep.* **2019**, *12*, 816–830. [CrossRef]

29. Forbes, B.E.; McCarthy, P.; Norton, R.S. Insulin-like growth factor binding proteins: A structural perspective. *Front. Endocrinol.* **2012**, *3*, 38. [CrossRef]

30. Brown, J.; Jones, E.Y.; Forbes, B.E. Interactions of IGF-II with the IGF2R/cation-independent mannose-6-phosphate receptor mechanism and biological outcomes. *Vitam. Horm.* **2009**, *80*, 699–719. [CrossRef]

31. Brown, J.; Delaine, C.; Zaccheo, O.J.; Siebold, C.; Gilbert, R.J.; Van Boxel, G.; Denley, A.; Wallace, J.C.; Hassan, A.B.; Forbes, B.E. Structure and functional analysis of the IGF-II/IGF2R interaction. *EMBO J.* **2008**, *27*, 265–276. [CrossRef]

32. Zha, J.; Lackner, M.R. Targeting the insulin-like growth factor receptor-1R pathway for cancer therapy. *Clin. Cancer Res.* **2010**, *16*, 2512–2517. [CrossRef] [PubMed]

33. Belfiore, A.; Malaguarnera, R. Insulin receptor and cancer. *Endocr. Relat. Cancer* **2011**, *18*, R125–R147. [CrossRef] [PubMed]

34. Livingstone, C. IGF2 and cancer. *Endocr. Relat. Cancer* **2013**, *20*, R321–R339. [CrossRef]

35. Pollak, M. Insulin and insulin-like growth factor signalling in neoplasia. *Nat. Rev. Cancer* **2008**, *8*, 915. [CrossRef] [PubMed]

36. Samani, A.A.; Yakar, S.; LeRoith, D.; Brodt, P. The role of the IGF system in cancer growth and metastasis: Overview and recent insights. *Endocr. Rev.* **2007**, *28*, 20–47. [CrossRef]

37. Sciacca, L.; Costantino, A.; Pandini, G.; Mineo, R.; Frasca, F.; Scalia, P.; Sbraccia, P.; Goldfine, I.D.; Vigneri, R.; Belfiore, A. Insulin receptor activation by IGF-II in breast cancers: Evidence for a new autocrine/paracrine mechanism. *Oncogene* **1999**, *18*, 2471–2479. [CrossRef]

38. Chao, W.; D'Amore, P.A. IGF2: Epigenetic regulation and role in development and disease. *Cytokine Growth Factor Rev.* **2008**, *19*, 111–120. [CrossRef]

39. Harris, L.K.; Westwood, M. Biology and significance of signalling pathways activated by IGF-II. *Growth Factors* **2012**, *30*, 1–12. [CrossRef]

40. Lui, J.C.; Baron, J. Evidence that Igf2 down-regulation in postnatal tissues and up-regulation in malignancies is driven by transcription factor E2f3. *Proc. Natl. Acad. Sci. USA* **2013**, *110*, 6181–6186. [CrossRef]

41. Arcaro, A. Targeting the insulin-like growth factor-1 receptor in human cancer. *Front. Pharmacol.* **2013**, *4*, 30. [CrossRef] [PubMed]

42. Gallagher, E.J.; LeRoith, D. The proliferating role of insulin and insulin-like growth factors in cancer. *Trends Endocrinol. Metab. TEM* **2010**, *21*, 610–618. [CrossRef] [PubMed]

43. Denley, A.; Wallace, J.C.; Cosgrove, L.J.; Forbes, B.E. The insulin receptor isoform exon 11- (IR-A) in cancer and other diseases: A review. *Horm. Metab. Res.* **2003**, *35*, 778–785. [CrossRef] [PubMed]

44. Giorgino, F.; Belfiore, A.; Milazzo, G.; Costantino, A.; Maddux, B.; Whittaker, J.; Goldfine, I.D.; Vigneri, R. Overexpression of insulin receptors in fibroblast and ovary cells induces a ligand-mediated transformed phenotype. *Mol. Endocrinol.* **1991**, *5*, 452. [CrossRef] [PubMed]

45. Vella, V.; Nicolosi, M.L.; Cantafio, P.; Massimino, M.; Lappano, R.; Vigneri, P.; Ciuni, R.; Gangemi, P.; Morrione, A.; Malaguarnera, R.; et al. DDR1 regulates thyroid cancer cell differentiation via IGF-2/IR-A autocrine signaling loop. *Endocr. Relat. Cancer* **2019**, *26*, 197–214. [CrossRef]

46. Tominaga, K.; Shimamura, T.; Kimura, N.; Murayama, T.; Matsubara, D.; Kanauchi, H.; Niida, A.; Shimizu, S.; Nishioka, K.; Tsuji, E.I.; et al. Addiction to the IGF2-ID1-IGF2 circuit for maintenance of the breast cancer stem-like cells. *Oncogene* **2017**, *36*, 1276–1286. [CrossRef]

47. Laviola, L.; Natalicchio, A.; Giorgino, F. The IGF-I signaling pathway. *Curr. Pharm. Des.* **2007**, *13*, 663–669. [CrossRef]

48. Siddle, K. Signalling by insulin and IGF receptors: Supporting acts and new players. *J. Mol. Endocrinol.* **2011**, *47*, R1–R10. [CrossRef]

49. Cabail, M.Z.; Li, S.; Lemmon, E.; Bowen, M.E.; Hubbard, S.R.; Miller, W.T. The insulin and IGF1 receptor kinase domains are functional dimers in the activated state. *Nat. Commun.* **2015**, *6*, 6406. [CrossRef]

50. Hubbard, S.R.; Wei, L.; Ellis, L.; Hendrickson, W.A. Crystal structure of the tyrosine kinase domain of the human insulin receptor. *Nature* **1994**, *372*, 746–754. [CrossRef]

51. De Meyts, P.; Whittaker, J. Structural biology of insulin and IGF1 receptors: Implications for drug design. *Nat. Rev. Drug Discov.* **2002**, *1*, 769–783. [CrossRef] [PubMed]

52. Hanke, S.; Mann, M. The phosphotyrosine interactome of the insulin receptor family and its substrates IRS-1 and IRS-2. *Mol. Cell. Proteom.* **2009**, *8*, 519–534. [CrossRef] [PubMed]

53. Taniguchi, C.M.; Emanuelli, B.; Kahn, C.R. Critical nodes in signalling pathways: Insights into insulin action. *Nat. Rev. Mol. Cell Biol.* **2006**, *7*, 85–96. [CrossRef] [PubMed]

54. Hakuno, F.; Takahashi, S.I. IGF1 receptor signaling pathways. *J. Mol. Endocrinol.* **2018**, *61*, T69–T86. [CrossRef] [PubMed]

55. Torres, A.M.; Forbes, B.E.; Aplin, S.E.; Wallace, J.C.; Francise, G.L.; Norton, R.S. Solution structure of human insulin-like growthfactor II. Relationship to receptor and binding protein interactions. *J. Mol. Biol.* **1995**, *248*, 385–401. [CrossRef]

56. Smith, G.D.; Pangborn, W.A.; Blessing, R.H. The structure of T6 human insulin at 1.0 Å resolution. *Acta Crystallogr. Sect. D Biol. Crystallogr.* **2003**, *59*, 474–482. [CrossRef] [PubMed]

57. McKern, N.M.; Lawrence, M.C.; Streltsov, V.A.; Lou, M.Z.; Adams, T.E.; Lovrecz, G.O.; Elleman, T.C.; Richards, K.M.; Bentley, J.D.; Pilling, P.A.; et al. Structure of the insulin receptor ectodomain reveals a folded-over conformation. *Nature* **2006**, *443*, 218–221. [CrossRef]

58. Xu, Y.; Kong, G.K.; Menting, J.G.; Margetts, M.B.; Delaine, C.A.; Jenkin, L.M.; Kiselyov, V.V.; De Meyts, P.; Forbes, B.E.; Lawrence, M.C. How ligand binds to the type 1 insulin-like growth factor receptor. *Nat. Commun.* **2018**, *9*, 821. [CrossRef]

59. Kavran, J.M.; McCabe, J.M.; Byrne, P.O.; Connacher, M.K.; Wang, Z.; Ramek, A.; Sarabipour, S.; Shan, Y.; Shaw, D.E.; Hristova, K. How IGF-1 activates its receptor. *Elife* **2014**, *3*, e03772. [CrossRef]

60. Lou, M.; Garrett, T.P.; McKern, N.M.; Hoyne, P.A.; Epa, V.C.; Bentley, J.D.; Lovrecz, G.O.; Cosgrove, L.J.; Frenkel, M.J.; Ward, C.W. The first three domains of the insulin receptor differ structurally from the insulin-like growth factor 1 receptor in the regions governing ligand specificity. *Proc. Natl. Acad. Sci. USA* **2006**, *103*, 12429–12434. [CrossRef]

61. Garrett, T.P.; McKern, N.M.; Lou, M.; Frenkel, M.J.; Bentley, J.D.; Lovrecz, G.O.; Elleman, T.C.; Cosgrove, L.J.; Ward, C.W. Crystal structure of the first three domains of the type-1 insulin-like growth factor receptor. *Nature* **1998**, *394*, 395–399. [CrossRef] [PubMed]

62. Weis, F.; Menting, J.G.; Margetts, M.B.; Chan, S.J.; Xu, Y.; Tennagels, N.; Wohlfart, P.; Langer, T.; Muller, C.W.; Dreyer, M.K.; et al. The signalling conformation of the insulin receptor ectodomain. *Nat. Commun.* **2018**, *9*, 4420. [CrossRef] [PubMed]

63. Xu, Y.; Kirk, N.S.; Venugopal, H.; Margetts, M.B.; Croll, T.I.; Sandow, J.J.; Webb, A.I.; Delaine, C.A.; Forbes, B.E.; Lawrence, M.C. How IGF-II Binds to the Human Type 1 Insulin-like Growth Factor Receptor. *Structure* **2020**. [CrossRef] [PubMed]

64. Li, J.; Choi, E.; Yu, H.; Bai, X.-c. Structural basis of the activation of type 1 insulin-like growth factor receptor. *Nat. Commun.* **2019**, *10*, 4567. [CrossRef] [PubMed]

65. De Meyts, P. Insulin/receptor binding: The last piece of the puzzle. *BioEssays* **2015**, *37*, 389–397. [CrossRef] [PubMed]

66. Alvino, C.L.; McNeil, K.A.; Ong, S.C.; Delaine, C.; Booker, G.W.; Wallace, J.C.; Whittaker, J.; Forbes, B.E. A novel approach to identify two distinct receptor binding surfaces of insulin-like growth factor II. *J. Biol. Chem.* **2009**, *284*, 7656–7664. [CrossRef]

67. Gauguin, L.; Delaine, C.; Alvino, C.L.; McNeil, K.A.; Wallace, J.C.; Forbes, B.E.; De Meyts, P. Alanine scanning of a putative receptor binding surface of insulin-like growth factor-I. *J. Biol. Chem.* **2008**, *283*, 20821–20829. [CrossRef] [PubMed]

68. Kristensen, C.; Kjeldsen, T.; Wiberg, F.C.; Schäffer, L.; Hach, M.; Havelund, S.; Bass, J.; Steiner, D.F.; Andersen, A.S. Alanine scanning mutagenesis of insulin. *J. Biol. Chem.* **1997**, *272*, 12978–12983. [CrossRef]

69. Gutmann, T.; Schafer, I.B.; Poojari, C.; Brankatschk, B.; Vattulainen, I.; Strauss, M.; Coskun, U. Cryo-EM structure of the complete and ligand-saturated insulin receptor ectodomain. *J. Cell Biol.* **2020**, *219*. [CrossRef]
70. Uchikawa, E.; Choi, E.; Shang, G.; Yu, H.; Bai, X.C. Activation mechanism of the insulin receptor revealed by cryo-EM structure of the fully liganded receptor-ligand complex. *Elife* **2019**, *8*. [CrossRef]
71. Kiselyov, V.V.; Versteyhe, S.; Gauguin, L.; De Meyts, P. Harmonic oscillator model of the insulin and IGF1 receptors' allosteric binding and activation. *Mol. Syst. Biol.* **2009**, *5*, 243. [CrossRef] [PubMed]
72. Belfiore, A.; Frasca, F.; Pandini, G.; Sciacca, L.; Vigneri, R. Insulin receptor isoforms and insulin receptor/ insulin-like growth factor receptor hybrids in physiology and disease. *Endocr. Rev.* **2009**, *30*, 586–623. [CrossRef] [PubMed]
73. Gualberto, A.; Pollak, M. Emerging role of insulin-like growth factor receptor inhibitors in oncology: Early clinical trial results and future directions. *Oncogene* **2009**, *28*, 3009–3021. [CrossRef] [PubMed]

Review

Mechanisms of IGF-1-Mediated Regulation of Skeletal Muscle Hypertrophy and Atrophy

Tadashi Yoshida [1,2,*] **and Patrice Delafontaine** [1,2,*]

1   Heart and Vascular Institute, John W. Deming Department of Medicine, Tulane University School of Medicine, 1430 Tulane Ave SL-48, New Orleans, LA 70112, USA
2   Department of Physiology, Tulane University School of Medicine, 1430 Tulane Ave, New Orleans, LA 70112, USA
*   Correspondence: tyoshida@tulane.edu (T.Y.); pdelafon@tulane.edu (P.D.)

Received: 1 July 2020; Accepted: 19 August 2020; Published: 26 August 2020

Abstract: Insulin-like growth factor-1 (IGF-1) is a key growth factor that regulates both anabolic and catabolic pathways in skeletal muscle. IGF-1 increases skeletal muscle protein synthesis via PI3K/Akt/mTOR and PI3K/Akt/GSK3β pathways. PI3K/Akt can also inhibit FoxOs and suppress transcription of E3 ubiquitin ligases that regulate ubiquitin proteasome system (UPS)-mediated protein degradation. Autophagy is likely inhibited by IGF-1 via mTOR and FoxO signaling, although the contribution of autophagy regulation in IGF-1-mediated inhibition of skeletal muscle atrophy remains to be determined. Evidence has suggested that IGF-1/Akt can inhibit muscle atrophy-inducing cytokine and myostatin signaling via inhibition of the NF-κB and Smad pathways, respectively. Several miRNAs have been found to regulate IGF-1 signaling in skeletal muscle, and these miRs are likely regulated in different pathological conditions and contribute to the development of muscle atrophy. IGF-1 also potentiates skeletal muscle regeneration via activation of skeletal muscle stem (satellite) cells, which may contribute to muscle hypertrophy and/or inhibit atrophy. Importantly, IGF-1 levels and IGF-1R downstream signaling are suppressed in many chronic disease conditions and likely result in muscle atrophy via the combined effects of altered protein synthesis, UPS activity, autophagy, and muscle regeneration.

Keywords: insulin-like growth factor-1; skeletal muscle; hypertrophy; atrophy; cachexia; muscle regeneration; autophagy

---

1. Introduction

Studies in various models in cell culture, animals, and humans have evaluated cytokines and growth factors that can regulate muscle growth. Insulin-like growth factor-1 (IGF-1) is one of the best-characterized growth factors, and it has been shown to modulate muscle size and play a critical role in regulating muscle function. IGF-1 is thought to mediate many of the beneficial outcomes of physical activity [1,2]. In a study analyzing healthy young subjects, circulating IGF-1 levels were negatively associated with body fat, body mass index (BMI), and total cholesterol and positively associated with aerobic fitness and muscular endurance parameters (VO$_2$ peak, sit-ups, push-ups, and repetitive squats) [3]. In contrast, lower IGF-1 levels were associated with various pathological conditions including chronic diseases, inflammation, and malnutrition [4,5]. Since skeletal muscle cells, or myofibers, are postmitotic, their size is determined by a balance between synthesis of new proteins and degradation of old proteins. Under physiological conditions, the rates of protein synthesis and degradation are balanced and the myofiber size is maintained. In cachectic conditions, on the contrary, myofiber protein degradation is accelerated and protein synthesis rate is suppressed, resulting in muscle weakness and fatigue. IGF-1 can regulate both protein synthesis and degradation pathways, and changes in IGF-1 signaling in skeletal muscle can greatly affect myofiber size and function.

This review summarizes and discusses different aspects of IGF-1-mediated protein synthetic and degradation pathways in skeletal muscle and its potential application to therapies to treat patients with reduced skeletal muscle function. The signaling pathways downstream of IGF-1 discussed in the following sections are summarized in Figure 1.

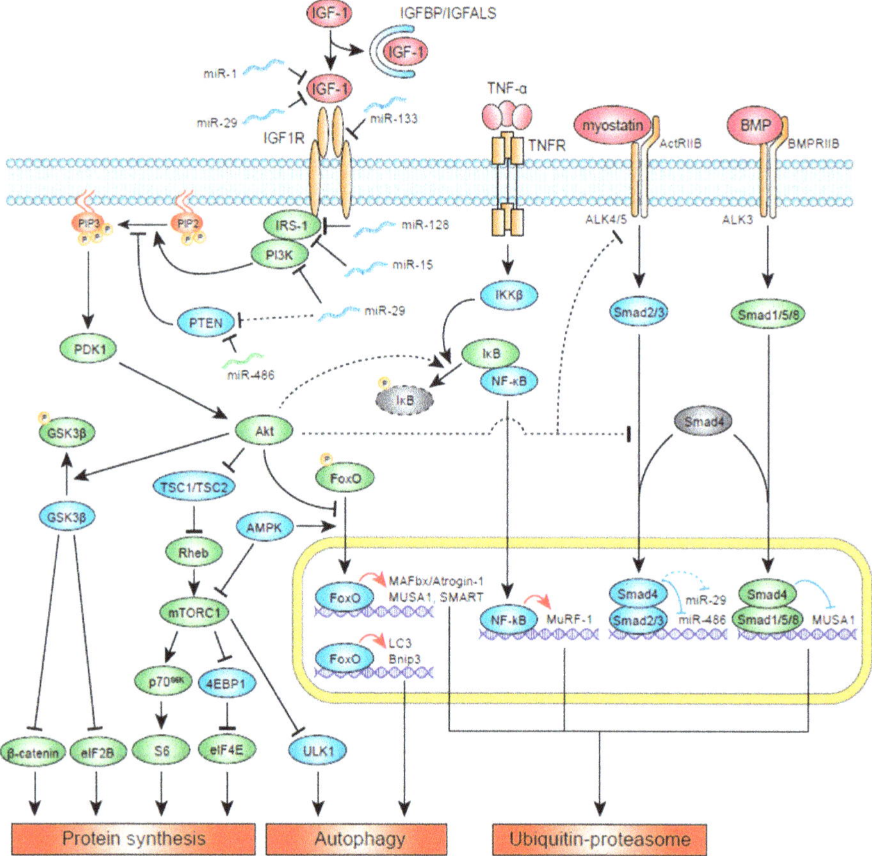

Figure 1. IGF-1 signaling pathways. In the figure, the signaling molecules and miRNAs that activate protein synthesis and/or inhibit protein degradation are shown in green, while the ones that inhibit protein synthesis and/or activate protein degradation are shown in blue. The majority of IGF-1 in the body are bound to IGFBP and IGFALS, and its activity is suppressed. Once IGF-1 binds to IGF-1R, IRS-1 and PI3K are recruited and activated. PI3K converts PIP2 to PIP3, which activates PDK1 and Akt. Akt activates protein synthesis via activation of ribosomal protein S6 and the translation initiation factor eIF4E downstream of mTORC1, and activation of β-catenin and eIF2B downstream of GSK3β. Akt can suppress UPS activity via inhibition of FoxO-mediated transcription of E3 ubiquitin ligases MAFbx/Atrogin-1, MUSA1, and SMART. MuRF1 expression is induced by cytokines such as TNF-α via NF-κB pathway. Akt could phosphorylate IκB and activate the NF-κB pathway, although it has not been shown in skeletal muscle and multiple studies have shown IGF-1 activation does not alter MuRF1 expression. Myostatin and BMP signaling compete against each other for their usage of Smad4. Activation of myostatin inhibits BMP-mediated Smad1/5/8 translocation to the nucleus, thus inhibiting

MUSA1-mediated UPS activity. Akt can also downregulate ActRIIB and inhibit ALK4/5 via unknown mechanisms. Although it has not been shown in skeletal muscle, Akt can interact directly with unphosphorylated Smad3 to sequester it outside the nucleus. Several miRNAs have been shown to regulate IGF-1 signaling. miR-486 is inhibited by the myostatin/Smad pathway, resulting in inhibition of IGF-1 signaling via PTEN increase. miR-1 and miR-133 target IGF-1 and IGF-1R, respectively, and their expression is reduced during muscle hypertrophy. IRS-1 could be inhibited by miR-128 and miR-15. LncIRS1 (not shown in the figure), which is upregulated in hypertrophic muscles, can act as sponge for miR-15, resulting in activation of IRS-1. Note that studies have shown conflicting evidence on miR-29's role in IGF-1 signaling in skeletal muscle, and it may potentiate or inhibit IGF-1 signaling. Pathways that are unclear and/or not shown in skeletal muscle are shown in dotted lines.

2. Muscle Protein Synthesis and IGF-1 Signaling

One of the most important function of IGF-1 is its regulation of protein synthesis in skeletal muscle and promotion of body growth. Upon binding to IGF-1, IGF-1 receptor (IGF-1R) phosphorylates an intracellular adaptor protein insulin receptor substrate-1 (IRS-1), which recruits and phosphorylates phosphoinositide 3-kinase (PI3K) followed by Akt phosphorylation. The PI3K/Akt pathway plays a critical role in myotube hypertrophy [6,7], and activation of Akt in rat muscle prevents denervation-induced atrophy [8,9]. Mammalian target of rapamycin (mTOR) is a downstream target of Akt, and in mammalian cells mTOR activity is tightly regulated by amino acid availability to the cells. As amino acids are necessary to build proteins, nucleic acid, glucose, and ATP in the body, mTOR activity is highly correlated with the anabolic/catabolic balance. The IGF-1/Akt/mTOR pathway has been shown to be indispensable in promoting muscle hypertrophy [10]. Akt phosphorylates and inhibits tuberous sclerosis 1 and 2 (TSC1/TSC2), resulting in activation of small G protein Ras homolog enriched in brain (Rheb) via its binding to GTP. GTP-bound Rheb activates mTOR complex-1 (mTORC1), resulting in phosphorylation of $p70^{S6K}$, which promotes protein synthesis by activating ribosomal protein S6, a component of the 40S ribosomal subunit. mTORC1 also phosphorylates 4EBP1, leading to its release from the inhibitory complex with the translation initiation factor eIF4E, the cap binding protein, permitting the binding of eIF4E to eIF4G to form the critical translation initiation complex [11]. When animals were treated with mTOR inhibitor rapamycin, phosphorylation of $p70^{S6K}$ and the release of 4EBP1 from eIF4E inhibitory complex were blocked, leading to inhibition of surgical overload-induced muscle hypertrophy [12]. Consistently, the Akt/mTOR pathway was inhibited during disuse (unloading)-induced atrophy and re-activated after reloading.

Besides mTOR, Akt-mediated phosphorylation of glycogen synthase kinase-3β (GSK3β) is another critical downstream pathway of IGF-1. In muscle hypertrophic conditions, GSK3β is phosphorylated and its activity is inhibited, leading to activation of eIF2B and transcriptional activator β-catenin [13,14]. In contrast, GSK3β activity is increased in a dexamethasone (Dex)-induced atrophy model. Local IGF-1 or constitutively active Akt gene transfer inhibited GSK3β, increased β-catenin levels, and prevented muscle atrophy [15].

In summary, IGF-1/Akt controls two protein synthetic pathways via mTORC1 and GSK3β. Although these pathways are decreased in various muscle atrophy conditions [16–19], the exact relationship and interaction between these pathways in skeletal muscle atrophy and hypertrophy conditions remain to be determined. IGF-1 also affects protein synthesis via myostatin signaling, and the mechanism is discussed below.

3. Muscle Protein Synthesis: Myostatin

Myostatin is a member of the transforming growth factor-β (TGF-β) superfamily, is secreted mainly from skeletal muscle, and negatively regulates muscle mass [20]. Myostatin has been found to be upregulated in cancer, heart disease, HIV, and aging, and systemic administration of myostatin caused cachexia in rodents. Studies have identified crosstalk between myostatin and IGF-1 signaling pathways. In cultured myotubes, myostatin inhibited Akt phosphorylation, resulting in decreased protein synthesis and reduced cell size [21–23]. In mice deficient in myostatin, total Akt expression was increased together with increased p70^{S6K} levels [21,24]. These data suggest that myostatin and IGF-1 signaling counteract each other. Indeed, IGF-1 treatment of cultured myotubes blocked myostatin-mediated downregulation of Akt and myotube diameter reduction [23]. Accordingly, the hypertrophic effect of IGF-1 was greater in the myostatin null background [25].

Myostatin signaling is mediated by activin type II receptors (ActRIIA and ActRIIB) and activin type I receptors (ALK4 and ALK5), leading to phosphorylation of Smad proteins (Smad2 and -3). Smad2/3 form a complex with Smad4, which is also a co-mediator of the bone morphogenic protein (BMP) signaling pathway. Therefore, when the myostatin signaling is low, Smad4 becomes more available to BMP signaling, leading to muscle hypertrophy [26]. Studies have suggested that IGF-1 and myostatin/Smad pathways cross-talk at different levels. Akt activation downregulated ActRIIB in denervated muscles, and blocked atrophy-inducing effects of constitutively active ALK4 and ALK5 [27]. Studies in cancer cells have demonstrated direct interaction of Akt and Smad3 to sequester Smad3 outside of the nucleus [28], although it remains to be determined whether the same mechanism exists in skeletal muscle. Although the entire picture of Akt-Smad interaction remains to be determined in skeletal muscle, these data suggest that the balance between competing IGF-1, myostatin, and BMP pathways are critical to maintain muscle mass.

4. Muscle Protein Degradation: UPS

The ubiquitin-proteasome system (UPS) is a crucial protein degradation system in eukaryotes, and studies have shown its importance in development of muscle atrophy [29,30]. Muscle atrophy F-box (MAFbx)/Atrogin-1 and muscle RING finger 1 (MuRF1) are the best characterized E3 ubiquitin ligases in skeletal muscle that mediate polyubiquitination of proteins and target them to degradation by the 26S proteasome. MAFbx/Atrogin-1 and MuRF-1 are shown to be increased in various muscle atrophy-inducing conditions, including disuse, denervation, inflammation, aging, glucocorticoid increase, high Ang II, and chronic diseases such as cancer, congestive heart failure, chronic kidney disease, chronic obstructive pulmonary disease (COPD), and AIDS [31,32]. Interestingly, studies have suggested that IGF-1 signaling is altered in many of these conditions, and signaling pathways that regulate MAFbx/Atrogin-1 and MuRF1 are in some part overlapping and regulated by IGF-1 signaling (Figure 1).

Various studies have shown that MAFbx/Atrogin-1 and MuRF1 expression is differentially regulated by FoxO and NF-κB pathways. Inhibition of FoxOs prevented MAFbx/Atrogin-1 increase and protected against muscle atrophy. Although FoxO1 activates both MAFbx/Atrogin-1 and MuRF1 expression in cultured myotubes, the ability of FoxO1 to induce MuRF1 expression is independent of its DNA binding [33]. Similarly, Senf et al. found that FoxO3a induced MAFbx/Atrogin-1 expression via promoter activation, whereas MuRF1 activation did not require FoxO3a DNA binding [34]. Transgenic overexpression of activated IκB kinase β (IKKβ) in skeletal muscle caused profound muscle wasting, with increased expression of MuRF1 but not MAFbx/Atrogin-1 [35]. In addition, promoter analysis revealed that MuRF1 expression is regulated by upstream NF-κB binding sites, but not FoxO sites in disuse atrophy [36].

It is of note that the IGF-1/PI3K/Akt pathway not only activates FoxO, but also NF-κB signaling via several mechanisms including stimulating p65 transactivation and activation of IKKβ [37]. However, it is not clear whether activation of IGF-1 in skeletal muscle alters NF-κB activation and MuRF-1 expression. In myofibers, IGF-1 rapidly and strongly reduced Dex-induced Atrogin-1 expression (~80% reduction after 6 h), whereas MuRF-1 mRNA reduction occurred more slowly (~30% reduction after 18 h) [38]. Importantly, changes in overall proteolysis with Dex and IGF-1 correlated tightly with changes in Atrogin-1 mRNA levels, but not with MuRF1. Consistently, systemic Ang II infusion increased both MAFbx/Atrogin-1 and MuRF-1, whereas IGF-1 inhibited expression and promoter activity of MAFbx/Atrogin-1, but not MuRF-1 [39].

Several substrates of MAFbx/Atrogin-1 and MuRF1 have been identified in skeletal muscle. By yeast two-hybrid screening, eukaryotic initiation factor 3 subunit 5 (eIF3-f) was identified as a target substrate of MAFbx/Atrogin-1 [40]. MAFbx/Atrogin-1 increased eIF3-f degradation in myotubes undergoing atrophy in vitro, and overexpression of eIF3-f caused hypertrophy both in vitro and in vivo. Immunoprecipitation of MAFbx/Atrogin-1 followed by LC-MS/MS analysis in myostatin-treated C2C12 myotubes identified desmin and vimentin as other targets of MAFbx/Atrogin-1 [41]. For MuRF1, myosin heavy chain (MYH) was identified as its target in Dex-induced myotube atrophy model [42]. In addition, by comparing the WT and transgenic mice expressing a RING deletion mutant of MuRF1, which binds but cannot ubiquitinate substrates, Cohen et al. found that atrophying muscles showed a loss of myosin-binding protein C (MyBP-C) and myosin light chains 1 and 2 (MyLC1 and MyLC2) from the myofibril, before loss of MYH [43]. MuRF1 also has been shown to associate with titin and stabilize the sarcomeric M-line [44]. Moreover, MuRF1 is suggested to regulate muscle energy metabolism by targeting creatine kinase [45,46]. However, changes in these target substrates in response to IGF-1 have not been determined in muscle atrophy models.

In addition to well-characterized MAFbx/Atrogin-1 and MuRF1, there are other E3 ubiquitin ligases that are involved in skeletal muscle protein breakdown and are potentially regulated by IGF-1. Milan et al. found that a group of ubiquitin ligases were upregulated in denervated or fasted skeletal muscle, and were blunted in FoxO1, -3, and -4 triple knockout mice (FoxO1,3,4$^{-/-}$) [47]. These ubiquitin ligases include muscle ubiquitin ligase of the SCF complex in atrophy-1 (MUSA1), Fbxo31, and Fbxo21 (also known as SMART). FoxO3 overexpression in myotubes was sufficient to induce MUSA1, but not other ubiquitin ligases. FoxO1 and FoxO3 bind to the promoter regions of MUSA1 and SMART, and the FoxO3 deletion completely blunted the induction of SMART, but not other ubiquitin ligases. These data suggest an overlapping and complex regulation of these ubiquitin ligases by FoxO1, -3, and -4, and therefore by IGF-1.

Nedd4 is a HECT domain ubiquitin ligase that is increased in skeletal muscles after denervation [48,49], unloading [48], and COPD [50]. Nedd4-null mice showed a reduction of IGF-1 and insulin signaling, delayed embryonic development, reduced growth and body weight, and neonatal lethality [51]. Furthermore, skeletal muscle-specific Nedd4 null mice were protected against denervation induced muscle atrophy [52].

Trim32 is a tripartite motif ubiquitin ligase that ubiquitinates and degrades the desmin cytoskeleton, thin filament (actin, tropomyosin, and troponins), and Z-band (α-actinin) [53]. Downregulation of Trim32 in hindlimb muscles reduced fasting-induced breakdown of these contractile and cytoskeletal proteins and muscle atrophy. Furthermore, downregulation of Trim32 in skeletal muscle increased PI3K/Akt/FoxO signaling, enhanced glucose uptake, and induced myofiber growth [54].

TNF receptor adaptor protein 6 (TRAF6) is a member of the TRAF family of adaptor proteins, with the unique property to have E3 ubiquitin ligase activity. TRAF6 is upregulated in skeletal muscle after denervation, starvation, and cancer cachexia development [55,56]. Interestingly, the induction of MAFbx/Atrogin-1 and MuRF1 was suppressed in TRAF6 null mice, suggesting that TRAF6 is an upstream regulator of these E3 ubiquitin ligases. Notably, TRAF6 directly ubiquitinates Akt and inhibits its activity [57]. Although the importance of the potential interaction between IGF-1 signaling and MUSA1, SMART, Nedd4, Trim32, and TRAF6 in skeletal muscle hypertrophy and atrophy remains to be determined, IGF-1 signaling pathway components could be novel targets to regulate these E3 ubiquitin ligase activities in skeletal muscle.

5. Muscle Protein Degradation: Autophagy

Another major proteolytic pathway in eukaryotic cells is the autophagy-lysosome system. Autophagy plays a critical role in removal of damaged organelles such as mitochondria, peroxisomes, nuclei and ribosomes, as well as in degradation of damaged or misfolded proteins. Another protective role of autophagy is to provide the degraded cellular components as an energy source to cells especially in the face of sustained starvation. Various skeletal muscle diseases that manifest atrophy and dystrophy such as Pompe disease and Danon disease are associated with lowered autophagic activity [58]. In addition, skeletal muscle-specific knockout mice for Atg7, which acts as an E1-like enzyme critical for autophagy regulation, showed profound muscle atrophy and age-dependent decline in muscle force [59].

Autophagy is regulated by two main pathways that overlap with IGF-1 signaling pathways: mTOR-mediated inhibitory phosphorylation of unc51-like kinase-1 (ULK1) and FoxO3-mediated induction of autophagy-related genes. Since IGF-1 activates mTOR (thus, inhibits ULK1) and inhibits FoxO (thus, inhibits autophagy-related gene expression), it is reasonable to assume IGF-1 inhibits autophagy, although some conflicting results have been reported on the relative importance of mTOR and FoxO pathways in regulation of skeletal muscle autophagy. A first group of studies suggested that mTOR-mediated regulation of autophagy only plays a minor role, at least in skeletal muscle. Only a small (10–15%) induction of autophagy was observed after rapamycin (mTOR inhibitor) treatment in cultured myotubes [60], and rapamycin administration or mTOR knockdown did not induce autophagy in skeletal muscle in vivo [61]. In contrast to these findings, skeletal muscle-specific TSC1-deficient mice (TSCmKO), which show sustained activation of mTORC1, developed a late-onset myopathy related to impaired autophagy [62].

Likely independent of mTOR, Akt activation blocked autophagy via inhibition of FoxO3 [60,61]. Blockade of FoxO3 inhibited the starvation-induced autophagy, and these effects are likely mediated by inhibition of FoxO3-mediated transcriptional activation of autophagy-related genes such as LC3, Bnip3, Beclin-1, Atg4, and Atg12 [63]. Interestingly, Zhao et al. showed that constitutively-active FoxO3 increased protein degradation in cultured myotubes, and, surprisingly, approximately 80% of the effect was mediated by autophagy [60].

These data suggest that both IGF-1/Akt/mTOR and IGF-1/Akt/FoxO pathways inhibit autophagy. However, few studies have extensively analyzed the effect of IGF-1 in skeletal muscle autophagy, and conflicting evidence has been presented. Nakashima et al. treated chicken myotubes with IGF-1 and found that LC3-I to LC3-II conversion, a critical step for autophagosome formation, was decreased [64]. In contrast, Ascenzi et al. showed that LC3-I to LC3-II conversion, which is normally decreased during aging, was increased in mice with skeletal muscle-specific overexpression of IGF-1 [65]. To understand these discrepancies, it is important to note that autophagy involves dynamic and complicated processes, and it has been a challenge in autophagy research to capture a dynamic process with static measurements [66]. Neither of the above studies measured the autophagic flux (i.e., dynamic process of autophagy), therefore more studies are required to understand the role of IGF-1 in regulation of autophagic flux in skeletal muscle in vivo. In other cell types, IGF-1 has been shown to inhibit autophagy. In human colorectal carcinoma drug-resistant cells, IGF-1 inhibited autophagy via Akt/mTOR pathway [67]. IGF-1 knockdown increased autophagy via reduction of Akt/mTOR in aged bone marrow mesenchymal stem cells (BM-MSCs) in hypoxic condition and protected cells against hypoxic injury [68]. This IGF-1-mediated autophagy reduction is suggested to be involved in cellular senescence and longevity. Long-term exposure of quiescent human fibroblasts to IGF-1 reduced viability and increased senescent cells, associated with reduced autophagy and dysfunctional mitochondria. These effects were reversed by rapamycin treatment (mTOR inhibition). Consistently, autophagy is increased in mouse fibroblasts in vivo with lowered IGF-1 levels [69].

In various muscle atrophy conditions such as disuse and denervation, autophagy has been shown to be activated [70]. Although IGF-1 has been used in attempts to prevent muscle atrophy in various models, careful evaluation of autophagy is not always conducted. In models such as cancer cachexia, in which UPS-mediated protein breakdown in increased, overall autophagic activity is likely decreased despite the observation of increased autophagy marker such as Beclin-1, p62, and LC3B [71]. Similarly, in the Ang II-induced muscle atrophy model, autophagy is reduced and likely caused accumulation of dysfunctional mitochondria and impaired skeletal muscle energy metabolism [72]. In both of these models, IGF-1 is reduced [73,74] and IGF-1 administration rescued muscle atrophy [39,75,76]. However, in C26 tumor-bearing mice, neither inhibition nor activation of autophagy rescued the muscle function, and both treatments worsened the outcome [77]. The IGF-1 pathway could still be a promising target to treat muscle atrophy where both autophagy and UPS are activated, and protein synthesis is decreased, as IGF-1 activation could theoretically normalize all of these pathways. However, more careful evaluation of IGF-1's effects on autophagy is necessary for the development of therapies, as both excessive activation and insufficiency of autophagy could be deleterious to skeletal muscle.

6. Muscle Energy Homeostasis: AMPK and IGF-1

5′-adenosine monophosphate-activated protein kinase (AMPK) is an intracellular sensor of ATP consumption and acts as a key regulator of skeletal muscle metabolism. When ATP level is low (thus AMP/ATP ratio is high), AMPK is activated and protein synthesis, which consumes ATP, inhibited. Furthermore, activated AMPK promotes ATP-producing catabolic processes including glucose and fat oxidation, UPS- and autophagy-mediated protein degradation [78]. Via these mechanisms, dominant-negative AMPK overexpression in skeletal muscle or skeletal muscle-specific AMPK gene deletion increased muscle mass [79–81]. Mechanistically, AMPK targets two major components of IGF-1 signaling: mTOR and FoxO. AMPK decreases protein translation via activation of mTORC1 and promotes protein breakdown via activation of FoxO1 and FoxO3, which in turn increase UPS and autophagy-related genes (Figure 1). Therefore, it is consistent with these mechanisms that pharmacological or genetic activation of AMPK blocked overloading-induced muscle hypertrophy [82,83]. However, the role of AMPK in muscle atrophy is unclear. In rodent muscle unloading-induced atrophy models, both increased and decreased AMPK activity has been reported [84–87]. In these models, genetic inactivation of AMPK prevented muscle atrophy [88,89]. In contrast, in Ang II-induced muscle atrophy model, AMPK activity is reduced, and pharmacological and genetic AMPK activation restored muscle mass [90,91]. The proposed mechanistic model is that elevated Ang II reduces ATP content in skeletal muscle, which is supposed to activate AMPK, while Ang II inhibits AMPK activation, causing severe ATP depletion and muscle atrophy. It is not clear whether muscle ATP content is altered in unloading muscle atrophy models and muscle atrophy is caused in a similar mechanism. Importantly, IGF-1 level is reduced in both of these atrophying conditions and Akt is inhibited in skeletal muscle, although the role of AMPK (which is known to inhibit Akt/mTOR and activate FoxO [78,92]) in relation to IGF-1 signaling in atrophying conditions is not clear.

7. Alternative Splicing of IGF-1 mRNA to Produce a Local Form

In addition to circulating IGF-1 secreted by the liver, peripheral tissues including skeletal muscle produce IGF-1. Interestingly, some studies suggest distinct roles between circulating and local IGF-1. The IGF-1 gene contains six exons that are differentially spliced to generate multiple transcript variants that result in different pre-pro-IGF-1s (Figure 2). Although the different pre-pro-IGF-1s eventually give rise to the same mature 70-amino acid IGF-1 molecule, it has been shown that these variants have different stabilities, binding partners, and activity. The first two exons are mutually exclusive for their use, and each exon has multiple transcription initiation sites, therefore generating different 5′-UTRs and N-terminal signal sequences. Transcripts containing exon 1 or 2 are referred to as Class 1 and 2, respectively. Exons 3 and 4 are used in all the variants and encode the B, C, A, and D domains, which are named based on their similarity to those in insulin. The 3′-end of IGF-1 gene generates three types of mRNAs with different termination codons, polyadenylation sites, and 3′-UTRs. The C-terminus of pre-pro-IGF-1, termed as E-peptide domain, thus has the greatest variability within the entire protein. The E-peptide domain includes part of exon 4 (16 amino acids), with differential inclusion of exon 5 and 6; Ea consists of exon 6 (19 amino acids) and Eb of exon 5 (61 amino acids). Due to alternative splicing, Ec consists of part of exon 5 (16 amino acids) and part of exon 6 (8 amino acids). Note that these are terminologies for human IGF-1; rodents' equivalent of human Ec is termed as Eb, as they do not express human Eb-equivalent form. Overall, this alternative splicing generates at least 6 pre-pro-IGF-1: Class 1-Ea, Eb, Ec, and Class 2-Ea, Eb, and Ec. Studies have suggested distinct functions among these different forms of proteins. For instance, Class 1 proteins have a longer signal peptide that is potentially myristoylated and may retain the protein in the ER during the translation process, whereas Class 2 mRNAs are highly expressed in the liver, the primary source of circulating IGF-1. Therefore, Class 1 peptides represent a locally-produced autocrine/paracrine form, and Class 2 peptides represent the circulating endocrine form in the body. Bikle et al. demonstrated that muscle atrophy is more pronounced after ablation of muscle IGF-1 production than when liver IGF-1 production is inhibited [93], suggesting that local IGF-1 is a crucial factor for muscle hypertrophy. However, Temmerman et al. demonstrated the deletion of exon 2 (thus Class 2 mRNAs) in mice did not affect viability, growth, and maintenance of circulating IGF-1 levels [94], and the exact physiological roles of Class 1 and Class 2 proteins remain to be determined. For the E peptide domain, Annibalini et al. identified a highly conserved N-glycosylaton site in the Ea domain, which regulated intracellular pro-IGF-1Ea level via prevention of proteasome-mediated degradation and subcellular localization [95]. Interestingly, Durzyńska et al. found that the predominant forms that are expressed in skeletal muscle are pro-IGF-1s, which contain E peptide, rather than mature IGF-1. Both glycosylated and non-glycosylated forms of pro-IGF-1 were expressed in skeletal muscle, whereas non-glycosylated pro-IGF-1 is more potent to activate IGF-1R [96]. Ascenzi et al. analyzed the effects of IGF-1-Ea and IGF-1-Eb in skeletal muscle and found that only IGF-1-Ea promoted a pronounced hypertrophic phenotype in young mice. Interestingly, however, both isoforms of IGF-1 were protective against age-related loss of muscle mass and force [65]. These data suggest that E domains regulate not only IGF-1 production and secretion but also its local activity.

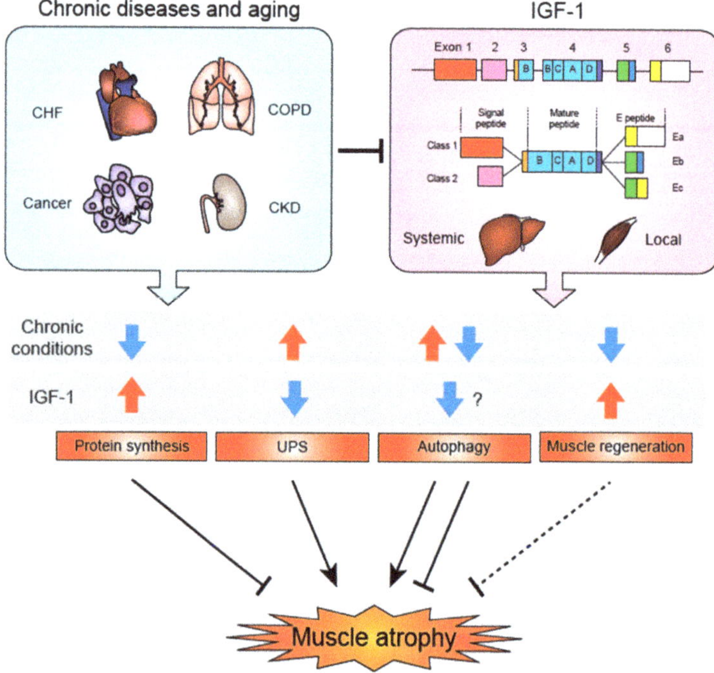

Figure 2. IGF-1 and skeletal muscle atrophy in chronic diseases and aging. In various chronic disease conditions, such as congestive heart failure (CHF), cancer, chronic obstructive pulmonary disease (COPD), and chronic kidney disease (CKD), and aging, muscle atrophy develops through various mechanisms: decreased protein synthesis, increased UPS, and lowered muscle regeneration. Depending

on the pathophysiological conditions, autophagy could be increased or decreased, and both excessive and defective autophagy could lead to muscle atrophy. IGF-1 is thought to decrease autophagy, but the role of IGF-1 regulation of autophagy in chronic disease-induced muscle atrophy is yet to be determined. IGF-1 stimulates skeletal muscle regeneration via activation of satellite cells. Systemic (circulating) IGF-1 is predominantly produced in the liver, whereas locally produced IGF-1 likely acts in a paracrine or autocrine manner. The first two exons of IGF-1 are mutually exclusive and generate different signal peptides, termed Class 1 (exon 1) and Class 2 (exon 2). The mature IGF-1 peptide is coded in exons 3 and 4 (B, C, A, and D domains). Three types of C-terminus E-peptides are generated by alternative splicing. Ea is from exon 6, Eb is from exon 5, and Ec is from part of exons 5 and 6. Class 2 IGF-1 is mainly expressed in the liver (considered to be the systemic isoform), and Class 1 IGF-1 is mainly expressed in peripheral tissues including skeletal muscle. Both systemic and local IGF-1 levels are decreased in various chronic disease conditions, and the combination of these reductions affect protein synthesis, UPS activity, autophagy, and muscle regeneration and regulate the development of muscle atrophy.

8. IGF-1 Binding Proteins in Skeletal Muscle

IGF-1's actions are regulated by six IGF-1-binding proteins (IGFBPs), which serve as IGF-1 transport proteins. Approximately 98% of IGF-1 exists as a bound form to one of the IGFBPs, with IGFBP3 accounting for 80% of all the binding. The binding of IGF-1 to IGFBPs is either in a binary complex (an IGF-1 and an IBFBP), or a ternary complex consisting of an IGF-1, an IGFBP and an IGF binding protein acid labile subunit (IGFALS). The binding of IGF-1 to IGFBPs and IGFALS significantly prolongs the half-life of IGF-1 in circulation. The half-lives of unbound IGF-1, IGF-1 in a binary complex, and IGF-1 in a ternary complex are less than 10 min, 25 min and more than 16 h, respectively [97–99]. Therefore, circulating levels of IGF-1 are greatly affected by IGFBPs and IGFALS. IGFBP3 gene deletion resulted in 40% decrease in serum IGF-1. IGFALS knockout mice showed 60% reduction in serum IGF-1, and also 90% reduction in IGFBP-3 [100]. As IGFBPs bind to IGF-1 with equal or greater affinity compared to IGF-1R, the binding of IGFBPs to IGF-1 is crucial for the regulation of IGF-1's availability to peripheral tissues. Another important function of IGFBPs is to prevent the potential interaction of IGF-1 with insulin receptor (IR). Since IGF-1R and IR are structurally similar and IGF-1 can bind to IR with lower affinity, IGF-1 could cause hypoglycemic effects if it can freely access to the IR [101,102]. IGFBP3 is expressed in the liver and peripheral tissues, and its hepatic expression is regulated by GH, allowing the coordinated regulation of circulating IGF-1 and IGFBP3 levels. When bound to IGF-1, IGFPB3 blocks its binding to IGF-1R, thereby impairing the downstream signaling. Furthermore, IGFBP3 has been shown to exhibit antiproliferative and proapoptotic actions via an IGF-1/IGF-1R-independent mechanism [103]. Studies suggest different roles of IGFBPs in regulation of skeletal muscle function depending on muscle type, age, and atrophy conditions. In a study analyzing the expression of mouse IGFBPs at different ages [104], IGFBP4 and -5 were found to be increased with age, whereas IGFBP3 and -6 were regulated differently between males and females: IGFBP-3 decreased with age in males but increased in females, while IGFBP-6 decreased with age in females and remained unchanged in males. Transgenic overexpression of IGF-1 did not alter expression of any of the IGFBPs. Huang et al. analyzed two independent datasets of gene profiles in pancreatic tumors, and found that IGFBP3 was dramatically increased in pancreatic ductal adenocarcinoma, which causes cancer cachexia with high prevalence. The conditioned medium from pancreatic cancer cells contained high IGFBP3 and caused significant myofiber wasting, which was prevented by IGFBP3 knockdown or neutralizing antibody [105]. These results indicate that IGFBPs inhibit IGF-1's action to induce muscle growth and hypertrophy. Consistently, global overexpression of IGFBP5 in mice caused a severe reduction in prenatal and postnatal growth, resulting in increased neonatal mortality and decreased skeletal muscle weight [106]. Similarly, AAV-mediated overexpression of IGFBP2 in skeletal muscle reduced muscle mass and induced a slower muscle phenotype [107]. On the other hand, mice lacking IGFBP3, -4, or -5 developed normally and only IGFBP4 deficient mice showed a modest (85–90% compared to wild type) growth retardation [108], suggesting that other IGFBPs compensate for the loss of IGFBP5. Indeed, triple knockout of IGFBP3, -4, and -5 had significantly smaller body and quadriceps weight (78% and 60% of wild type, respectively). The triple knockout mice showed lower circulating levels of IGF-1 (45% of wild type) and had lower IGF-1 activity measured by IGF-1R phosphorylation in the cells treated with the serum of the animals (37% of wild type). Interestingly, ERK/MAPK phosphorylation was decreased in the skeletal muscle of triple knockout mice, whereas Akt phosphorylation was not altered [108]. Although these studies indicate that IGFBPs inhibit IGF-1 signaling locally, whether or how IGFBPs affect the outcome of IGF-1 signaling, such as protein synthesis, protein degradation, and autophagy, remains to be elucidated.

9. Skeletal Muscle-Specific IGF-1/IGF-1R Gene Deletion Studies

Liver is the major source of circulating IGF-1, and liver-specific IGF-1 gene deletion resulted in 70–80% reduction in serum IGF-1 levels [109,110]. These studies showed normal growth of the animals and threw into question the requirement of circulating IGF-1 for postnatal body growth. However, a later genetic study using a mouse strain with conditional liver-specific IGF-1 expression in IGF-1 null background demonstrated that IGF-1 from the liver contributes approximately 30% of the adult body size [111]. These studies indicate that liver-derived circulating IGF-1 certainly plays a significant role in growth of animals, although it cannot explain all of IGF-1's growth promoting function in the body.

Transgene, AAV, or electroporation-mediated overexpression of a locally-acting isoform of IGF-1 in skeletal muscle increased muscle mass, myofiber cross sectional area (CSA), and maximum isometric force [112–114]. These animals were protected against aging-associated loss of muscle mass [113], Dex-induced atrophy [115], and Ang II-induced atrophy [39,76], whereas disuse atrophy was not prevented [116].

To define the roles of growth hormone (GH) and IGF-1 signaling in skeletal muscle, Mavalli et al. treated primary myoblasts with GH and IGF-1 [117]. Utilizing GH receptor (GHR) and IGF-1R deficient myoblasts, the authors found that, although both GH and IGF-1 induced myoblast proliferation and fusion, the effect was primarily mediated by IGF-1. Both skeletal muscle-specific GHR and IGF-1R knockout mice exhibited reduced myofiber size and number, and impaired muscle force, which are associated with diminished myoblast fusion. Interestingly, muscle-specific GHR deficient mice developed marked peripheral adiposity, insulin resistance, and glucose intolerance, none of which were observed in muscle IGF-1R knockout mice. These data suggest that GH's action to promote muscle development is mainly mediated by IGF-1, whereas GH facilitates normal insulin action in skeletal muscle independently from IGF-1, leading to changes in global nutrient metabolism. While the study by Mavalli et al. used a cre strain driven by the mef-2c-73k promoter, which is active from an embryonic stage, O'Neill et al. generated skeletal muscle-specific IGF-1R-null mice using the skeletal muscle actin promoter, which is active in differentiated muscle cells, and found that these mice did not show altered body weight or muscle mass [118] In the same study, O'Neill et al. generated mice with muscle-specific double knockout of IGF-1R and IR (MIGIRKO). These animals showed a marked decrease in skeletal muscle mass and fiber size and died earlier (between 15 and 25 weeks), likely due to respiratory failure. Surprisingly, however, glucose and insulin tolerance were not affected in MIGIRKO mice, instead these animals showed increased basal glucose uptake in muscle.

10. IGF-1, Satellite Cells and Skeletal Muscle Regeneration

Skeletal muscle stem cells, or satellite cells (SCs), are normally quiescent and located between the basal lamina and sarcolemma of the myofiber. During growth and after muscle damage, a myogenic program of SCs is activated, and SCs self-renew to maintain their pool and/or differentiate to form myoblasts and eventually myofibers.

IGF-1 has been shown to increase both proliferation and differentiation of cultured myoblasts [119]. When cells are in the proliferative stage, IGF-1 increased the expression of cell-cycle progression factors, whereas IGF-1 promoted myoblast differentiation when cells are withdrawn from the cell cycle by myogenic regulatory factors such as myogenin. L6E9 cell line is a subclone of the parental rat myoblast cell line L6, and does not express IGF-1 whereas IGF-1R expression is intact. Utilizing these cells, Musaro et al. demonstrated that IGF-1 overexpression in differentiated L6E9 cells resulted in pronounced myotube hypertrophy and myogenin induction [120]. PI3K/Akt and MAPK pathways have been shown to mediate downstream signaling of IGF-1 in these cells, although the relative importance of these pathways seems to differ depending on the model systems analyzed. Blockade of MAPK inhibited IGF-1-mediated L6A1 myoblast (another subclone of rat neonatal myoblast cell line L6) proliferation, whereas blockade of PI3K or mTOR abolished myoblast differentiation [121]. In contrast, SCs isolated from muscle-specific IGF-1 transgenic mice showed enhanced proliferative capacity in vitro, and the effect was mediated by activation of PI3K/Akt, independent of MAPK, and downregulation of the cyclin-dependent kinase inhibitor $p27^{Kip1}$, supporting the role of IGF-1 in regulation of the cell cycle in SCs [122].

In addition to the above-mentioned in vitro studies, a series of in vivo studies have shown the importance of IGF-1 signaling in SC function. Barton-Davis et al. proposed that the increase in skeletal muscle mass and strength in mice that overexpress IGF-1 specifically in skeletal muscle is primarily due to the activation of SCs and increased regeneration [112]. In mice treated with hindlimb gamma-irradiation to prevent SC proliferation, approximately half of IGF-1's hypertrophic effect was prevented. However, a following study by Heslop et al. showed hindlimb gamma-irradiation does not completely abolish SC function [123], questioning whether the observation in the study by Barton-Davis et al. is due to depletion of SCs. More recent studies presented conflicting evidence whether SCs are required for muscle hypertrophy [124,125], indicating the importance of careful evaluation and selection of appropriate animal model to address the in vivo contribution of SCs to muscle hypertrophy and the role of IGF-1.

Another consideration needs to be given when analyzing IGF-1's role in SCs is the potential isoform-specific effects of IGF-1. By differential screening, IGF-1 mRNA with the Ec form of E peptide domain (see Section 7) was identified as the transcript that is increased in exercised muscle compared to the resting state, and named mechano-growth factor (MGF). MGF has been shown to stimulate SCs to re-enter the cell cycle and proliferate, facilitating new myofibers to replace damaged myofibers [126] In addition, impairments of IGF-1 splicing to produce MGF were observed during muscle wasting and age-related decline of muscle regeneration [127–129]. Attention needs to be drawn to the usage of the MGF terminology, as some studies use it in referring to the Ec portion of the peptide alone, not including the IGF-1 mature peptide (to avoid any confusion, it is called the Ec peptide in this article). Yang et al. showed that, unlike mature IGF-1, the Ec peptide inhibited C2C12 myoblast terminal differentiation, while increasing proliferation in IGF-1R-independent manner [130]. Furthermore, the Ec peptide increased the proliferative lifespan and delayed senescence of SCs isolated from healthy human subjects [131], and increased the number of primary cultured muscle progenitor cells isolated from patients with muscular dystrophies (CMD, FSHD) and amyotrophic lateral sclerosis (ALS) [132]. However, a contradictory study has been reported [133], in which investigators failed to show any effect of the Ec peptide on C2C12 or primary human myoblasts. A study investigating another IGF-1 isoform class 2 IGF-1-Ea showed that this isoform exerts its hypertrophic effect only when the muscles are in growing status (e.g., during postnatal development or during regeneration) [130]. These studies suggest that IGF-1's effects on SCs differ between isoforms, but no study has been conducted to compare the isoform-specific effects of IGF-1 on muscle regeneration and atrophy in vivo.

IGF-1 seems to regulate SCs in concert with other myogenic factors. The morphogenic factor sonic Hedgehog (Shh) has been reported to be expressed in adult myoblasts and to promote their proliferation and differentiation [134,135]. Both Shh and IGF-1 enhanced Akt and MAPK phosphorylation and myogenic factor expression levels in C2 myoblasts in a dose-responsive manner, having additive effects. In cultured myoblasts isolated from mice with a muscle-specific knockout of Smoothened (Smo), a component of the Shh receptor, IGF-1-induced Akt and MAPK phosphorylation and myogenic differentiation were significantly blocked. Interestingly, Smo physically associates with the IGF-1R, the p85 regulatory subunit of PI3K, and IRS1 in a Shh and IGF-1 dose-responsive manner, indicating that mutual regulation of Shh and IGF-1 occurs at the receptor complex level [136].

Another potential mechanism whereby IGF-1 affects SC function is via regulation of autophagy. Zecchini et al. showed that autophagy is required for neonatal myogenesis and muscle development [137]. Atg7 is an E1-like activating enzyme that regulates fusion of peroxisomal and vacuolar membranes during autophagy, and Atg7 knockdown in SCs caused severe reduction in neonatal myogenesis. Interestingly, the expression of GHR and IGF-1 were reduced in the skeletal muscle of these animals. In primary cultures of neonatal SCs, the defective autophagy decreased proliferation and differentiation, and GH's action to promote myotube growth was completely abolished. As discussed in Section 5, IGF-1 likely reduces autophagy in skeletal muscle. In addition, IGF-1 is known to be reduced in various muscle atrophy conditions. However, it is not clear whether reduced IGF-1 results in an increased autophagy in these conditions, or whether altered autophagy affects the SC functions in these atrophy conditions.

11. Atrophy-Related miRs and Their Potential Regulation of IGF-1 Signaling

Various non-coding RNAs have been proposed to regulate IGF-1 signaling in skeletal muscle. Using miRNA arrays, Li et al. found miR-29b as the only miRNA whose expression was increased in five different in vivo murine muscle atrophy models (denervation, Dex-treatment, fasting, cancer cachexia, and aging) as well as three in vitro atrophy-inducing cell culture models (C2C12 myotubes treated with dexamethasone, TNF-α, and H_2O_2) [138]. miR-29b overexpression promoted muscle atrophy, while miR-29b inhibition prevented denervation-induced muscle atrophy. Importantly, the authors found that miR-29b targets two members of the IGF-1/Akt/mTOR pathway, IGF-1 and PI3K (p85α). miR-29b agomir decreased Akt activity and activated FoxO3A, as well as decreased mTORC1 and p70^{S6K} both in vitro and in vivo. However, conflicting evidence was presented by Goodman et al., showing that Smad3 gene transfer to skeletal muscle decreased miR-29 promoter activity, whereas Akt/mTOR activity was decreased and skeletal muscle atrophy was induced [139]. Furthermore, in a mouse model of CKD-induced muscle atrophy, miR-29 was decreased in skeletal muscle [140], and exosome-mediated miR-29 transfer prevented muscle atrophy [141]. In these studies, phosphatase and tensin homolog (PTEN), which suppresses IGF-1 pathway, and transcriptional repressor Yin Yang 1, which suppresses IGF-1 transcription [142], were shown to be targets of miR-29. It is not clear the potential reasons of these discrepancies, although it is interesting that multiple IGF-1 signaling pathway molecules are potentially targeted by one miRNA, and studies are required to investigate its relationship with other miRNAs discussed below.

During myogenesis, the expression of miR-1 and miR-133 are greatly induced [143], whereas these miRs are reduced during muscle hypertrophy [144]. In C2C12 myoblasts, miR-1 and miR-133 are shown to inhibit the IGF-1 pathway by targeting IGF-1, IGF-1R and HSP70 [145–147], although their roles in skeletal muscle remain to be determined.

miR-128a is highly expressed in brain and skeletal muscle, and it has been shown to target IRS1 [148]. Inhibition of miR-128a in C2C12 myotubes increased IRS1 protein and Akt activity, resulting in increased the size of the myotubes. Furthermore, administration of antisense miR-128a caused skeletal muscle hypertrophy in mice.

miR-486 is encoded in the intron of the Ank1.5 gene, which functions to connect sarcomeres to the sarcoplasmic reticulum [149,150], and is co-expressed with Ank1.5 mRNA [151] miR-486 is found to target PTEN and FoxO1. PTEN dephosphorylates PIP3, and thus inhibits PI3K's activity to phosphorylate PIP2 to produce PIP3, resulting in inhibition of Akt. It is suggested that myostatin inhibits miR-486; overexpression of miR-486 induced myotube hypertrophy via activation of Akt [152] and restored Akt activity and muscle mass in CKD-induced muscle atrophy model [153].

Long noncoding RNAs (lncRNAs) are novel class of regulatory RNAs, which are involved in numerous biological processes via interaction with mRNAs and miRNAs, such as miRNA and lncRNA competition for the same mRNA target, and lncRNAs acting as decoys (or sponges) for miRNAs [154]. By RNA sequencing of hypertrophic and leaner broilers, Li et al. identified a novel lncRNA, termed lncIRS1, is upregulated in hypertrophic muscles. LncIRS1 promoted proliferation and differentiation of myoblasts in vitro, and muscle mass and myofiber size in vivo [155]. Mechanistically, lncIRS1 acts as a molecular sponge for miR-15a, miR-15b-5p, and miR-15c-p, all of which interact with IRS1 mRNA. Increased lncIRS1 inhibits the activity of these miRs, leading to activation of IRS1 and muscle hypertrophy.

These studies strongly suggest the involvement of different miRs in IGF-1-mediated hypertrophy and atrophy prevention. However, some conflicting studies have been published as in the case of miR-29, and further studies are required analyzing these miRs in specific hypertrophy and atrophy models, especially in human patients.

12. IGF-1 Changes in Chronic Conditions and Aging-Associated Sarcopenia

Local overexpression of IGF-1 has successfully rescued muscles in various chronic and experimental muscle atrophy models including Dex injected rats [115], age-related muscle atrophy [113], hindlimb suspension [156], and Ang II infusion in rodents [39,74,76], as well as in the mouse models of ALS [157] and muscular dystrophy [158–160]. Rheumatoid arthritis (RA) is associated with low muscle mass and density, and skeletal muscle of RA patients have been shown to have lower levels of IGF-1, which were associated with the severity of the disease, low appendicular lean mass, and lower myofiber CSA [161]. In a rat RA model, both circulating and skeletal muscle IGF-1 were decreased, the animals showed lower muscle mass, and subcutaneous injection of IGF-1 (100 µg/kg; twice daily for 12 days) increased body and hindlimb muscle weight without changing arthritis. RA increased skeletal muscle MAFbx/Atrogin-1, MuRF1, IGFBP3, and IGFBP5 expression, and IGF-1 treatment attenuated the increase of MAFbx/Atrogin-1, MuRF1, and IGFBP3, but not IGFBP5 [162]. Although a decrease in circulating and skeletal muscle IGF-1 has been reported in various chronic conditions, including cancer, congestive heart failure, chronic kidney disease, and COPD, and aging [5,163] (Figure 2), more studies are required to determine whether IGF-1 administration could be a therapeutic approach to treat muscle atrophy in these patients (discussed below).

In rats bearing AH-130 hepatomas, IGF-1 mRNA expression in hindlimb muscles progressively decreased, whereas that of IGF-1R and IR increased [73]. Circulating and hepatic IGF-1 levels were also decreased in this model, and these changes were associated with increased MAFbx/Atrogin-1 and MuRF1 expression in skeletal muscle. In the Apc$^{Min/+}$ mice, a model of colorectal cancer that develops cachexia, muscle IGF-1 mRNA expression was decreased with suppressed mTOR targets [164]. Similar results were observed in humans, as muscle IGF-1 mRNA was decreased in gastric cancer patients [165]. Interestingly, the reduction of IGF-1 was observed irrespective of the weight loss, suggesting that IGF-1 downregulation precedes cachexia development. In the rat AH-130 hepatoma model, subcutaneous injection of IGF-1 for 16 days attenuated the loss of lean mass at low-dose (0.3 mg/kg/day) and high-dose (3 mg/kg/day), with improvement of spontaneous activity, food intake, and mortality at low-dose treatment [75]. However, in the same animal model, the parenteral administration of IGF-1 did not alter E3 ubiquitin ligase expression or muscle atrophy [73]. The same group of authors also found that phosphorylation of Akt was comparable or increased in skeletal muscle of mice bearing AH-130 hepatomas or C26 colon adenocarcinomas, with hyperphosphorylation of GSK3β, p70^{S6K}, and FoxO1 and reduced eIF2α phosphorylation. Electroporation-mediated IGF-1 gene transfer to the hindlimbs of these animals did not alter myofiber size and muscle mass [166]. These data suggest that IGF-1's effect to treat cancer-induced muscle atrophy may depend on the cancer type, animal species, and/or the route and dose of administration.

Low circulating IGF-1 levels have been associated with an increased risk and worse prognosis of cardiovascular diseases in human patients [167–169]. Deficiency in liver-derived IGF-1 caused impaired contractility of cardiac myocytes and compensatory hypertrophic response [170,171]. Importantly, skeletal muscle atrophy is a hallmark of rodent myocardial infarction models of congestive heart failure. In skeletal muscles of these animals, Akt/mTOR/p70^{S6K} signaling is decreased and

MAFbx/Atrogin-1 is increased [76,172], and transgenic overexpression of IGF-1 inhibited muscle atrophy [167]. Interestingly, skeletal muscle-specific Akt activation decreased cardiac myocyte hypertrophy, decreased interstitial fibrosis, and restored contractile function in the heart, suggesting skeletal muscle to cardiac communication [173].

In experimental models of CKD, UPS-mediated protein degradation is increased with impaired insulin and IGF-1 signaling [174,175]. Interestingly, SC proliferation and differentiation are impaired in CKD mouse model, and Akt activity was decreased. Kido et al. proposed that advanced glycation end-products (AGEs), which is accumulated in patients with CKD, increases fibroblast growth factor 23 (FGF23) and its receptor Klotho-mediated suppression of insulin/IGF-1, leading to inhibition of S differentiation. In addition to UPS activation and SC inhibition, CKD was associated with increased autophagy in skeletal muscles of human CKD patients [176]. These pathways can be initiated by complications associated with CKD, such as metabolic acidosis, defective insulin and IGF-1 signaling, inflammation, increased angiotensin II levels, abnormal appetite regulation, and impaired microRNA responses [175]. Whether IGF-1 administration can rescue CKD-mediated skeletal muscle atrophy remains to be determined.

Muscle dysfunction is one of the most relevant systemic manifestations of patients with COPD, and lower limb muscle atrophy is frequently observed in COPD patients [177]. Survival in patients with COPD is negatively associated with skeletal muscle dysfunction and lower mass, and COPD exacerbations rapidly induce loss of muscle mass and function [178–181]. As in other cases of muscle atrophy, UPS-mediated protein degradation is activated in COPD skeletal muscles [50,182]. In addition, SC senescence and reduced regenerative capacity were reported in SCs isolated from COPD patients [183,184], suggesting the lower SC funciton contributes to muscle atrophy in COPD. However, the contribution of IGF-1 in COPD patients is not clear. Circulating levels of IGF-1 were reported to be unchanged in COPD patients [185], and in cachectic vs. non-cachectic patients with COPD [186]. However, IGF-1 levels were decreased during periods of acute exacerbation [187], which is known to result in muscle atrophy. More careful evaluation of IGF-1 levels and signaling is necessary for these patients.

Aging-associated decline in skeletal muscle mass, quality, and strength mostly occurs in type 2 (fast-twitch) muscle fibers and is associated with marked infiltration of fibrous and adipose tissues in the muscle [188]. Both circulating and local IGF-1 levels are reduced in aging [189], with decreased Akt/mTOR/p70^{S6K} in skeletal muscle [9,189,190]. AAV-mediated IGF-1 gene transfer prevented aging-related muscle changes in old mice [191], and, conversely, deletion of liver specific IGF-1 at one year of age dramatically impaired health span of the mice [192]. Furthermore, the age-related reduction in IGF-1 levels are accompanied by increased IGFBP levels, further decreasing IGF-1 availability to peripheral tissues. Contrary to these findings, Sandri et al. reported only modest to no changes in IGF-1/Akt/mTOR pathway in old human subjects [193].

13. Conclusions

We review the role of IGF-1 and its downstream signaling in skeletal muscle atrophy associated with various chronic diseases and aging. IGF-1 regulates skeletal muscle protein synthesis and protein degradation via the UPS and autophagy, and multiple pathways and mechanisms have been identified (Figure 1). IGF-1 has also been shown to activate satellite cell proliferation, although the involvement of these cells in atrophy development in in vivo animal models and human patients remains to be elucidated. One of the difficulties in IGF-1 research in skeletal muscle is that IGF-1 regulates numerous biological pathways, and these pathways likely interact with each other. For instance, growing evidence suggests the involvement of different miRNAs in IGF-1 signaling, and, considering that each miRNA can target multiple mRNAs, careful examination of changes and biological functions of these miRNAs will be required. Furthermore, it is possible that delivering a specific isoform of IGF-1 may be required to have effective activation of downstream signaling. The role and relative importance of IGF-1 signaling likely differs between muscle atrophy models, and further studies are required to develop effective strategies to apply IGF-1 to treat muscle atrophy in human patients.

Funding: This study was funded by NIH/NHLBI (R01HL080682 and R01HL070241), NIH/NIGMS (P30GM103337, U54GM104940 and 20GM103629), and American Heart Association (15SDG25240022 and 19TPA34850165).

Conflicts of Interest: The authors declare no conflict of interest.

References

1. Stein, A.M.; Silva, T.M.V.; Coelho, F.G.M.; Arantes, F.J.; Costa, J.L.R.; Teodoro, E.; Santos-Galduroz, R.F. Physical exercise, IGF-1 and cognition A systematic review of experimental studies in the elderly. *Dement. Neuropsychol.* **2018**, *12*, 114–122. [CrossRef]
2. Majorczyk, M.; Smolag, D. Effect of physical activity on IGF-1 and IGFBP levels in the context of civilization diseases prevention. *Rocz. Panstw. Zakl. Hig.* **2016**, *67*, 105–111.
3. Nindl, B.C.; Santtila, M.; Vaara, J.; Hakkinen, K.; Kyrolainen, H. Circulating IGF-I is associated with fitness and health outcomes in a population of 846 young healthy men. *Growth Horm. IGF Res.* **2011**, *21*, 124–128. [CrossRef]
4. Maggio, M.; De Vita, F.; Lauretani, F.; Butto, V.; Bondi, G.; Cattabiani, C.; Nouvenne, A.; Meschi, T.; Dall'Aglio, E.; Ceda, G.P. IGF-1, the cross road of the nutritional, inflammatory and hormonal pathways to frailty. *Nutrients* **2013**, *5*, 4184–4205. [CrossRef]
5. Puche, J.E.; Castilla-Cortazar, I. Human conditions of insulin-like growth factor-I (IGF-I) deficiency. *J. Transl. Med.* **2012**, *10*, 224. [CrossRef]
6. Rommel, C.; Bodine, S.C.; Clarke, B.A.; Rossman, R.; Nunez, L.; Stitt, T.N.; Yancopoulos, G.D.; Glass, D.J. Mediation of IGF-1-induced skeletal myotube hypertrophy by PI(3)K/Akt/mTOR and PI(3)K/Akt/GSK3 pathways. *Nat. Cell Biol.* **2001**, *3*, 1009–1013. [CrossRef]
7. Peng, X.D.; Xu, P.Z.; Chen, M.L.; Hahn-Windgassen, A.; Skeen, J.; Jacobs, J.; Sundararajan, D.; Chen, W.S.; Crawford, S.E.; Coleman, K.G.; et al. Dwarfism, impaired skin development, skeletal muscle atrophy, delayed bone development, and impeded adipogenesis in mice lacking Akt1 and Akt2. *Genes Dev.* **2003**, *17*, 1352–1365. [CrossRef] [PubMed]
8. Bodine, S.C.; Latres, E.; Baumhueter, S.; Lai, V.K.; Nunez, L.; Clarke, B.A.; Poueymirou, W.T.; Panaro, F.J.; Na, E.; Dharmarajan, K.; et al. Identification of ubiquitin ligases required for skeletal muscle atrophy. *Science* **2001**, *294*, 1704–1708. [CrossRef]
9. Pallafacchina, G.; Calabria, E.; Serrano, A.L.; Kalhovde, J.M.; Schiaffino, S. A protein kinase B-dependent and rapamycin-sensitive pathway controls skeletal muscle growth but not fiber type specification. *Proc. Natl. Acad. Sci. USA* **2002**, *99*, 9213–9218. [CrossRef]
10. Glass, D.J. Molecular mechanisms modulating muscle mass. *Trends Mol. Med.* **2003**, *9*, 344–350. [CrossRef]
11. Gruner, S.; Peter, D.; Weber, R.; Wohlbold, L.; Chung, M.Y.; Weichenrieder, O.; Valkov, E.; Igreja, C.; Izaurralde, E. The structures of eIF4E-eIF4G complexes reveal an extended interface to regulate translation initiation. *Mol. Cell* **2016**, *64*, 467–479. [CrossRef] [PubMed]
12. Bodine, S.C.; Stitt, T.N.; Gonzalez, M.; Kline, W.O.; Stover, G.L.; Bauerlein, R.; Zlotchenko, E.; Scrimgeour, A.; Lawrence, J.C.; Glass, D.J.; et al. Akt/mTOR pathway is a crucial regulator of skeletal muscle hypertrophy and can prevent muscle atrophy in vivo. *Nat. Cell Biol.* **2001**, *3*, 1014–1019. [CrossRef]
13. Desbois-Mouthon, C.; Cadoret, A.; Blivet-Van Eggelpoel, M.J.; Bertrand, F.; Cherqui, G.; Perret, C.; Capeau, J. Insulin and IGF-1 stimulate the beta-catenin pathway through two signalling cascades involving GSK-3beta inhibition and Ras activation. *Oncogene* **2001**, *20*, 252–259. [CrossRef] [PubMed]
14. Armstrong, D.D.; Esser, K.A. Wnt/beta-catenin signaling activates growth-control genes during overload-induced skeletal muscle hypertrophy. *Am. J. Physiol. Cell Physiol.* **2005**, *289*, C853–C859. [CrossRef]
15. Schakman, O.; Kalista, S.; Bertrand, L.; Lause, P.; Verniers, J.; Ketelslegers, J.M.; Thissen, J.P. Role of Akt/GSK-3beta/beta-catenin transduction pathway in the muscle anti-atrophy action of insulin-like growth factor-I in glucocorticoid-treated rats. *Endocrinology* **2008**, *149*, 3900–3908. [CrossRef] [PubMed]
16. Yoon, M.S. MTOR as a key regulator in maintaining skeletal muscle mass. *Front. Physiol.* **2017**, *8*, 788. [CrossRef]

17.	Verhees, K.J.; Schols, A.M.; Kelders, M.C.; Op den Kamp, C.M.; van der Velden, J.L.; Langen, R.C. Glycogen synthase kinase-3beta is required for the induction of skeletal muscle atrophy. *Am. J. Physiol. Cell Physiol.* **2011**, *301*, C995–C1007. [CrossRef]

18.	Pansters, N.A.; Schols, A.M.; Verhees, K.J.; de Theije, C.C.; Snepvangers, F.J.; Kelders, M.C.; Ubags, N.D.; Haegens, A.; Langen, R.C. Muscle-specific GSK-3beta ablation accelerates regeneration of disuse-atrophied skeletal muscle. *Biochim. Biophys. Acta* **2015**, *1852*, 490–506. [CrossRef]

19.	Verhees, K.J.; Pansters, N.A.; Baarsma, H.A.; Remels, A.H.; Haegens, A.; de Theije, C.C.; Schols, A.M.; Gosens, R.; Langen, R.C. Pharmacological inhibition of GSK-3 in a guinea pig model of LPS-induced pulmonary inflammation: II. Effects on skeletal muscle atrophy. *Respir. Res.* **2013**, *14*, 117. [CrossRef]

20.	McPherron, A.C.; Lawler, A.M.; Lee, S.J. Regulation of skeletal muscle mass in mice by a new TGF-beta superfamily member. *Nature* **1997**, *387*, 83–90. [CrossRef]

21.	Morissette, M.R.; Cook, S.A.; Buranasombati, C.; Rosenberg, M.A.; Rosenzweig, A. Myostatin inhibits IGF-I-induced myotube hypertrophy through Akt. *Am. J. Physiol. Cell Physiol.* **2009**, *297*, C1124–C1132. [CrossRef]

22.	Taylor, W.E.; Bhasin, S.; Artaza, J.; Byhower, F.; Azam, M.; Willard, D.H., Jr.; Kull, F.C., Jr.; Gonzalez-Cadavid, N. Myostatin inhibits cell proliferation and protein synthesis in C2C12 muscle cells. *Am. J. Physiol. Endocrinol. Metab.* **2001**, *280*, E221–E228. [CrossRef] [PubMed]

23.	Trendelenburg, A.U.; Meyer, A.; Rohner, D.; Boyle, J.; Hatakeyama, S.; Glass, D.J. Myostatin reduces Akt/TORC1/p70S6K signaling, inhibiting myoblast differentiation and myotube size. *Am. J. Physiol. Cell Physiol.* **2009**, *296*, C1258–C1270. [CrossRef] [PubMed]

24.	Lipina, C.; Kendall, H.; McPherron, A.C.; Taylor, P.M.; Hundal, H.S. Mechanisms involved in the enhancement of mammalian target of rapamycin signalling and hypertrophy in skeletal muscle of myostatin-deficient mice. *FEBS Lett.* **2010**, *584*, 2403–2408. [CrossRef]

25.	Hennebry, A.; Oldham, J.; Shavlakadze, T.; Grounds, M.D.; Sheard, P.; Fiorotto, M.L.; Falconer, S.; Smith, H.K.; Berry, C.; Jeanplong, F.; et al. IGF1 stimulates greater muscle hypertrophy in the absence of myostatin in male mice. *J. Endocrinol.* **2017**, *234*, 187–200. [CrossRef]

26.	Sartori, R.; Schirwis, E.; Blaauw, B.; Bortolanza, S.; Zhao, J.; Enzo, E.; Stantzou, A.; Mouisel, E.; Toniolo, L.; Ferry, A.; et al. BMP signaling controls muscle mass. *Nat. Genet.* **2013**, *45*, 1309–1318. [CrossRef] [PubMed]

27.	Sartori, R.; Milan, G.; Patron, M.; Mammucari, C.; Blaauw, B.; Abraham, R.; Sandri, M. Smad2 and 3 transcription factors control muscle mass in adulthood. *Am. J. Physiol. Cell Physiol.* **2009**, *296*, C1248–C1257. [CrossRef]

28.	Zhang, L.; Zhou, F.; Ten Dijke, P. Signaling interplay between transforming growth factor-beta receptor and PI3K/AKT pathways in cancer. *Trends Biochem. Sci.* **2013**, *38*, 612–620. [CrossRef]

29.	Bilodeau, P.A.; Coyne, E.S.; Wing, S.S. The ubiquitin proteasome system in atrophying skeletal muscle: Roles and regulation. *Am. J. Physiol. Cell Physiol.* **2016**, *311*, C392–C403. [CrossRef]

30.	Attaix, D.; Ventadour, S.; Codran, A.; Bechet, D.; Taillandier, D.; Combaret, L. The ubiquitin-proteasome system and skeletal muscle wasting. *Essays Biochem.* **2005**, *41*, 173–186. [CrossRef]

31.	Bodine, S.C.; Baehr, L.M. Skeletal muscle atrophy and the E3 ubiquitin ligases MuRF1 and MAFbx/atrogin-1. *Am. J. Physiol. Endocrinol. Metab.* **2014**, *307*, E469–E484. [CrossRef] [PubMed]

32.	Yoshida, T.; Delafontaine, P. Mechanisms of cachexia in chronic disease states. *Am. J. Med. Sci.* **2015**, *350*, 250–256. [CrossRef]

33.	McLoughlin, T.J.; Smith, S.M.; DeLong, A.D.; Wang, H.; Unterman, T.G.; Esser, K.A. FoxO1 induces apoptosis in skeletal myotubes in a DNA-binding-dependent manner. *Am. J. Physiol. Cell Physiol.* **2009**, *297*, C548–C555. [CrossRef] [PubMed]

34.	Senf, S.M.; Dodd, S.L.; Judge, A.R. FOXO signaling is required for disuse muscle atrophy and is directly regulated by Hsp70. *Am. J. Physiol. Cell Physiol.* **2010**, *298*, C38–C45. [CrossRef] [PubMed]

35.	Cai, D.; Frantz, J.D.; Tawa, N.E., Jr.; Melendez, P.A.; Oh, B.C.; Lidov, H.G.; Hasselgren, P.O.; Frontera, W.R.; Lee, J.; Glass, D.J.; et al. IKKbeta/NF-kappaB activation causes severe muscle wasting in mice. *Cell* **2004**, *119*, 285–298. [CrossRef]

36.	Wu, C.L.; Cornwell, E.W.; Jackman, R.W.; Kandarian, S.C. NF-kappaB but not FoxO sites in the MuRF1 promoter are required for transcriptional activation in disuse muscle atrophy. *Am. J. Physiol. Cell Physiol.* **2014**, *306*, C762–C767. [CrossRef]

37. Tilstra, J.S.; Clauson, C.L.; Niedernhofer, L.J.; Robbins, P.D. NF-kappaB in aging and disease. *Aging Dis.* **2011**, *2*, 449–465.

38. Sacheck, J.M.; Ohtsuka, A.; McLary, S.C.; Goldberg, A.L. IGF-I stimulates muscle growth by suppressing protein breakdown and expression of atrophy-related ubiquitin ligases, atrogin-1 and MuRF1. *Am. J. Physiol. Endocrinol. Metab.* **2004**, *287*, E591–E601. [CrossRef]

39. Yoshida, T.; Semprun-Prieto, L.; Sukhanov, S.; Delafontaine, P. IGF-1 prevents ANG II-induced skeletal muscle atrophy via Akt- and Foxo-dependent inhibition of the ubiquitin ligase atrogin-1 expression. *Am. J. Physiol. Heart Circ. Physiol.* **2010**, *298*, H1565–H1570. [CrossRef]

40. Lagirand-Cantaloube, J.; Offner, N.; Csibi, A.; Leibovitch, M.P.; Batonnet-Pichon, S.; Tintignac, L.A.; Segura, C.T.; Leibovitch, S.A. The initiation factor eIF3-f is a major target for atrogin1/MAFbx function in skeletal muscle atrophy. *EMBO J.* **2008**, *27*, 1266–1276. [CrossRef]

41. Lokireddy, S.; Wijesoma, I.W.; Sze, S.K.; McFarlane, C.; Kambadur, R.; Sharma, M. Identification of atrogin-1-targeted proteins during the myostatin-induced skeletal muscle wasting. *Am. J. Physiol. Cell Physiol.* **2012**, *303*, C512–C529. [CrossRef] [PubMed]

42. Clarke, B.A.; Drujan, D.; Willis, M.S.; Murphy, L.O.; Corpina, R.A.; Burova, E.; Rakhilin, S.V.; Stitt, T.N.; Patterson, C.; Latres, E.; et al. The E3 ligase MuRF1 degrades myosin heavy chain protein in dexamethasone-treated skeletal muscle. *Cell Metab.* **2007**, *6*, 376–385. [CrossRef] [PubMed]

43. Cohen, S.; Brault, J.J.; Gygi, S.P.; Glass, D.J.; Valenzuela, D.M.; Gartner, C.; Latres, E.; Goldberg, A.L. During muscle atrophy, thick, but not thin, filament components are degraded by MuRF1-dependent ubiquitylation. *J. Cell Biol.* **2009**, *185*, 1083–1095. [CrossRef]

44. McElhinny, A.S.; Kakinuma, K.; Sorimachi, H.; Labeit, S.; Gregorio, C.C. Muscle-specific RING finger-1 interacts with titin to regulate sarcomeric M-line and thick filament structure and may have nuclear functions via its interaction with glucocorticoid modulatory element binding protein-1. *J. Cell Biol.* **2002**, *157*, 125–136. [CrossRef] [PubMed]

45. Koyama, S.; Hata, S.; Witt, C.C.; Ono, Y.; Lerche, S.; Ojima, K.; Chiba, T.; Doi, N.; Kitamura, F.; Tanaka, K.; et al. Muscle RING-finger protein-1 (MuRF1) as a connector of muscle energy metabolism and protein synthesis. *J. Mol. Biol.* **2008**, *376*, 1224–1236. [CrossRef]

46. Hirner, S.; Krohne, C.; Schuster, A.; Hoffmann, S.; Witt, S.; Erber, R.; Sticht, C.; Gasch, A.; Labeit, S.; Labeit, D. MuRF1-dependent regulation of systemic carbohydrate metabolism as revealed from transgenic mouse studies. *J. Mol. Biol.* **2008**, *379*, 666–677. [CrossRef]

47. Milan, G.; Romanello, V.; Pescatore, F.; Armani, A.; Paik, J.H.; Frasson, L.; Seydel, A.; Zhao, J.; Abraham, R.; Goldberg, A.L.; et al. Regulation of autophagy and the ubiquitin-proteasome system by the FoxO transcriptional network during muscle atrophy. *Nat. Commun.* **2015**, *6*, 6670. [CrossRef]

48. Koncarevic, A.; Jackman, R.W.; Kandarian, S.C. The ubiquitin-protein ligase Nedd4 targets Notch1 in skeletal muscle and distinguishes the subset of atrophies caused by reduced muscle tension. *FASEB J.* **2007**, *21*, 427–437. [CrossRef]

49. Batt, J.; Bain, J.; Goncalves, J.; Michalski, B.; Plant, P.; Fahnestock, M.; Woodgett, J. Differential gene expression profiling of short and long term denervated muscle. *FASEB J.* **2006**, *20*, 115–117. [CrossRef]

50. Plant, P.J.; Brooks, D.; Faughnan, M.; Bayley, T.; Bain, J.; Singer, L.; Correa, J.; Pearce, D.; Binnie, M.; Batt, J. Cellular markers of muscle atrophy in chronic obstructive pulmonary disease. *Am. J. Respir. Cell Mol. Biol.* **2010**, *42*, 461–471. [CrossRef]

51. Cao, X.R.; Lill, N.L.; Boase, N.; Shi, P.P.; Croucher, D.R.; Shan, H.; Qu, J.; Sweezer, E.M.; Place, T.; Kirby, P.A.; et al. Nedd4 controls animal growth by regulating IGF-1 signaling. *Sci. Signal.* **2008**, *1*, ra5. [CrossRef] [PubMed]

52. Nagpal, P.; Plant, P.J.; Correa, J.; Bain, A.; Takeda, M.; Kawabe, H.; Rotin, D.; Bain, J.R.; Batt, J.A. The ubiquitin ligase Nedd4-1 participates in denervation-induced skeletal muscle atrophy in mice. *PLoS ONE* **2012**, *7*, e46427. [CrossRef] [PubMed]

53. Cohen, S.; Zhai, B.; Gygi, S.P.; Goldberg, A.L. Ubiquitylation by Trim32 causes coupled loss of desmin, Z-bands, and thin filaments in muscle atrophy. *J. Cell Biol.* **2012**, *198*, 575–589. [CrossRef] [PubMed]

54. Cohen, S.; Lee, D.; Zhai, B.; Gygi, S.P.; Goldberg, A.L. Trim32 reduces PI3K-Akt-FoxO signaling in muscle atrophy by promoting plakoglobin-PI3K dissociation. *J. Cell Biol.* **2014**, *204*, 747–758. [CrossRef]

55. Paul, P.K.; Gupta, S.K.; Bhatnagar, S.; Panguluri, S.K.; Darnay, B.G.; Choi, Y.; Kumar, A. Targeted ablation of TRAF6 inhibits skeletal muscle wasting in mice. *J. Cell Biol.* **2010**, *191*, 1395–1411. [CrossRef] [PubMed]

56. Paul, P.K.; Bhatnagar, S.; Mishra, V.; Srivastava, S.; Darnay, B.G.; Choi, Y.; Kumar, A. The E3 ubiquitin ligase TRAF6 intercedes in starvation-induced skeletal muscle atrophy through multiple mechanisms. *Mol. Cell Biol.* **2012**, *32*, 1248–1259. [CrossRef]

57. Yang, W.L.; Wang, J.; Chan, C.H.; Lee, S.W.; Campos, A.D.; Lamothe, B.; Hur, L.; Grabiner, B.C.; Lin, X.; Darnay, B.G.; et al. The E3 ligase TRAF6 regulates Akt ubiquitination and activation. *Science* **2009**, *325*, 1134–1138. [CrossRef]

58. Malicdan, M.C.; Noguchi, S.; Nonaka, I.; Saftig, P.; Nishino, I. Lysosomal myopathies: An excessive build-up in autophagosomes is too much to handle. *Neuromuscul. Disord.* **2008**, *18*, 521–529. [CrossRef]

59. Masiero, E.; Agatea, L.; Mammucari, C.; Blaauw, B.; Loro, E.; Komatsu, M.; Metzger, D.; Reggiani, C.; Schiaffino, S.; Sandri, M. Autophagy is required to maintain muscle mass. *Cell Metab.* **2009**, *10*, 507–515. [CrossRef]

60. Zhao, J.; Brault, J.J.; Schild, A.; Cao, P.; Sandri, M.; Schiaffino, S.; Lecker, S.H.; Goldberg, A.L. FoxO3 coordinately activates protein degradation by the autophagic/lysosomal and proteasomal pathways in atrophying muscle cells. *Cell Metab.* **2007**, *6*, 472–483. [CrossRef]

61. Mammucari, C.; Milan, G.; Romanello, V.; Masiero, E.; Rudolf, R.; Del Piccolo, P.; Burden, S.J.; Di Lisi, R.; Sandri, C.; Zhao, J.; et al. FoxO3 controls autophagy in skeletal muscle in vivo. *Cell Metab.* **2007**, *6*, 458–471. [CrossRef] [PubMed]

62. Castets, P.; Lin, S.; Rion, N.; Fulvio, S.D.; Romanino, K.; Guridi, M.; Frank, S.; Tintignac, L.A.; Sinnreich, M.; Rüegg, M.A. Sustained activation of mTORC1 in skeletal muscle inhibits constitutive and starvation-induced autophagy and causes a severe, late-onset myopathy. *Cell Metab.* **2013**, *17*, 731–744. [CrossRef] [PubMed]

63. Fullgrabe, J.; Ghislat, G.; Cho, D.H.; Rubinsztein, D.C. transcriptional regulation of mammalian autophagy at a glance. *J. Cell Sci.* **2016**, *129*, 3059–3066. [CrossRef] [PubMed]

64. Nakashima, K.; Ishida, A. Regulation of autophagy in chick myotubes: Effects of insulin, insulin-like growth factor-I., and amino acids. *J. Poult. Sci.* **2018**, *55*, 257–262. [CrossRef]

65. Ascenzi, F.; Barberi, L.; Dobrowolny, G.; Villa Nova Bacurau, A.; Nicoletti, C.; Rizzuto, E.; Rosenthal, N.; Scicchitano, B.M.; Musaro, A. Effects of IGF-1 isoforms on muscle growth and sarcopenia. *Aging Cell* **2019**, *18*, e12954. [CrossRef]

66. Mizushima, N.; Yoshimori, T.; Levine, B. Methods in mammalian autophagy research. *Cell* **2010**, *140*, 313–326. [CrossRef]

67. Wang, S.; Gu, K. Insulin-like growth factor 1 inhibits autophagy of human colorectal carcinoma drug-resistant cells via the protein kinase B/mammalian target of rapamycin signaling pathway. *Mol. Med. Rep.* **2018**, *17*, 2952–2956. [CrossRef]

68. Yang, M.; Wen, T.; Chen, H.; Deng, J.; Yang, C.; Zhang, Z. Knockdown of insulin-like growth factor 1 exerts a protective effect on hypoxic injury of aged BM-MSCs: Role of autophagy. *Stem Cell Res. Ther.* **2018**, *9*, 284. [CrossRef]

69. Bitto, A.; Lerner, C.; Torres, C.; Roell, M.; Malaguti, M.; Perez, V.; Lorenzini, A.; Hrelia, S.; Ikeno, Y.; Matzko, M.E.; et al. Long-term IGF-I exposure decreases autophagy and cell viability. *PLoS ONE* **2010**, *5*, e12592. [CrossRef]

70. Sandri, M. Protein breakdown in muscle wasting: Role of autophagy-lysosome and ubiquitin-proteasome. *Int. J. Biochem. Cell Biol.* **2013**, *45*, 2121–2129. [CrossRef]

71. Aversa, Z.; Pin, F.; Lucia, S.; Penna, F.; Verzaro, R.; Fazi, M.; Colasante, G.; Tirone, A.; Rossi Fanelli, F.; Ramaccini, C.; et al. Autophagy is induced in the skeletal muscle of cachectic cancer patients. *Sci. Rep.* **2016**, *6*, 30340. [CrossRef] [PubMed]

72. Silva, K.A.S.; Ghiarone, T.; Schreiber, K.; Grant, D.; White, T.; Frisard, M.I.; Sukhanov, S.; Chandrasekar, B.; Delafontaine, P.; Yoshida, T. Angiotensin II suppresses autophagy and disrupts ultrastructural morphology and function of mitochondria in mouse skeletal muscle. *J. Appl. Physiol.* **2019**, *126*, 1550–1562. [CrossRef] [PubMed]

73. Costelli, P.; Muscaritoli, M.; Bossola, M.; Penna, F.; Reffo, P.; Bonetto, A.; Busquets, S.; Bonelli, G.; Lopez-Soriano, F.J.; Doglietto, G.B.; et al. IGF-1 is downregulated in experimental cancer cachexia. *Am. J. Physiol. Regul. Integr. Comp. Physiol.* **2006**, *291*, R674–R683. [CrossRef] [PubMed]

74. Brink, M.; Wellen, J.; Delafontaine, P. Angiotensin II causes weight loss and decreases circulating insulin-like growth factor I in rats through a pressor-independent mechanism. *J. Clin. Investig.* **1996**, *97*, 2509–2516. [CrossRef]

75. Schmidt, K.; von Haehling, S.; Doehner, W.; Palus, S.; Anker, S.D.; Springer, J. IGF-1 treatment reduces weight loss and improves outcome in a rat model of cancer cachexia. *J. Cachexia Sarcopenia Muscle* **2011**, *2*, 105–109. [CrossRef]

76. Song, Y.H.; Li, Y.; Du, J.; Mitch, W.E.; Rosenthal, N.; Delafontaine, P. Muscle-specific expression of IGF-1 blocks angiotensin II-induced skeletal muscle wasting. *J. Clin. Investig.* **2005**, *115*, 451–458. [CrossRef]

77. Penna, F.; Ballaro, R.; Martinez-Cristobal, P.; Sala, D.; Sebastian, D.; Busquets, S.; Muscaritoli, M.; Argiles, J.M.; Costelli, P.; Zorzano, A. Autophagy exacerbates muscle wasting in cancer cachexia and impairs mitochondrial function. *J. Mol. Biol.* **2019**, *431*, 2674–2686. [CrossRef]

78. Thomson, D.M. The role of AMPK in the regulation of skeletal muscle size, hypertrophy, and regeneration. *Int. J. Mol. Sci.* **2018**, *19*, 3125. [CrossRef]

79. Lantier, L.; Mounier, R.; Leclerc, J.; Pende, M.; Foretz, M.; Viollet, B. Coordinated maintenance of muscle cell size control by AMP-activated protein kinase. *FASEB J.* **2010**, *24*, 3555–3561. [CrossRef]

80. Mu, J.; Barton, E.R.; Birnbaum, M.J. Selective suppression of AMP-activated protein kinase in skeletal muscle: Update on 'lazy mice'. *Biochem. Soc. Trans.* **2003**, *31*, 236–241. [CrossRef]

81. O'Neill, H.M.; Maarbjerg, S.J.; Crane, J.D.; Jeppesen, J.; Jorgensen, S.B.; Schertzer, J.D.; Shyroka, O.; Kiens, B.; van Denderen, B.J.; Tarnopolsky, M.A.; et al. AMP-activated protein kinase (AMPK) beta1beta2 muscle null mice reveal an essential role for AMPK in maintaining mitochondrial content and glucose uptake during exercise. *Proc. Natl. Acad. Sci. USA* **2011**, *108*, 16092–16097. [CrossRef] [PubMed]

82. Gordon, S.E.; Lake, J.A.; Westerkamp, C.M.; Thomson, D.M. Does AMP-activated protein kinase negatively mediate aged fast-twitch skeletal muscle mass? *Exerc. Sport Sci. Rev.* **2008**, *36*, 179–186. [CrossRef]

83. Mounier, R.; Lantier, L.; Leclerc, J.; Sotiropoulos, A.; Pende, M.; Daegelen, D.; Sakamoto, K.; Foretz, M.; Viollet, B. Important role for AMPKalpha1 in limiting skeletal muscle cell hypertrophy. *FASEB J.* **2009**, *23*, 2264–2273. [CrossRef] [PubMed]

84. Hilder, T.L.; Baer, L.A.; Fuller, P.M.; Fuller, C.A.; Grindeland, R.E.; Wade, C.E.; Graves, L.M. Insulin-independent pathways mediating glucose uptake in hindlimb-suspended skeletal muscle. *J. Appl. Physiol.* **2005**, *99*, 2181–2188. [CrossRef]

85. Han, B.; Zhu, M.J.; Ma, C.; Du, M. Rat hindlimb unloading down-regulates insulin like growth factor-1 signaling and AMP-activated protein kinase, and leads to severe atrophy of the soleus muscle. *Appl. Physiol. Nutr. Metab.* **2007**, *32*, 1115–1123. [CrossRef]

86. Liu, J.; Peng, Y.; Cui, Z.; Wu, Z.; Qian, A.; Shang, P.; Qu, L.; Li, Y.; Liu, J.; Long, J. Depressed mitochondrial biogenesis and dynamic remodeling in mouse tibialis anterior and gastrocnemius induced by 4-week hindlimb unloading. *IUBMB Life* **2012**, *64*, 901–910. [CrossRef]

87. Cannavino, J.; Brocca, L.; Sandri, M.; Grassi, B.; Bottinelli, R.; Pellegrino, M.A. The role of alterations in mitochondrial dynamics and PGC-1alpha over-expression in fast muscle atrophy following hindlimb unloading. *J. Physiol.* **2015**, *593*, 1981–1995. [CrossRef]

88. Egawa, T.; Goto, A.; Ohno, Y.; Yokoyama, S.; Ikuta, A.; Suzuki, M.; Sugiura, T.; Ohira, Y.; Yoshioka, T.; Hayashi, T.; et al. Involvement of AMPK in regulating slow-twitch muscle atrophy during hindlimb unloading in mice. *Am. J. Physiol. Endocrinol. Metab.* **2015**, *309*, E651–E662. [CrossRef]

89. Guo, Y.; Meng, J.; Tang, Y.; Wang, T.; Wei, B.; Feng, R.; Gong, B.; Wang, H.; Ji, G.; Lu, Z. AMP-activated kinase alpha2 deficiency protects mice from denervation-induced skeletal muscle atrophy. *Arch. Biochem. Biophys.* **2016**, *600*, 56–60. [CrossRef]

90. Tabony, A.M.; Yoshida, T.; Galvez, S.; Higashi, Y.; Sukhanov, S.; Chandrasekar, B.; Mitch, W.E.; Delafontaine, P. Angiotensin II upregulates protein phosphatase 2Calpha and inhibits AMP-activated protein kinase signaling and energy balance leading to skeletal muscle wasting. *Hypertension* **2011**, *58*, 643–649. [CrossRef]

91. Tabony, A.M.; Yoshida, T.; Sukhanov, S.; Delafontaine, P. Protein phosphatase 2C-alpha knockdown reduces angiotensin II-mediated skeletal muscle wasting via restoration of mitochondrial recycling and function. *Skelet. Muscle* **2014**, *4*, 20. [CrossRef] [PubMed]

92. Sanchez, A.M.; Candau, R.; Bernardi, H. Recent data on cellular component turnover: Focus on adaptations to physical exercise. *Cells* **2019**, *8*, 542. [CrossRef]

93. Bikle, D.D.; Tahimic, C.; Chang, W.; Wang, Y.; Philippou, A.; Barton, E.R. Role of IGF-I signaling in muscle bone interactions. *Bone* **2015**, *80*, 79–88. [CrossRef] [PubMed]

94. Temmerman, L.; Slonimsky, E.; Rosenthal, N. Class 2 IGF-1 isoforms are dispensable for viability, growth and maintenance of IGF-1 serum levels. *Growth Horm. IGF Res.* **2010**, *20*, 255–263. [CrossRef] [PubMed]

95. Annibalini, G.; Contarelli, S.; De Santi, M.; Saltarelli, R.; Di Patria, L.; Guescini, M.; Villarini, A.; Brandi, G.; Stocchi, V.; Barbieri, E. The intrinsically disordered E-domains regulate the IGF-1 prohormones stability, subcellular localisation and secretion. *Sci. Rep.* **2018**, *8*, 8919. [CrossRef] [PubMed]

96. Durzynska, J.; Philippou, A.; Brisson, B.K.; Nguyen-McCarty, M.; Barton, E.R. The pro-forms of insulin-like growth factor I (IGF-I) are predominant in skeletal muscle and alter IGF-I receptor activation. *Endocrinology* **2013**, *154*, 1215–1224. [CrossRef]

97. Guler, H.P.; Zapf, J.; Schmid, C.; Froesch, E.R. Insulin-like growth factors I and II in healthy man. Estimations of half-lives and production rates. *Acta Endocrinol.* **1989**, *121*, 753–758. [CrossRef]

98. Rajaram, S.; Baylink, D.J.; Mohan, S. Insulin-like growth factor-binding proteins in serum and other biological fluids: Regulation and functions. *Endocr. Rev.* **1997**, *18*, 801–831. [CrossRef]

99. Lewitt, M.S.; Saunders, H.; Phuyal, J.L.; Baxter, R.C. Complex formation by human insulin-like growth factor-binding protein-3 and human acid-labile subunit in growth hormone-deficient rats. *Endocrinology* **1994**, *134*, 2404–2409. [CrossRef]

100. Yakar, S.; Rosen, C.J.; Bouxsein, M.L.; Sun, H.; Mejia, W.; Kawashima, Y.; Wu, Y.; Emerton, K.; Williams, V.; Jepsen, K.; et al. Serum complexes of insulin-like growth factor-1 modulate skeletal integrity and carbohydrate metabolism. *FASEB J.* **2009**, *23*, 709–719. [CrossRef]

101. Jones, J.I.; Clemmons, D.R. Insulin-like growth factors and their binding proteins: Biological actions. *Endocr. Rev.* **1995**, *16*, 3–34. [CrossRef] [PubMed]

102. Daughaday, W.H.; Rotwein, P. Insulin-like growth factors I. and II. Peptide, messenger ribonucleic acid and gene structures, serum, and tissue concentrations. *Endocr. Rev.* **1989**, *10*, 68–91. [CrossRef] [PubMed]

103. Jogie-Brahim, S.; Feldman, D.; Oh, Y. Unraveling insulin-like growth factor binding protein-3 actions in human disease. *Endocr. Rev.* **2009**, *30*, 417–437. [CrossRef] [PubMed]

104. Oliver, W.T.; Rosenberger, J.; Lopez, R.; Gomez, A.; Cummings, K.K.; Fiorotto, M.L. The local expression and abundance of insulin-like growth factor (IGF) binding proteins in skeletal muscle are regulated by age and gender but not local IGF-I in vivo. *Endocrinology* **2005**, *146*, 5455–5462. [CrossRef]

105. Huang, X.Y.; Huang, Z.L.; Yang, J.H.; Xu, Y.H.; Sun, J.S.; Zheng, Q.; Wei, C.; Song, W.; Yuan, Z. Pancreatic cancer cell-derived IGFBP-3 contributes to muscle wasting. *J. Exp. Clin. Cancer Res.* **2016**, *35*, 46. [CrossRef]

106. Salih, D.A.; Tripathi, G.; Holding, C.; Szestak, T.A.; Gonzalez, M.I.; Carter, E.J.; Cobb, L.J.; Eisemann, J.E.; Pell, J.M. Insulin-like growth factor-binding protein 5 (Igfbp5) compromises survival, growth, muscle development, and fertility in mice. *Proc. Natl. Acad. Sci. USA* **2004**, *101*, 4314–4319. [CrossRef]

107. Swiderski, K.; Martins, K.J.; Chee, A.; Trieu, J.; Naim, T.; Gehrig, S.M.; Baum, D.M.; Brenmoehl, J.; Chau, L.; Koopman, R.; et al. Skeletal muscle-specific overexpression of IGFBP-2 promotes a slower muscle phenotype in healthy but not dystrophic mdx mice and does not affect the dystrophic pathology. *Growth Horm. IGF Res.* **2016**, *30–31*, 1–10. [CrossRef]

108. Ning, Y.; Schuller, A.G.; Bradshaw, S.; Rotwein, P.; Ludwig, T.; Frystyk, J.; Pintar, J.E. Diminished growth and enhanced glucose metabolism in triple knockout mice containing mutations of insulin-like growth factor binding protein-3, -4, and -5. *Mol. Endocrinol.* **2006**, *20*, 2173–2186. [CrossRef]

109. Yakar, S.; Liu, J.L.; Stannard, B.; Butler, A.; Accili, D.; Sauer, B.; LeRoith, D. Normal growth and development in the absence of hepatic insulin-like growth factor I. *Proc. Natl. Acad. Sci. USA* **1999**, *96*, 7324–7329. [CrossRef]

110. Sjogren, K.; Liu, J.L.; Blad, K.; Skrtic, S.; Vidal, O.; Wallenius, V.; LeRoith, D.; Tornell, J.; Isaksson, O.G.; Jansson, J.O.; et al. Liver-derived insulin-like growth factor I (IGF-I) is the principal source of IGF-I in blood but is not required for postnatal body growth in mice. *Proc. Natl. Acad. Sci. USA* **1999**, *96*, 7088–7092. [CrossRef]

111. Stratikopoulos, E.; Szabolcs, M.; Dragatsis, I.; Klinakis, A.; Efstratiadis, A. The hormonal action of IGF1 in postnatal mouse growth. *Proc. Natl. Acad. Sci. USA* **2008**, *105*, 19378–19383. [CrossRef] [PubMed]

112. Barton-Davis, E.R.; Shoturma, D.I.; Sweeney, H.L. Contribution of satellite cells to IGF-I induced hypertrophy of skeletal muscle. *Acta Physiol. Scand.* **1999**, *167*, 301–305. [CrossRef] [PubMed]

113. Musaro, A.; McCullagh, K.; Paul, A.; Houghton, L.; Dobrowolny, G.; Molinaro, M.; Barton, E.R.; Sweeney, H.L.; Rosenthal, N. Localized Igf-1 transgene expression sustains hypertrophy and regeneration in senescent skeletal muscle. *Nat. Genet.* **2001**, *27*, 195–200. [CrossRef] [PubMed]

114. Coleman, M.E.; DeMayo, F.; Yin, K.C.; Lee, H.M.; Geske, R.; Montgomery, C.; Schwartz, R.J. Myogenic vector expression of insulin-like growth factor I stimulates muscle cell differentiation and myofiber hypertrophy in transgenic mice. *J. Biol. Chem.* **1995**, *270*, 12109–12116. [CrossRef]

115. Schakman, O.; Gilson, H.; de Coninck, V.; Lause, P.; Verniers, J.; Havaux, X.; Ketelslegers, J.M.; Thissen, J.P. Insulin-like growth factor-I gene transfer by electroporation prevents skeletal muscle atrophy in glucocorticoid-treated rats. *Endocrinology* **2005**, *146*, 1789–1797. [CrossRef]

116. Criswell, D.S.; Booth, F.W.; DeMayo, F.; Schwartz, R.J.; Gordon, S.E.; Fiorotto, M.L. Overexpression of IGF-I in skeletal muscle of transgenic mice does not prevent unloading-induced atrophy. *Am. J. Physiol.* **1998**, *275*, E373–E379. [CrossRef]

117. Mavalli, M.D.; DiGirolamo, D.J.; Fan, Y.; Riddle, R.C.; Campbell, K.S.; van Groen, T.; Frank, S.J.; Sperling, M.A.; Esser, K.A.; Bamman, M.M.; et al. Distinct growth hormone receptor signaling modes regulate skeletal muscle development and insulin sensitivity in mice. *J. Clin. Investig.* **2010**, *120*, 4007–4020. [CrossRef]

118. O'Neill, B.T.; Lauritzen, H.P.; Hirshman, M.F.; Smyth, G.; Goodyear, L.J.; Kahn, C.R. Differential role of insulin/IGF-1 receptor signaling in muscle growth and glucose homeostasis. *Cell Rep.* **2015**, *11*, 1220–1235. [CrossRef]

119. Engert, J.C.; Berglund, E.B.; Rosenthal, N. Proliferation precedes differentiation in IGF-I-stimulated myogenesis. *J. Cell Biol.* **1996**, *135*, 431–440. [CrossRef]

120. Musaro, A.; Rosenthal, N. Maturation of the myogenic program is induced by postmitotic expression of insulin-like growth factor I. *Mol. Cell Biol.* **1999**, *19*, 3115–3124. [CrossRef]

121. Coolican, S.A.; Samuel, D.S.; Ewton, D.Z.; McWade, F.J.; Florini, J.R. The mitogenic and myogenic actions of insulin-like growth factors utilize distinct signaling pathways. *J. Biol. Chem.* **1997**, *272*, 6653–6662. [CrossRef] [PubMed]

122. Chakravarthy, M.V.; Abraha, T.W.; Schwartz, R.J.; Fiorotto, M.L.; Booth, F.W. Insulin-like growth factor-I extends in vitro replicative life span of skeletal muscle satellite cells by enhancing G1/S cell cycle progression via the activation of phosphatidylinositol 3′-kinase/Akt signaling pathway. *J. Biol. Chem.* **2000**, *275*, 35942–35952. [CrossRef] [PubMed]

123. Heslop, L.; Morgan, J.E.; Partridge, T.A. Evidence for a myogenic stem cell that is exhausted in dystrophic muscle. *J. Cell Sci.* **2000**, *113 Pt 12*, 2299–2308.

124. Egner, I.M.; Bruusgaard, J.C.; Gundersen, K. Satellite cell depletion prevents fiber hypertrophy in skeletal muscle. *Development* **2016**, *143*, 2898–2906. [CrossRef]

125. McCarthy, J.J.; Mula, J.; Miyazaki, M.; Erfani, R.; Garrison, K.; Farooqui, A.B.; Srikuea, R.; Lawson, B.A.; Grimes, B.; Keller, C.; et al. Effective fiber hypertrophy in satellite cell-depleted skeletal muscle. *Development* **2011**, *138*, 3657–3666. [CrossRef]

126. Ates, K.; Yang, S.Y.; Orrell, R.W.; Sinanan, A.C.; Simons, P.; Solomon, A.; Beech, S.; Goldspink, G.; Lewis, M.P. The IGF-I splice variant MGF increases progenitor cells in ALS, dystrophic, and normal muscle. *FEBS Lett.* **2007**, *581*, 2727–2732. [CrossRef]

127. Goldspink, G. Impairment of IGF-I gene splicing and MGF expression associated with muscle wasting. *Int. J. Biochem. Cell Biol.* **2006**, *38*, 481–489. [CrossRef]

128. Hameed, M.; Orrell, R.W.; Cobbold, M.; Goldspink, G.; Harridge, S.D. Expression of IGF-I splice variants in young and old human skeletal muscle after high resistance exercise. *J Physiol* **2003**, *547*, 247–254. [CrossRef]

129. Owino, V.; Yang, S.Y.; Goldspink, G. Age-related loss of skeletal muscle function and the inability to express the autocrine form of insulin-like growth factor-1 (MGF) in response to mechanical overload. *Febs Lett* **2001**, *505*, 259–263. [CrossRef]

130. Yang, S.Y.; Goldspink, G. Different roles of the IGF-I Ec peptide (MGF) and mature IGF-I in myoblast proliferation and differentiation. *FEBS Lett.* **2002**, *522*, 156–160. [CrossRef]

131. Kandalla, P.K.; Goldspink, G.; Butler-Browne, G.; Mouly, V. Mechano Growth Factor E peptide (MGF-E), derived from an isoform of IGF-1, activates human muscle progenitor cells and induces an increase in their fusion potential at different ages. *Mech. Ageing Dev.* **2011**, *132*, 154–162. [CrossRef]

132. Fornaro, M.; Hinken, A.C.; Needle, S.; Hu, E.; Trendelenburg, A.U.; Mayer, A.; Rosenstiel, A.; Chang, C.; Meier, V.; Billin, A.N.; et al. Mechano-growth factor peptide, the COOH terminus of unprocessed insulin-like growth factor 1, has no apparent effect on myoblasts or primary muscle stem cells. *Am. J. Physiol. Endocrinol. Metab.* **2014**, *306*, E150–E156. [CrossRef] [PubMed]

133. Shavlakadze, T.; Chai, J.; Maley, K.; Cozens, G.; Grounds, G.; Winn, N.; Rosenthal, N.; Grounds, M.D. A growth stimulus is needed for IGF-1 to induce skeletal muscle hypertrophy in vivo. *J. Cell Sci.* **2010**, *123*, 960–971. [CrossRef] [PubMed]

134. Elia, D.; Madhala, D.; Ardon, E.; Reshef, R.; Halevy, O. Sonic hedgehog promotes proliferation and differentiation of adult muscle cells: Involvement of MAPK/ERK and PI3K/Akt pathways. *Biochim. Biophys. Acta* **2007**, *1773*, 1438–1446. [CrossRef] [PubMed]

135. Koleva, M.; Kappler, R.; Vogler, M.; Herwig, A.; Fulda, S.; Hahn, H. Pleiotropic effects of sonic hedgehog on muscle satellite cells. *Cell Mol. Life Sci.* **2005**, *62*, 1863–1870. [CrossRef]

136. Madhala-Levy, D.; Williams, V.C.; Hughes, S.M.; Reshef, R.; Halevy, O. Cooperation between Shh and IGF-I in promoting myogenic proliferation and differentiation via the MAPK/ERK and PI3K/Akt pathways requires Smo activity. *J. Cell Physiol.* **2012**, *227*, 1455–1464. [CrossRef]

137. Zecchini, S.; Giovarelli, M.; Perrotta, C.; Morisi, F.; Touvier, T.; Di Renzo, I.; Moscheni, C.; Bassi, M.T.; Cervia, D.; Sandri, M.; et al. Autophagy controls neonatal myogenesis by regulating the GH-IGF1 system through a NFE2L2- and DDIT3-mediated mechanism. *Autophagy* **2019**, *15*, 58–77. [CrossRef]

138. Li, J.; Chan, M.C.; Yu, Y.; Bei, Y.; Chen, P.; Zhou, Q.; Cheng, L.; Chen, L.; Ziegler, O.; Rowe, G.C.; et al. miR-29b contributes to multiple types of muscle atrophy. *Nat. Commun.* **2017**, *8*, 15201. [CrossRef]

139. Goodman, C.A.; McNally, R.M.; Hoffmann, F.M.; Hornberger, T.A. Smad3 induces atrogin-1, inhibits mTOR and protein synthesis, and promotes muscle atrophy in vivo. *Mol. Endocrinol.* **2013**, *27*, 1946–1957. [CrossRef]

140. Wang, X.H.; Hu, Z.; Klein, J.D.; Zhang, L.; Fang, F.; Mitch, W.E. Decreased miR-29 suppresses myogenesis in CKD. *J. Am. Soc. Nephrol.* **2011**, *22*, 2068–2076. [CrossRef]

141. Wang, H.; Wang, B.; Zhang, A.; Hassounah, F.; Seow, Y.; Wood, M.; Ma, F.; Klein, J.D.; Price, S.R.; Wang, X.H. Exosome-mediated miR-29 transfer reduces muscle atrophy and kidney fibrosis in mice. *Mol. Ther.* **2019**, *27*, 571–583. [CrossRef] [PubMed]

142. Blattler, S.M.; Cunningham, J.T.; Verdeguer, F.; Chim, H.; Haas, W.; Liu, H.; Romanino, K.; Ruegg, M.A.; Gygi, S.P.; Shi, Y.; et al. Yin Yang 1 deficiency in skeletal muscle protects against rapamycin-induced diabetic-like symptoms through activation of insulin/IGF signaling. *Cell Metab.* **2012**, *15*, 505–517. [CrossRef] [PubMed]

143. Ge, Y.; Chen, J. MicroRNAs in skeletal myogenesis. *Cell Cycle* **2011**, *10*, 441–448. [CrossRef]

144. McCarthy, J.J.; Esser, K.A. MicroRNA-1 and microRNA-133a expression are decreased during skeletal muscle hypertrophy. *J. Appl. Physiol.* **2007**, *102*, 306–313. [CrossRef]

145. Elia, L.; Contu, R.; Quintavalle, M.; Varrone, F.; Chimenti, C.; Russo, M.A.; Cimino, V.; De Marinis, L.; Frustaci, A.; Catalucci, D.; et al. Reciprocal regulation of microRNA-1 and insulin-like growth factor-1 signal transduction cascade in cardiac and skeletal muscle in physiological and pathological conditions. *Circulation* **2009**, *120*, 2377–2385. [CrossRef]

146. Huang, M.B.; Xu, H.; Xie, S.J.; Zhou, H.; Qu, L.H. Insulin-like growth factor-1 receptor is regulated by microRNA-133 during skeletal myogenesis. *PLoS ONE* **2011**, *6*, e29173. [CrossRef]

147. Kukreti, H.; Amuthavalli, K.; Harikumar, A.; Sathiyamoorthy, S.; Feng, P.Z.; Anantharaj, R.; Tan, S.L.; Lokireddy, S.; Bonala, S.; Sriram, S.; et al. Muscle-specific microRNA1 (miR1) targets heat shock protein 70 (HSP70) during dexamethasone-mediated atrophy. *J. Biol. Chem.* **2013**, *288*, 6663–6678. [CrossRef]

148. Motohashi, N.; Alexander, M.S.; Shimizu-Motohashi, Y.; Myers, J.A.; Kawahara, G.; Kunkel, L.M. Regulation of IRS1/Akt insulin signaling by microRNA-128a during myogenesis. *J. Cell Sci.* **2013**, *126*, 2678–2691. [CrossRef]

149. Porter, N.C.; Resneck, W.G.; O'Neill, A.; Van Rossum, D.B.; Stone, M.R.; Bloch, R.J. Association of small ankyrin 1 with the sarcoplasmic reticulum. *Mol. Membr. Biol.* **2005**, *22*, 421–432. [CrossRef]

150. Zhou, L.; Wang, L.; Lu, L.; Jiang, P.; Sun, H.; Wang, H. Inhibition of miR-29 by TGF-beta-Smad3 signaling through dual mechanisms promotes transdifferentiation of mouse myoblasts into myofibroblasts. *PLoS ONE* **2012**, *7*, e33766. [CrossRef]

151. Small, E.M.; O'Rourke, J.R.; Moresi, V.; Sutherland, L.B.; McAnally, J.; Gerard, R.D.; Richardson, J.A.; Olson, E.N. Regulation of PI3-kinase/Akt signaling by muscle-enriched microRNA-486. *Proc. Natl. Acad. Sci. USA* **2010**, *107*, 4218–4223. [CrossRef] [PubMed]

152. Hitachi, K.; Nakatani, M.; Tsuchida, K. Myostatin signaling regulates Akt activity via the regulation of miR-486 expression. *Int. J. Biochem. Cell Biol.* **2014**, *47*, 93–103. [CrossRef] [PubMed]

153. Xu, J.; Li, R.; Workeneh, B.; Dong, Y.; Wang, X.; Hu, Z. Transcription factor FoxO1, the dominant mediator of muscle wasting in chronic kidney disease, is inhibited by microRNA-486. *Kidney Int.* **2012**, *82*, 401–411. [CrossRef]
154. Lopez-Urrutia, E.; Bustamante Montes, L.P.; Ladron de Guevara Cervantes, D.; Perez-Plasencia, C.; Campos-Parra, A.D. Crosstalk between long non-coding RNAs, Micro-RNAs and mRNAs: Deciphering molecular mechanisms of master regulators in cancer. *Front. Oncol.* **2019**, *9*, 669. [CrossRef]
155. Li, Z.; Cai, B.; Abdalla, B.A.; Zhu, X.; Zheng, M.; Han, P.; Nie, Q.; Zhang, X. LncIRS1 controls muscle atrophy via sponging miR-15 family to activate IGF1-PI3K/AKT pathway. *J. Cachexia Sarcopenia Muscle* **2019**, *10*, 391–410. [CrossRef]
156. Park, S.; Brisson, B.K.; Liu, M.; Spinazzola, J.M.; Barton, E.R. Mature IGF-I excels in promoting functional muscle recovery from disuse atrophy compared with pro-IGF-IA. *J. Appl. Physiol.* **2014**, *116*, 797–806. [CrossRef]
157. Dobrowolny, G.; Giacinti, C.; Pelosi, L.; Nicoletti, C.; Winn, N.; Barberi, L.; Molinaro, M.; Rosenthal, N.; Musaro, A. Muscle expression of a local Igf-1 isoform protects motor neurons in an ALS mouse model. *J. Cell Biol.* **2005**, *168*, 193–199. [CrossRef]
158. Barton, E.R.; Morris, L.; Musaro, A.; Rosenthal, N.; Sweeney, H.L. Muscle-specific expression of insulin-like growth factor I counters muscle decline in mdx mice. *J. Cell Biol.* **2002**, *157*, 137–148. [CrossRef]
159. Kumar, A.; Yamauchi, J.; Girgenrath, T.; Girgenrath, M. Muscle-specific expression of insulin-like growth factor 1 improves outcome in Lama2Dy-w mice, a model for congenital muscular dystrophy type 1A. *Hum. Mol. Genet.* **2011**, *20*, 2333–2343. [CrossRef]
160. Shavlakadze, T.; White, J.; Hoh, J.F.; Rosenthal, N.; Grounds, M.D. Targeted expression of insulin-like growth factor-I reduces early myofiber necrosis in dystrophic mdx mice. *Mol. Ther.* **2004**, *10*, 829–843. [CrossRef]
161. Baker, J.F.; Von Feldt, J.M.; Mostoufi-Moab, S.; Kim, W.; Taratuta, E.; Leonard, M.B. Insulin-like Growth Factor 1 and Adiponectin and Associations with Muscle Deficits, Disease Characteristics, and Treatments in Rheumatoid Arthritis. *J. Rheumatol.* **2015**, *42*, 2038–2045. [CrossRef] [PubMed]
162. Lopez-Menduina, M.; Martin, A.I.; Castillero, E.; Villanua, M.A.; Lopez-Calderon, A. Systemic IGF-I administration attenuates the inhibitory effect of chronic arthritis on gastrocnemius mass and decreases atrogin-1 and IGFBP-3. *Am. J. Physiol. Regul. Integr. Comp. Physiol.* **2010**, *299*, R541–R551. [CrossRef] [PubMed]
163. AsghariHanjani, N.; Vafa, M. The role of IGF-1 in obesity, cardiovascular disease, and cancer. *Med. J. Islam. Repub. Iran* **2019**, *33*, 56. [CrossRef] [PubMed]
164. White, J.P.; Baynes, J.W.; Welle, S.L.; Kostek, M.C.; Matesic, L.E.; Sato, S.; Carson, J.A. The Regulation of Skeletal Muscle Protein Turnover during the Progression of Cancer Cachexia in the ApcMin/ + Mouse. *PLoS ONE* **2011**, *6*, e24650. [CrossRef] [PubMed]
165. Bonetto, A.; Penna, F.; Aversa, Z.; Mercantini, P.; Baccino, F.M.; Costelli, P.; Ziparo, V.; Lucia, S.; Rossi Fanelli, F.; Muscaritoli, M. Early changes of muscle insulin-like growth factor-1 and myostatin gene expression in gastric cancer patients. *Muscle Nerve* **2013**, *48*, 387–392. [CrossRef]
166. Penna, F.; Bonetto, A.; Muscaritoli, M.; Costamagna, D.; Minero, V.G.; Bonelli, G.; Fanelli, F.R.; Baccino, F.M.; Costelli, P. Muscle atrophy in experimental cancer cachexia: Is the IGF-1 signaling pathway involved? *Int. J. Cancer* **2010**, *127*, 1706–1717. [CrossRef]
167. Ungvari, Z.; Csiszar, A. The emerging role of IGF-1 deficiency in cardiovascular aging: Recent advances. *J. Gerontol. A Biol. Sci. Med. Sci.* **2012**, *67*, 599–610. [CrossRef]
168. Laughlin, G.A.; Barrett-Connor, E.; Criqui, M.H.; Kritz-Silverstein, D. The prospective association of serum insulin-like growth factor I (IGF-I) and IGF-binding protein-1 levels with all cause and cardiovascular disease mortality in older adults: The rancho bernardo study. *J. Clin. Endocrinol. Metab.* **2004**, *89*, 114–120. [CrossRef]
169. Conti, E.; Carrozza, C.; Capoluongo, E.; Volpe, M.; Crea, F.; Zuppi, C.; Andreotti, F. Insulin-like growth factor-1 as a vascular protective factor. *Circulation* **2004**, *110*, 2260–2265. [CrossRef]
170. Bailey-Downs, L.C.; Mitschelen, M.; Sosnowska, D.; Toth, P.; Pinto, J.T.; Ballabh, P.; Valcarcel-Ares, M.N.; Farley, J.; Koller, A.; Henthorn, J.C.; et al. Liver-specific knockdown of IGF-1 decreases vascular oxidative stress resistance by impairing the Nrf2-dependent antioxidant response: A novel model of vascular aging. *J. Gerontol. A Biol. Sci. Med. Sci.* **2012**, *67*, 313–329. [CrossRef] [PubMed]
171. Li, Q.; Ceylan-Isik, A.F.; Li, J.; Ren, J. Deficiency of insulin-like growth factor 1 reduces sensitivity to aging-associated cardiomyocyte dysfunction. *Rejuvenation Res.* **2008**, *11*, 725–733. [CrossRef]

172. Schulze, P.C.; Fang, J.; Kassik, K.A.; Gannon, J.; Cupesi, M.; MacGillivray, C.; Lee, R.T.; Rosenthal, N. Transgenic overexpression of locally acting insulin-like growth factor-1 inhibits ubiquitin-mediated muscle atrophy in chronic left-ventricular dysfunction. *Circ. Res.* **2005**, *97*, 418–426. [CrossRef] [PubMed]

173. Araki, S.; Izumiya, Y.; Hanatani, S.; Rokutanda, T.; Usuku, H.; Akasaki, Y.; Takeo, T.; Nakagata, N.; Walsh, K.; Ogawa, H. Akt1-mediated skeletal muscle growth attenuates cardiac dysfunction and remodeling after experimental myocardial infarction. *Circ. Heart Fail.* **2012**, *5*, 116–125. [CrossRef] [PubMed]

174. Price, S.R.; Gooch, J.L.; Donaldson, S.K.; Roberts-Wilson, T.K. Muscle atrophy in chronic kidney disease results from abnormalities in insulin signaling. *J. Ren. Nutr.* **2010**, *20*, S24–S28. [CrossRef]

175. Wang, X.H.; Mitch, W.E. Mechanisms of muscle wasting in chronic kidney disease. *Nat. Rev. Nephrol.* **2014**, *10*, 504–516. [CrossRef]

176. Zhang, Y.Y.; Gu, L.J.; Huang, J.; Cai, M.C.; Yu, H.L.; Zhang, W.; Bao, J.F.; Yuan, W.J. CKD autophagy activation and skeletal muscle atrophy-a preliminary study of mitophagy and inflammation. *Eur. J. Clin. Nutr.* **2019**, *73*, 950–960. [CrossRef]

177. Barreiro, E.; Jaitovich, A. Muscle atrophy in chronic obstructive pulmonary disease: Molecular basis and potential therapeutic targets. *J. Thorac. Dis.* **2018**, *10*, S1415–S1424. [CrossRef]

178. Maltais, F.; Decramer, M.; Casaburi, R.; Barreiro, E.; Burelle, Y.; Debigare, R.; Dekhuijzen, P.N.; Franssen, F.; Gayan-Ramirez, G.; Gea, J.; et al. An official american thoracic society/european respiratory society statement: Update on limb muscle dysfunction in chronic obstructive pulmonary disease. *Am. J. Respir. Crit. Care Med.* **2014**, *189*, e15–e62. [CrossRef]

179. Marquis, K.; Debigare, R.; Lacasse, Y.; LeBlanc, P.; Jobin, J.; Carrier, G.; Maltais, F. Midthigh muscle cross-sectional area is a better predictor of mortality than body mass index in patients with chronic obstructive pulmonary disease. *Am. J. Respir. Crit. Care Med.* **2002**, *166*, 809–813. [CrossRef]

180. Shrikrishna, D.; Patel, M.; Tanner, R.J.; Seymour, J.M.; Connolly, B.A.; Puthucheary, Z.A.; Walsh, S.L.; Bloch, S.A.; Sidhu, P.S.; Hart, N.; et al. Quadriceps wasting and physical inactivity in patients with COPD. *Eur. Respir. J.* **2012**, *40*, 1115–1122. [CrossRef]

181. Swallow, E.B.; Reyes, D.; Hopkinson, N.S.; Man, W.D.; Porcher, R.; Cetti, E.J.; Moore, A.J.; Moxham, J.; Polkey, M.I. Quadriceps strength predicts mortality in patients with moderate to severe chronic obstructive pulmonary disease. *Thorax* **2007**, *62*, 115–120. [CrossRef] [PubMed]

182. Doucet, M.; Russell, A.P.; Leger, B.; Debigare, R.; Joanisse, D.R.; Caron, M.A.; LeBlanc, P.; Maltais, F. Muscle atrophy and hypertrophy signaling in patients with chronic obstructive pulmonary disease. *Am. J. Respir. Crit. Care Med.* **2007**, *176*, 261–269. [CrossRef] [PubMed]

183. Theriault, M.E.; Pare, M.E.; Maltais, F.; Debigare, R. Satellite cells senescence in limb muscle of severe patients with COPD. *PLoS ONE* **2012**, *7*, e39124. [CrossRef] [PubMed]

184. Theriault, M.E.; Pare, M.E.; Lemire, B.B.; Maltais, F.; Debigare, R. Regenerative defect in vastus lateralis muscle of patients with chronic obstructive pulmonary disease. *Respir. Res.* **2014**, *15*, 35. [CrossRef]

185. Piehl-Aulin, K.; Jones, I.; Lindvall, B.; Magnuson, A.; Abdel-Halim, S.M. Increased serum inflammatory markers in the absence of clinical and skeletal muscle inflammation in patients with chronic obstructive pulmonary disease. *Respiration* **2009**, *78*, 191–196. [CrossRef]

186. Debigare, R.; Marquis, K.; Cote, C.H.; Tremblay, R.R.; Michaud, A.; LeBlanc, P.; Maltais, F. Catabolic/anabolic balance and muscle wasting in patients with COPD. *Chest* **2003**, *124*, 83–89. [CrossRef]

187. Crul, T.; Spruit, M.A.; Gayan-Ramirez, G.; Quarck, R.; Gosselink, R.; Troosters, T.; Pitta, F.; Decramer, M. Markers of inflammation and disuse in vastus lateralis of chronic obstructive pulmonary disease patients. *Eur. J. Clin. Investig.* **2007**, *37*, 897–904. [CrossRef]

188. Walston, J.D. Sarcopenia in older adults. *Curr. Opin. Rheumatol.* **2012**, *24*, 623–627. [CrossRef]

189. Leger, B.; Derave, W.; De Bock, K.; Hespel, P.; Russell, A.P. Human sarcopenia reveals an increase in SOCS-3 and myostatin and a reduced efficiency of Akt phosphorylation. *Rejuvenation Res.* **2008**, *11*, 163–175. [CrossRef]

190. Cuthbertson, D.; Smith, K.; Babraj, J.; Leese, G.; Waddell, T.; Atherton, P.; Wackerhage, H.; Taylor, P.M.; Rennie, M.J. Anabolic signaling deficits underlie amino acid resistance of wasting, aging muscle. *FASEB J.* **2005**, *19*, 422–424. [CrossRef]

191. Barton-Davis, E.R.; Shoturma, D.I.; Musaro, A.; Rosenthal, N.; Sweeney, H.L. Viral mediated expression of insulin-like growth factor I blocks the aging-related loss of skeletal muscle function. *Proc. Natl. Acad. Sci. USA* **1998**, *95*, 15603–15607. [CrossRef] [PubMed]

192. Gong, Z.; Kennedy, O.; Sun, H.; Wu, Y.; Williams, G.A.; Klein, L.; Cardoso, L.; Matheny, R.W., Jr.; Hubbard, G.B.; Ikeno, Y.; et al. Reductions in serum IGF-1 during aging impair health span. *Aging Cell* **2014**, *13*, 408–418. [CrossRef] [PubMed]

193. Sandri, M.; Barberi, L.; Bijlsma, A.Y.; Blaauw, B.; Dyar, K.A.; Milan, G.; Mammucari, C.; Meskers, C.G.; Pallafacchina, G.; Paoli, A.; et al. Signalling pathways regulating muscle mass in ageing skeletal muscle: The role of the IGF1-Akt-mTOR-FoxO pathway. *Biogerontology* **2013**, *14*, 303–323. [CrossRef] [PubMed]

Review

Glucose-Regulated Protein 94 (GRP94): A Novel Regulator of Insulin-Like Growth Factor Production

Yair Argon [1,2,*], Sophie E. Bresson [3], Michal T. Marzec [3] and Adda Grimberg [4,5]

[1] Department of Pathology and Laboratory Medicine, Perelman School of Medicine, University of Pennsylvania, Philadelphia, PA 19104, USA

[2] Division of Cell Pathology, Children's Hospital of Philadelphia, Philadelphia, PA 19104, USA

[3] Department of Biomedical Sciences, University of Copenhagen, Blegdamsvej 3B, 2200 Copenhagen N, Denmark; sophie.bresson@sund.ku.dk (S.E.B.); michal@sund.ku.dk (M.T.M.)

[4] Department of Pediatrics, Perelman School of Medicine, University of Pennsylvania, Philadelphia, PA 19104, USA; grimberg@email.chop.edu

[5] Division of Endocrinology and Diabetes, Children's Hospital of Philadelphia, Philadelphia, PA 19104, USA

* Correspondence: yargon@pennmedicine.upenn.edu; Tel.: +01-267-426-5131

Received: 17 July 2020; Accepted: 4 August 2020; Published: 6 August 2020

Abstract: Mammals have two insulin-like growth factors (IGF) that are key mediators of somatic growth, tissue differentiation, and cellular responses to stress. Thus, the mechanisms that regulate the bioavailability of IGFs are important in both normal and aberrant development. IGF-I levels are primarily controlled via the growth hormone-IGF axis, in response to nutritional status, and also reflect metabolic diseases and cancer. One mechanism that controls IGF bioavailablity is the binding of circulating IGF to a number of binding proteins that keep IGF in a stable, but receptor non-binding state. However, even before IGF is released from the cells that produce it, it undergoes an obligatory association with a ubiquitous chaperone protein, GRP94. This binding is required for secretion of a properly folded, mature IGF. This chapter reviews the known aspects of the interaction and highlights the specificity issues yet to be determined. The IGF–GRP94 interaction provides a potential novel mechanism of idiopathic short stature, involving the obligatory chaperone and not just IGF gene expression. It also provides a novel target for cancer treatment, as GRP94 activity can be either inhibited or enhanced.

Keywords: glucose regulated protein (GRP) 94; insulin-like growth factor; obligate chaperone

1. Introduction

As primary drivers of growth and proliferation at the somatic, tissue and cellular levels, the insulin-like growth factors (IGFs) must have tightly regulated activity—in terms of amount, timing, and spatial specificity and coordination. The "somatomedin hypothsies", the original framework of understanding IGF-I production and action, has undergone considerable development with time, as additional layers of complexity and nuance have been elucidated [1]. This review summarizes traditionally recognized regulators of IGF production and action in health and disease, and adds yet another: glucose regulated protein 94 (GPR94). GRP94, a ubiquitously expressed chaperone in the endoplasmic reticulum, is required for the proper folding and secretion of both IGFs. Although insulin, the other member of the IGF hormone family, shares similarities with the IGFs structurally (including approximately 50% amino acid homology to the IGFs), in their receptors and signaling, and is itself chaperoned by GRP94, this review will focus on the IGFs. By providing a novel nexus of regulating IGF production, GRP94 and its alterations serve as a potentially new mechanism of dysregulated growth, such as idiopathic short stature and cancer, and thereby may lead to new therapeutic interventions.

2. Regulators of IGF Levels Clinically

Insulin-like growth factor (IGF)-I is made throughout the body, though ~70% of circulating levels are of hepatic origin. Clinically, measurement of circulating IGF-I concentration is used most commonly as a marker of growth hormone (GH) bioactivity. Due to the GH dependence of transcription of the genes encoding IGF-I and its principal circulating partner, IGF binding protein (IGFBP)-3, normal levels of IGF-I and IGFBP-3 generally reflect normal GH activity (i.e., exclude GH deficiency) [2,3]. Further, because serum levels of these molecules do not fluctuate diurnally as does the pulsatile secretion of GH, they serve as convenient indicators of GH secretion that are measurable on a random blood sample. Indeed, circulating concentration of IGF-I was shown in 114 healthy children and adolescents to both correlate with height and reflect spontaneous GH secretion [4]. IGF-I levels are monitored during treatment with exogenous GH to assess adherence and inform dose adjustments [2,3,5]. Conversely, IGF-I levels are also employed in diagnosing and evaluating treatment efficacy for acromegaly, the state of excessive GH secretion [6,7].

Altered nutritional status represents the major non-GH, clinically relevant regulator of IGF-I levels [8]. Inadequate nutrition causes hepatic GH insensitivity, with a primary reduction in IGF-I production despite normal or even elevated levels of GH from loss of the normal IGF-I negative feedback on the pituitary gland and hypothalamus. This has been characterized in patients with anorexia nervosa [9,10], but also can be seen due to malnutrition from food insecurity (e.g., marasmus and kwashiorkor), milder dietary intake inadequacy that fails to meet daily demands, or specific micronutrient deficit like zinc deficiency [8]. Gastrointestinal diseases like Crohn's disease, celiac disease, cystic fibrosis, and gastroesophageal reflux disease also can lower circulating IGF-I levels even in the absence of gastrointestinal symptoms [8]. In the other extreme, obesity blunts GH secretion, yet circulating IGF-I concentrations in obese individuals often are maintained or even higher than in non-obese controls. Although obesity did not associate with higher total IGF-I levels in multiple studies (indeed, an inverse U-shaped association between IGF-I z-score and body mass index (BMI) was shown in a population-based study of more than 6000 adults [11]), IGF-I bioavailability is increased [8].

Circulating IGF-I concentrations are dependent on age, gender, and gonadal status (pubertal status in adolescence, reproductive function in adults, and even whether estrogen replacement in women is administered orally versus transdermally) [12]. Normal IGF-I concentrations rise from infancy through childhood, peak during puberty (the pubertal growth spurt), and gradually decline throughout adulthood [13,14]. Alterations in circulating IGF-I concentration can also result from hepatic disease, renal dysfunction, and diabetes mellitus. Even controlling for these factors, various IGF-I assays often produce discordant results [15], leading to calls for harmonization of IGF-I assays to prevent diagnostic misclassifications and to allow meaningful inter-study comparisons of results in the literature [2,3,16].

Although closely related to IGF-I, IGF-II shows a different pattern. In healthy newborns, IGF-II concentrations are highest in the fetus, about half that of adult levels, which are reached by 1 year of age and persist through life [17]. In contrast, rodent IGF-II expression declines early postnatally, such that rodent models cannot serve to elucidate the physiologic function of IGF-II persistence in humans. Nonetheless, it is clear IGF-II plays an important role, especially in prenatal growth. In humans, the *IGF2* gene is imprinted and paternally expressed. DNA hypomethylation in the region of the *IGF2* gene that reduces paternal *IGF2* expression presents clinically with Silver–Russell syndrome, which is characterized by both prenatal and postnatal growth failure, often with body segment asymmetry [18]. *IGF2* overexpression (biallelic expression from relaxation or loss of imprinting) can result in Beckwith–Wiedemann syndrome, an overgrowth syndrome that also affects both pre- and postnatal growth, can include disproportionate growth (such as macroglossia and hemihypertrophy), and is associated with increased risk of embryonal tumors [19].

3. Molecular Regulation of IGF-I Production

IGF action can be modulated at the level of hormone (or autocrine/paracrine) production, hormone bioavailability, or receptor density and activity. Transcription of the *Igf1* gene, on chromosome *12q23.2*, is regulated by GH, in a mechanistic axis termed the "somatomedin hypothesis", which has undergone considerable refinement over the years [20,21]. The GH receptor is a paradigmal cytokine receptor, whose activation recruits the tyrosine kinase Janus kinase 2 (JAK2). This, in turn, activates the signal transducers and activators of transcription, especially STAT5b [22], a transcription factor that stimulates transcription of the IGF genes. Apart from JAK2, the GH receptor also directly activates the Src tyrosine kinase pathway, the MAP kinase pathway, the PI3K/Akt pathway, and the mTOR pathway [23–25]. Naturally occurring and experimentally induced mutations have shed light on the specificity of second messenger recruitment and the specificity of outcomes conferred by them. Mutations in the GH receptor or in STAT5b are known to impair IGF production and lead to patients with primary IGF-deficient growth failure [26,27].

The *Igf1* gene encodes the 7.6 kD, single chain 70 amino acid polypeptide, that is cross-linked by disulfide bridges [28]. The *Igf2* gene, on chromosome 11p15.5, encodes the single chain 67 amino acid polypeptide [29] and is primarily regulated by imprinting. As discussed below, IGF-I and IGF-II production is determined not only by their transcriptional regulation, but also by interactions with dedicated molecular chaperones.

IGF bioavailability is primarily regulated via a family of six high-affinity IGF binding proteins (IGFBPs). Additional lower-affinity IGF binding proteins (named IGFBP-related proteins (IGFBPrPs)) were found by in silico searches for homology to the known IGFBPs; many of these molecules were previously known in other contexts, serving roles in normal or neoplastic growth [30]. The IGFBPs prolong the circulating half-life of IGF, transport the IGFs to target cells, and modulate the interaction of the IGFs with their surface membrane receptors via competitive inhibition. Local proteases, such as metalloproteinase pregnancy associated plasma protein A2 (PAPPA2), cleave the IGFBP, releasing the IGF for binding and activation of its receptor [31]. Of note, the IGFBPs have been found to perform various IGF-independent functions as well [30].

The actions of both IGF-I and IGF-II is mediated via the type 1 IGF receptor (IGF1R), an α2β2 transmembrane tyrosine kinase receptor that upon ligand binding, autophosphorylates and phosphorylates signaling pathways such as MAPK and PI3K/Akt [32]. IGF1R bears a high degree of homology to the insulin receptor, and αβ-hemireceptors of the two can form functioning hybrid receptors [33]. IGF1R signaling is regulated by internalization of bound receptors into clathrin-coated pits [34]. Phosphatases like SHP2 also can limit IGF1R signaling [35]. In contrast to IGF1R, the type 2 IGF receptor binds only IGF-II with high affinity, does not possess any recognizable signal transduction mechanism, and is identical to the cation-independent mannose-6-phosphate (CIM6P) receptor, a protein involved in intracellular lysosomal targeting [36]. Given the complexities of the system, an IGF-IR kinase receptor activation assay has been developed to measure IGF-1R stimulating activity (phosphorylation of tyrosine residues of the IGFIR) as a means of assessing the net effects of the system's multiple players in various conditions [37].

Whereas the transcriptional and translational regulation of IGF production follow usual paradigms, the post-translational regulation of IGFs has unique features. First, as discussed below, maturation of IGF-I depends on the activity of GRP94, and without it IGF-I does not complete its biosynthesis and is not secreted [38]. This chaperone interaction provides a new element of regulation outside the "standard" GH system. Second, as mentioned above, the complexes of IGF-I with the binding proteins are important for IGF-I function.

4. GRP94

GRP94 is a glucose-regulated protein of 94 kDa molecular size, encoded by the gene *HSP90B1* (*OMIM* #191175), whose chromosomal location is remarkably close to the IGF-I gene. Its expression is ubiquitous and its transcription is upregulated by low glucose tension [39], among other conditions.

GRP94 also is commonly known as gp96, ERp99, or endoplasmin [40], referring to its extensive glycosylation and its abundance in the endoplasmic reticulum. It has the domain structure typical of the heat shock 90 (HSP90) family of proteins, including a C-terminal domain that mediates the constitutive dimerization of GRP94 (Figure 1). Like all family members, the N-terminal domain of GRP94 is a typical ATP-binding domain [41] that affects the dimerization of GRP94 and its action cycle [42]. The N-terminal domain also mediates binding of antigenic peptides [43] through which GRP94 activates T cells, the basis for the immunological function of GRP94 [43,44]. The protein chaperone function of GRP94 also requires the N-terminal ATPase domain [45], but the protein binding site is thought to reside in the C-terminal domain, around residues 652–678 [46].

GRP94 is an essential chaperone for multiple receptors and secreted proteins [47] (Table 1). Protein interaction data using GRP94-sufficient and -deficient cells show physical interactions with ~200 proteins and effects on expression levels of ~500 proteins [48], including some of the verified substrates listed in Table 1. Much of this interactome remains to be characterized. For some of the substrates (e.g., IGFs) there is genetic evidence that GRP94 is essential, whereas others can be expressed properly (albeit at lower abundance) even absent GRP94. For many of the susbstrates (e.g., thyroglobulin), data only show physical association without a physiological conseqence.

As can be gleaned from this non-exhaustive listing, GRP94 substrates (also called "clients") are found in a variety of tissues and cell types. These substrates share no common structural motif that would predict their association with GRP94, nor do they share a common protein fold or a characteristic post-translational modification, aside from internal disulfide bonds (Table 1). The only obvious common denominator is that the substrates are secreted or membrane-bound proteins that are made in the endoplasmic reticulum. Importantly, even in cases of verified GRP94-substrate interactions, there can be exceptional isoforms or family members that do not interact, for example, TLR3 vs. most other TLRs [49].

Table 1. Protein substrates of GRP94.

Protein Substrate	Refs	Major Expression	Notable Structural Features
Immunoglobulin L chain H chain	[50,51]	B lineage cells	Immunoglobulin fold Non-glycosylated secreted Glycosylated secreted or membrane-spanning
Toll-like receptor	[52,53]	Ubiquitous, predominantly leukocytes	Leucine-rich repeats; Membrane-spanning proteins
Integrins	[48]	Ubiquitous	Immunoglobulin superfamily membrane-spanning heterodimers
LRP6	[54,55]		EGF-like repeats β-propeller motifs Interacts indirectly via MesD
Glycoprotein Ib-IX-V complex	[56]	Platelets	
Insulin-like proteins IGF-I IGF-II Insulin	[38,57,58]	Ubiquitous Pancreatic β cells	
Thyroglobulin	[59,60]	Thyrocytes	Large disulfide-bonded protease-type repeats
GARP	[61]	Treg cells; Platelets	Membrane-spanning leucine-rich repeats domains Tregs and platelets

No GRP94–susbstrate complex has been purified and analysed so far, so the exact mode of interaction currently can only be simulated, as shown in Figure 1 with human IGF-I [62] and GRP94 [63], using the ZDOCK algorithm [64]. Furthermore, as GRP94, like all chaperones, binds substrates that have not yet reached their final three-dimensional structure, the precise interaction is only approximated based on known mutations in the interacting proteins. This is an implicit limitation of docking studies such as that shown in Figure 1.

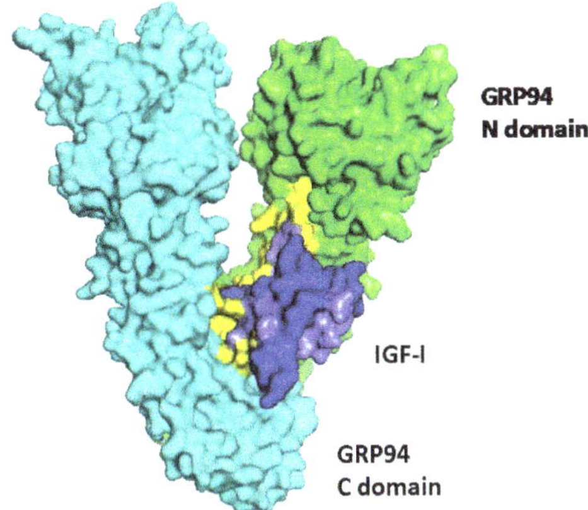

Figure 1. A predicted complex between GRP94 and IGF-I. The crystal structure of human IGF-1 (1IMX [63]) was docked onto the crystal structure of GRP94 (2O1U [64]) with the ZDOCK algorithm (version 3.0.2 [65]). The two monomers of GRP94, are shown in cyan and green, with the N-terminal and C-terminal domains indicated. The interacting amino acids are colored yellow. The complex shown is the highest scoring predicted complex, and eight other complexes out of the 10 highest scoring ones overlap with it, predicting the same topology of binding. The GRP94 interacting residues are from from the internal face of the Middle domain and the C-terminal domain of the chaperone. The IGF-I interacting residues are mostly derived from its N-terminal 32 amino acids, colored light blue, while the C-terminal 28 residues (deep blue) are mostly predicted as non-interacting.

Along with molecular specificity, another enigmatic feature of GRP94 is the paucity of co-chaperones. Its cytosolic homolog, HSP90, has a well-known set of auxiliary proteins that form transient complexes and impact the quality and/or speed of enhanced folding of the respective substrates [65]. Even other types of ER chaperones have well known co-chaperones, some general for all substrates and some substrate-dedicated [66]. In contrast, GRP94 is currently known to work with only one co-chaperone, CNPY3 [49] (see below). As far as the insulin/IGF substrates, GRP94 co-chaperones are presumably yet to be characterized, because genetic data show that ASNA-1 is an evolutionarily conserved ATPase that is important for insulin/IGF maturation in both worms and mammals [67] (see the next section). Further characterization of co-chaperones will no doubt explain many of the unresolved details about the action cycle of the GRP94 chaperone machine.

GRP94 differs from the cytosolic HSP90 orthologs in inherent, functionally-relevant structural properties [42] such as the nucleotide-dependent conformational changes of the N-terminal domain [63] as well as the interactions mediated by the charged linker domain. These differences lead to a different action cycle of this protein [68,69] and probably also to its ability to chaperone folding of client proteins without the many co-chaperones that are required for activity of the cytosolic HSP90 orthologs [70].

5. GRP94 as an Obligate Chaperone for IGF-I and IGF-II Production

The dependence of IGF-I maturation and secretion on GRP94 is a property also exhibited by IGF-II [38,71] and insulin [58] (Table 1), and even by the insulin-like proteins of the nematode *C. elegans*, some of which have only weak primary sequence similarity to IGF-I [72], showing that it is evolutionarily conserved. In contrast, within the TLR family of substrates, TLR3 is exceptional in its refractiveness to GRP94, showing the selectivity of substrate selection by GRP94 [49]. The chaperone dependence of the IGFs is based on physical association of pro-IGFs (or pro-insulin) with GRP94, an association that is transient and occurs early during biosynthesis [38,73]. The precise amino acids of the pro-insulins that interact with GRP94 have not been mapped, but some experiments plus molecular modelling indicate that the pro-insulins do not bind at the site of GRP94 that is responsible for binding of antigenic peptides [43], but rather bind at a more distal site encompassing the middle and C-terminal domains of GRP94 [58]. Apparently, their binding site overlaps residues 652–678 [58], the region that was identified for binding integrins and TLRs [46]. Nonetheless, despite such overlap, there is more complex specificity built into client selection, for example, the pair Met658/Met662 residues are essential for integrin folding but not TLRs [46].

Folding of client proteins often involves not just a chaperone protein, but also recruitment of additional proteins dedicated for the client, which serve as co-chaperones. In the case of GRP94, the ER luminal protein CNPY3 binds to GRP94 when it is engaged in biosynthesis of Toll-like receptors, but not other clients. CNPY3 and GRP94 interact with each other and with the TLR client in nucleotide-dependent manner [49]. Similar complexes have not been defined for the IGF/insulin proteins, but they likely exist; ASNA1, for example, is an ATPase expressed in insulin/IGF-producing cells in both worms and humans which regulates insulin secretion [67].

The biological importance of the IGF–GRP94 interaction is highlighted by the discovery of a hypomorphic variant of human GRP94, P300L, that affects the IGF chaperone activity and limits IGF biosynthesis [74]. Only four homozygotes have been identified so far, a lower frequency than expected from genetic principles [74], and heterozygous carriers of P300L are a noncommon single nucleotide polymorphism with frequencies of 1–4% in various populations. Carriers have 9% lower circulating IGF-1 concentration. In cell models of P300L heterozygosity, half as much IGF was secreted relative to wild type GRP94 [74]. It should be noted that the marked dependence of IGFs on GRP94 activity is unusual—depletion of the chaperone has much milder effects on the expression of some GRP94 client proteins compared to the secretion of the insulin family clients [52].

Why does the insulin/IGF structure require GRP94? At present, this question is not properly answered, and the available data only provide hints. The insulin-like family of proteins is unusual in that they are made initially as small (less than 100 amino acids) pro-proteins, that are processed proteolytically [72,75]. Furthermore, most of these sequences encode for three disulfides [72] that need to be bonded in a precise order within a small molecular space, a considerable folding challenge [76,77]. The surprising finding that at least one IGF-I variant has alternative folded states [78] underscores the folding difficulty, which is one likely reason for the need for molecular chaperones. As GRP94 has been found to interact with PDIs [60,79] it may act as a scaffolding protein in the recruitment of PDIs during the folding of the substrates [80].

The essential chaperoning role of GRP94 towards IGFs has implications for cell growth, for normal tissue differentiation and for cancer progression. A common cellular stress situation is the withdrawal of growth factors from cells, many of which respond to such stress by autocrine production of the growth factors [81]. However, cells with mutated or drug-inhibited GRP94 cannot produce these growth factors [38], leading to arrested growth/differentiation and, in extreme cases, cell death. The requirement for functional GRP94 in development is illustrated by the dramatic impact of tissue-specific GRP94 depletion on striated muscle [57], where myotube fusion and expression of contractile proteins downstream of the master MyoD transcription program are inhibited, coincident with the known need for synergistic input from growth factor signaling [73]. In cancer, elevated expression of GRP94 is observed in melanoma, ovarian cancer, multiple myeloma, lung cancer, and inflammation-associated colon cancer. GRP94 expression in cancer cells is closely linked to cancer growth and metastasis through a number of its clients, as listed above [82]. In part, this is due to response of the GRP94 promoter to some aspects of the tumor microenvironment that may include low glucose level [83], but is distinct from hypoxia [84]. The increase in GRP94 expression in tumors is tightly linked to their increased cellular proliferation rate and migration capacities and to their increased production of growth factors [85].

Constitutive overexpression of GRP94 is a common survival pathway that is usually used during oxidative stress [86], reflecting the many pathways that involve GRP94. The above three examples highlight situations that upregulate GRP94 more specifically, because of subsets of interacting proteins.

Association with GRP94 is by no means the only protein–protein interaction that IGF-I undergoes. Circulating IGF-I is secreted mainly by the liver and circulates bound to IGF-binding proteins (IGFBPs), either as binary complexes or ternary complexes primarily with IGFBP-3 or IGFBP-5 and an acid-labile subunit (ALS). The components of these circulating complexes are produced by different cells and the complexes assemble after secretion to the circulation [87]. The complexes are important for the stability of circulating IGFs and also for their signaling function; in the absence of IGFBPs, there is much lower level of serum IGF-1, but surprisingly, this neither predicted growth potential or skeletal integrity nor defined GH secretion or metabolic abnormalities [88].

Each IGF-I associated protein appears to play a distinct role in determining musculoskeletal phenotype, with different effects on cortical and trabecular bone compartments and the striated muscles [88,89]. The differential effects of hepatic vs. autocrine/paracrine IGF-I is likewise attributable to different complexes, either due to differential assembly or to different proteases at the target tissue that cleave the IGFBP to release IGF-I to interact with IGF1R, IGF1R and insulin receptor density, etc. [90]. Similarly, when skeletal muscle deletion of GRP94 is used to limit production of IGF-I, endocrine and paracrine IGF-I are shown to regulate both tissue growth and body plan [57,88].

6. Conclusions

6.1. Implications for Novel Mechanisms of Idiopathic Short Stature

The novel association of IGFs with GRP94 that modulates production of IGFs has two implications for idiopathic short stature and other growth deficiencies. First, as allelic variations of the chaperone are likely to be new determinants of stature, there are now new target genes that can be screened to explain clinical observations. Second, based on other interacting proteins like ASNA1, we expect that the production of multiple insulin-related proteins will be sensitive to the activity of these proteins, in addition to the quality of the insulin-related protein itself. The chaperone machinery can be modulated with small molecules, so either GRP94 itself or its interacting proteins provide a novel way to manipulate both IGF deficiency and excessive production.

6.2. Implications for Cancer Treatment

The IGF-GRP94 interaction has similar implications for cancer, suggesting a potential role for both genetic screening for and pharmacological agents against the GRP94 machinery. Tumors often conscript IGF system overactivity as a means of furthering the neoplastic process. Autocrine/paracrine IGF overexpression by tumor cells or supporting stromal cells serves to stimulate cancer progression. As an obligate chaperone for secretion of both IGF-I and IGF-II, GRP94 may become a novel target for anti-neoplastic therapy. This may be particularly important for cancers like breast and prostate that become IGF-dependent when they become sex hormone-independent. It is conceivable that differences in the association of IGF-I and IGF-II with GRP94 can be exploited for selective tissue targeting of compounds and it is also possible that distinct, tissue-specific auxiliary proteins are involved in complex formation in different cells and therefore can be targeted selectively.

Funding: This research was funded in part by NIH grant AG-18001 (to Y.A. and A.G.) and by European Foundation for the Study of Diabetes/Lilly European Diabetes Research Programme (to M.T.M.).

Conflicts of Interest: A.G. and Y.A. declare that they have interest in intellectual property in growth control, held by CHOP, related to the subject of this work.

References

1. Le, R.D.; Bondy, C.; Yakar, S.; Liu, J.L.; Butler, A. The somatomedin hypothesis: 2001. *Endocr. Rev.* **2001**, *22*, 53–74.
2. Grimberg, A.; Divall, S.A.; Polychronakos, C.; Allen, D.B.; Cohen, L.E.; Quintos, J.B.; Rossi, W.C.; Feudtner, C.; Murad, M.H. On Behalf Of The Drug And Therapeutics Committee And Ethics Committee Of The Pediatric Endocrine Society Guidelines for Growth Hormone and Insulin-Like Growth Factor-I Treatment in Children and Adolescents: Growth Hormone Deficiency, Idiopathic Short Stature, and Primary Insulin-Like Growth Factor-I Deficiency. *Horm. Res. Paediatr.* **2016**, *86*, 361–397. [CrossRef] [PubMed]
3. Collett-Solberg, P.F.; Ambler, G.; Backeljauw, P.F.; Bidlingmaier, M.; Biller, B.M.; Boguszewski, M.C.; Cheung, P.T.; Choong, C.S.Y.; Cohen, L.E.; Cohen, P.; et al. Diagnosis, Genetics, and Therapy of Short Stature in Children: A Growth Hormone Research Society International Perspective. *Horm. Res. Paediatr.* **2019**, *92*, 1–14. [CrossRef] [PubMed]
4. Blum, W.F.; Albertsson-Wikland, K.; Rosberg, S.; Ranke, M.B. Serum levels of IGF-I and IGF binding protein 3 reflect spontaneous growth hormone secretion. *J. Clin. Endocrinol. Metab.* **1993**, *76*, 1610–1616. [PubMed]
5. Allen, D.B.; Backeljauw, P.; Bidlingmaier, M.; Biller, B.M.K.; Boguszewski, M.; Burman, P.; Butler, G.; Chihara, K.; Christiansen, J.S.; Cianfarani, S.; et al. GH safety workshop position paper: A critical appraisal of recombinant human GH therapy in children and adults. *Eur. J. Endocrinol.* **2015**, *174*, P1–P9. [CrossRef]
6. Katznelson, L.; Laws, E.R.; Melmed, S.; Molitch, M.; Murad, M.H.; Utz, A.; Wass, J.A.H. Acromegaly: An Endocrine Society Clinical Practice Guideline. *J. Clin. Endocrinol. Metab.* **2014**, *99*, 3933–3951. [CrossRef]
7. Qiao, N.; He, M.; Shen, M.; Zhang, Q.; Zhang, Z.; Shou, X.; Wang, Y.; Zhao, Y.; Tritos, N.A. Comparative Efficacy Of Medical Treatment For Acromegaly: A Systemic Review And Network Meta-Analysis Of Integrated Randomized Trials And Observational Studies. *Endocr. Pr.* **2020**, *26*, 454–462. [CrossRef]
8. Hawkes, C.P.; Grimberg, A. Insulin-Like Growth Factor-I is a Marker for the Nutritional State. *Pediatr. Endocrinol. Rev.* **2015**, *13*, 499–511.
9. Misra, M.; Klibanski, A. Endocrine consequences of anorexia nervosa. *Lancet Diabetes Endocrinol.* **2014**, *2*, 581–592. [CrossRef]
10. Schorr, M.; Miller, K.K. The endocrine manifestations of anorexia nervosa: Mechanisms and management. *Nat. Rev. Endocrinol.* **2016**, *13*, 174–186. [CrossRef]
11. Schneider, H.J.; Saller, B.; Klotsche, J.; Erwa, W.; Wittchen, H.-U.; März, W.; Stalla, G.K. Opposite associations of age-dependent insulin-like growth factor-I standard deviation scores with nutritional state in normal weight and obese subjects. *Eur. J. Endocrinol.* **2006**, *154*, 699–706. [CrossRef] [PubMed]
12. Van Der Klaauw, A.; Biermasz, N.R.; Zelissen, P.M.J.; Pereira, A.M.; Lentjes, E.G.W.M.; Smit, J.W.; Van Thiel, S.W.; Romijn, J.A.; Roelfsema, F. Administration route dependent effects of estrogens on IGF-I levels during fixed GH replacement in women with hypopituitarism. *Eur. J. Endocrinol.* **2007**, *157*, 709–716. [CrossRef] [PubMed]
13. Lofqvist, C.; Andersson, E.; Gelander, L.; Rosberg, S.; Blum, W.F.; Wikland, A.K. Reference values for IGF-I throughout childhood and adolescence: A model that accounts simultaneously for the effect of gender, age, and puberty. *J. Clin. Endocrinol. Metab.* **2001**, *86*, 5870–5876. [CrossRef] [PubMed]
14. Yamamoto, H.; Sohmiya, M.; Oka, N.; Kato, Y. Effects of aging and sex on plasma insulin-like growth factor I (IGF-I) levels in normal adults. *Eur. J. Endocrinol.* **1991**, *124*, 497–500. [CrossRef]
15. Chanson, P.; Arnoux, A.; Mavromati, M.; Brailly-Tabard, S.; Massart, C.; Young, J.; Piketty, M.-L.; Souberbielle, J.-C. VARIETE Investigators Reference Values for IGF-I Serum Concentrations: Comparison of Six Immunoassays. *J. Clin. Endocrinol. Metab.* **2016**, *101*, 3450–3458. [CrossRef]
16. Clemmons, D.R.; Participants, O.B.O.T.C. Consensus Statement on the Standardization and Evaluation of Growth Hormone and Insulin-like Growth Factor Assays. *Clin. Chem.* **2011**, *57*, 555–559. [CrossRef]
17. Bergman, D.; Halje, M.; Nordin, M.; Engström, W. Insulin-Like Growth Factor 2 in Development and Disease: A Mini-Review. *Gerontology* **2013**, *59*, 240–249. [CrossRef]
18. Wakeling, E.L.; Brioude, F.; Lokulo-Sodipe, O.; O'Connell, S.M.; Salem, J.; Bliek, J.; Canton, A.P.M.; Chrzanowska, K.H.; Davies, J.H.; Dias, R.P.; et al. Diagnosis and management of Silver–Russell syndrome: First international consensus statement. *Nat. Rev. Endocrinol.* **2016**, *13*, 105–124. [CrossRef]
19. Brioude, F.; Kalish, J.M.; Mussa, A.; Foster, A.C.; Bliek, J.; Ferrero, G.B.; Boonen, S.E.; Cole, T.; Baker, R.; Bertoletti, M.; et al. Expert consensus document: Clinical and molecular diagnosis, screening and management

of Beckwith-Wiedemann syndrome: An international consensus statement. *Yearb. Paediatr. Endocrinol.* **2018**, *14*, 229–249. [CrossRef]

20. Camacho-Hubner, C. Normal Physiology of Growth Hormone and Insulin-Like Growth Factors in Childhood. Available online: https://www.ncbi.nlm.nih.gov/sites/books/NBK279164/ (accessed on 6 August 2020).

21. Yakar, S.; Werner, H.; Rosen, C.J.; Rosen, C. 40 YEARS OF IGF1: Insulin-like growth factors: Actions on the skeleton. *J. Mol. Endocrinol.* **2018**, *61*, T115–T137. [CrossRef]

22. Teglund, S.; McKay, C.; Schuetz, E.; Van Deursen, J.M.; Stravopodis, D.J.; Wang, D.; Brown, M.; Bodner, S.; Grosveld, G.; Ihle, J.N. Stat5a and Stat5b Proteins Have Essential and Nonessential, or Redundant, Roles in Cytokine Responses. *Cell* **1998**, *93*, 841–850. [CrossRef]

23. Hayashi, A.A.; Proud, C.G. The rapid activation of protein synthesis by growth hormone requires signaling through mTOR. *Am. J. Physiol. Metab.* **2007**, *292*, E1647–E1655. [CrossRef] [PubMed]

24. Li, C.-J.; Elsasser, T.; Kahl, S. AKT/eNOS signaling module functions as a potential feedback loop in the growth hormone signaling pathway. *J. Mol. Signal.* **2009**, *4*, 1. [CrossRef]

25. Jin, H.; Lanning, N.J.; Carter-Su, C. JAK2, But Not Src Family Kinases, Is Required for STAT, ERK, and Akt Signaling in Response to Growth Hormone in Preadipocytes and Hepatoma Cells. *Mol. Endocrinol.* **2008**, *22*, 1825–1841. [CrossRef] [PubMed]

26. Hwa, V. STAT5B deficiency: Impacts on human growth and immunity. *Growth Horm. IGF Res.* **2016**, *28*, 16–20. [CrossRef] [PubMed]

27. Kofoed, E.M.; Hwa, V.; Little, B.; Woods, K.A.; Buckway, C.K.; Tsubaki, J.; Pratt, K.L.; Bezrodnik, L.; Jasper, H.; Tepper, A.; et al. Growth Hormone Insensitivity Associated with aSTAT5bMutation. *N. Engl. J. Med.* **2003**, *349*, 1139–1147. [CrossRef]

28. Rinderknecht, E.; Humbel, R.E. The amino acid sequence of human insulin-like growth factor I and its structural homology with proinsulin. *J. Boil. Chem.* **1978**, *253*, 2769–2776.

29. Tricoli, J.V.; Rall, L.B.; Scott, J.; Bell, G.I.; Shows, T.B. Localization of insulin-like growth factor genes to human chromosomes 11 and 12. *Nature* **1984**, *310*, 784–786. [CrossRef]

30. Hwa, V.; Oh, Y.; Rosenfeld, G.R. The insulin-like growth factor-binding protein (IGFBP) superfamily. *Endocr. Rev.* **1999**, *20*, 761–787. [CrossRef]

31. Dauber, A.; Muñoz-Calvo, M.T.; Barrios, V.; Domené, H.M.; Kloverpris, S.; Serra-Juhé, C.; Desikan, V.; Pozo, J.; Muzumdar, R.; Martos-Moreno, G.Á.; et al. Mutations in pregnancy-associated plasma protein A2 cause short stature due to low IGF -I availability. *EMBO Mol. Med.* **2016**, *8*, 363–374. [CrossRef]

32. Li, J.; Choi, E.; Yu, H.; Bai, X.-C. Structural basis of the activation of type 1 insulin-like growth factor receptor. *Nat. Commun.* **2019**, *10*, 1–11. [CrossRef] [PubMed]

33. Slaaby, R.; Schäffer, L.; Lautrup-Larsen, I.; Andersen, A.S.; Shaw, A.C.; Mathiasen, I.S.; Brandt, J. Hybrid Receptors Formed by Insulin Receptor (IR) and Insulin-like Growth Factor I Receptor (IGF-IR) Have Low Insulin and High IGF-1 Affinity Irrespective of the IR Splice Variant. *J. Boil. Chem.* **2006**, *281*, 25869–25874. [CrossRef] [PubMed]

34. Foti, M.; Moukil, M.A.; Dudognon, P.; Carpentier, J.-L. Insulin and IGF-1 receptor trafficking and signalling. *Novartis Found. Symp.* **2004**, *262*, 125–147. [CrossRef]

35. Maile, L.A.; Clemmons, D.R. The $\alpha V\beta 3$ Integrin Regulates Insulin-Like Growth Factor I (IGF-I) Receptor Phosphorylation by Altering the Rate of Recruitment of the Src-Homology 2-Containing Phosphotyrosine Phosphatase-2 to the Activated IGF-I Receptor. *Endocrinology* **2002**, *143*, 4259–4264. [CrossRef] [PubMed]

36. Brown, J.; Jones, E.Y.; Forbes, B.E.; Jones, E.Y. Keeping IGF-II under control: Lessons from the IGF-II–IGF2R crystal structure. *Trends Biochem. Sci.* **2009**, *34*, 612–619. [CrossRef] [PubMed]

37. Janssen, J.A.; Varewijck, A.J.; Brugts, M.P. The insulin-like growth factor-I receptor stimulating activity (IRSA) in health and disease. *Growth Horm. IGF Res.* **2019**, 16–28. [CrossRef]

38. Ostrovsky, O.; Ahmed, N.T.; Argon, Y. The Chaperone Activity of GRP94 Toward Insulin-like Growth Factor II Is Necessary for the Stress Response to Serum Deprivation. *Mol. Boil. Cell* **2009**, *20*, 1855–1864. [CrossRef]

39. Lee, A.S. Glucose-regulated proteins in cancer: Molecular mechanisms and therapeutic potential. *Nat. Rev. Cancer* **2014**, *14*, 263–276. [CrossRef]

40. Koch, G.L.; Macer, D.R.; Wooding, F.B. Endoplasmin is a reticuloplasmin. *J. Cell Sci.* **1988**, *90*, 485–491.

41. Soldano, K.L.; Jivan, A.; Nicchitta, C.V.; Gewirth, D.T. Structure of the N-terminal Domain of GRP94. *J. Boil. Chem.* **2003**, *278*, 48330–48338. [CrossRef]

42. Jahn, M.; Tych, K.; Girstmair, H.; Steinmaßl, M.; Hugel, T.; Buchner, J.; Rief, M. Folding and Domain Interactions of Three Orthologs of Hsp90 Studied by Single-Molecule Force Spectroscopy. *Structure* **2018**, *26*, 96–105. [CrossRef] [PubMed]

43. Biswas, C.; Sriram, U.; Ciric, B.; Ostrovsky, O.; Gallucci, S.; Argon, Y. The N-terminal fragment of GRP94 is sufficient for peptide presentation via professional antigen-presenting cells. *Int. Immunol.* **2006**, *18*, 1147–1157. [CrossRef] [PubMed]

44. Blachere, N.E.; Srivastava, P.K. Heat shock protein-based cancer vaccines and related thoughts on immunogenicity of human tumors. *Semin. Cancer Boil.* **1995**, *6*, 349–355. [CrossRef]

45. Ostrovsky, O.; Makarewich, C.A.; Snapp, E.L.; Argon, Y. An essential role for ATP binding and hydrolysis in the chaperone activity of GRP94 in cells. *Proc. Natl. Acad. Sci. USA* **2009**, *106*, 11600–11605. [CrossRef]

46. Wu, S.; Hong, F.; Gewirth, D.; Guo, B.; Liu, B.; Li, H.Z. The molecular chaperone gp96/GRP94 interacts with Toll-like receptors and integrins via its C-terminal hydrophobic domain. *J. Biol. Chem.* **2012**, *287*, 6735–6742. [CrossRef] [PubMed]

47. Ansa-Addo, E.A.; Thaxton, J.; Hong, F.; Wu, X.B.; Zhang, Y.L.; Fufle, W.C.; Metelli, A.; Riesenberg, B.; Williams, K.; Gewirth, T.D.; et al. Clients and Oncogenic Roles of Molecular Chaperone gp96/grp94. *Curr. Top. Med. Chem.* **2016**, *16*, 2765–2778. [CrossRef]

48. Hong, F.; Liu, B.; Chiosis, G.; Gewirth, D.T.; Li, Z. α7 Helix Region of αI Domain Is Crucial for Integrin Binding to Endoplasmic Reticulum Chaperone gp96. *J. Boil. Chem.* **2013**, *288*, 18243–18248. [CrossRef]

49. Liu, B.; Yang, Y.; Qiu, Z.; Staron, M.; Hong, F.; Li, Y.; Wu, S.; Li, Y.; Hao, B.; Bona, R.; et al. Folding of Toll-like receptors by the HSP90 paralogue gp96 requires a substrate-specific cochaperone. *Nat. Commun.* **2010**, *1*, 79. [CrossRef]

50. Melnick, J.; Dul, J.L.; Argon, Y. Sequential interaction of the chaperones BiP and GRP94 with immunoglobulin chains in the endoplasmic reticulum. *Nature* **1994**, *370*, 373–375. [CrossRef]

51. Gidalevitz, T.; Simen, B.B.; Ostrovsky, O.; Vogen, S.M.; Dul, J.L.; Argon, Y. GRP94 activity is necessary for completion of immunoglobulin light chain folding. In Revision. 2011.

52. Liu, B.; Li, Z. Endoplasmic reticulum HSP90b1 (gp96, grp94) optimizes B-cell function via chaperoning integrin and TLR but not immunoglobulin. *Blood* **2008**, *112*, 1223–1230. [CrossRef]

53. Staron, M.; Yang, Y.; Liu, B.; Li, J.; Shen, Y.; Zuniga-Pflucker, J.C.; Aguila, H.L.; Goldschneider, I.; Li, Z. gp96, an endoplasmic reticulum master chaperone for integrins and Toll-like receptors, selectively regulates early T and B lymphopoiesis. *Blood* **2010**, *115*, 2380–2390. [CrossRef] [PubMed]

54. Weekes, M.P.; Antrobus, R.; Talbot, S.; Hör, S.; Simecek, N.; Smith, D.L.; Bloor, S.; Randow, F.; Lehner, P.J. Proteomic Plasma Membrane Profiling Reveals an Essential Role for gp96 in the Cell Surface Expression of LDLR Family Members, Including the LDL Receptor and LRP6. *J. Proteome Res.* **2012**, *11*, 1475–1484. [CrossRef] [PubMed]

55. Liu, B.; Staron, M.; Hong, F.; Wu, B.X.; Sun, S.; Morales, C.; Crosson, C.E.; Tomlinson, S.; Kim, I.; Wu, D.; et al. Essential roles of grp94 in gut homeostasis via chaperoning canonical Wnt pathway. *Proc. Natl. Acad. Sci. USA* **2013**, *110*, 6877–6882. [CrossRef] [PubMed]

56. Staron, M.; Wu, S.; Hong, F.; Stojanovic, A.; Du, P.X.; Bona, R.; Liu, B.; Li, L.Z. Heat-shock protein gp96/grp94 is an essential chaperone for the platelet glycoprotein Ib-IX-V complex. *Blood* **2011**, *117*, 7136–7144. [CrossRef] [PubMed]

57. Barton, E.R.; Park, S.; James, J.K.; Makarewich, C.A.; Philippou, A.; Eletto, D.; Lei, H.; Brisson, B.; Ostrovsky, O.; Li, Z.; et al. Deletion of muscle GRP94 impairs both muscle and body growth by inhibiting local IGF production. *FASEB J.* **2012**, *26*, 3691–3702. [CrossRef]

58. Khilji, M.S.; Bresson, S.E.; Verstappen, D.; Pihl, C.; Andersen, P.A.K.; Agergaard, J.B.; Dahlby, T.; Bryde, T.H.; Klindt, K.; Nielsen, C.K.; et al. The inducible beta5i proteasome subunit contributes to proinsulin degradation in GRP94-deficient beta-cells and is overexpressed in type 2 diabetes pancreatic islets. *Am. J. Physiol. Endocrinol. Metab.* **2020**, *318*, E892–E900. [CrossRef]

59. Kuznetsov, G.; Chen, L.B.; Nigam, S.K. Several endoplasmic reticulum stress proteins, including ERp72, interact with thyroglobulin during its maturation. *J. Boil. Chem.* **1994**, *269*, 22990–22995.

60. Di Jeso, B.; Park, Y.-N.; Ulianich, L.; Treglia, A.S.; Urbanas, M.L.; High, S.; Arvan, P. Mixed-Disulfide Folding Intermediates between Thyroglobulin and Endoplasmic Reticulum Resident Oxidoreductases ERp57 and Protein Disulfide Isomerase. *Mol. Cell. Boil.* **2005**, *25*, 9793–9805. [CrossRef]

61. Zhang, Y.; Wu, B.X.; Metelli, A.; Thaxton, J.E.; Hong, F.; Rachidi, S.; Ansa-Addo, E.; Sun, S.; Vasu, C.; Yang, Y.; et al. GP96 is a GARP chaperone and controls regulatory T cell functions. *J. Clin. Investig.* **2015**, *125*, 859–869. [CrossRef]

62. Vajdos, F.F.; Ultsch, M.; Schaffer, M.L.; Deshayes, K.D.; Liu, J.; Skelton, N.J.; De Vos, A.M. Crystal structure of human insulin-like growth factor-1: Detergent binding inhibits binding protein interactions. *Biochemistry* **2001**, *40*, 11022–11029. [CrossRef]

63. Dollins, D.E.; Warren, J.J.; Immormino, R.M.; Gewirth, D.T. Structures of GRP94-Nucleotide Complexes Reveal Mechanistic Differences between the hsp90 Chaperones. *Mol. Cell* **2007**, *28*, 41–56. [CrossRef] [PubMed]

64. Pierce, B.G.; Hourai, Y.; Weng, Z. Accelerating Protein Docking in ZDOCK Using an Advanced 3D Convolution Library. *PLoS ONE* **2011**, *6*, e24657. [CrossRef] [PubMed]

65. Zuehlke, A.D.; Johnson, J.L. Hsp90 and co-chaperones twist the functions of diverse client proteins. *Biopolymers* **2010**, *93*, 211–217. [CrossRef] [PubMed]

66. King, F.W.; Wawrzynów, A.; Höhfeld, J.; Zylicz, M. Co-chaperones Bag-1, Hop and Hsp40 regulate Hsc70 and Hsp90 interactions with wild-type or mutant p53. *EMBO J.* **2001**, *20*, 6297–6305. [CrossRef] [PubMed]

67. Kao, G.; Nordenson, C.; Still, M.; Rönnlund, A.; Tuck, S.; Naredi, P.L.J. ASNA-1 Positively Regulates Insulin Secretion in C. elegans and Mammalian Cells. *Cell* **2007**, *128*, 577–587. [CrossRef] [PubMed]

68. Wandinger, S.K.; Richter, K.; Buchner, J. The Hsp90 Chaperone Machinery. *J. Boil. Chem.* **2008**, *283*, 18473–18477. [CrossRef]

69. Hoter, A.; El-Sabban, M.; Naim, H.Y. The HSP90 Family: Structure, Regulation, Function, and Implications in Health and Disease. *Int. J. Mol. Sci.* **2018**, *19*, 2560. [CrossRef]

70. Marzec, M.; Eletto, D.; Argon, Y. GRP94: An HSP90-like protein specialized for protein folding and quality control in the endoplasmic reticulum. *Biochim. Biophys. Acta* **2012**, *1823*, 774–787. [CrossRef]

71. Wanderling, S.; Simen, B.B.; Ostrovsky, O.; Ahmed, N.T.; Vogen, S.M.; Gidalevitz, T.; Argon, Y. GRP94 Is Essential for Mesoderm Induction and Muscle Development Because It Regulates Insulin-like Growth Factor Secretion. *Mol. Boil. Cell* **2007**, *18*, 3764–3775. [CrossRef]

72. Li, W.; Kennedy, S.; Ruvkun, G. daf-28 encodes a C. elegans insulin superfamily member that is regulated by environmental cues and acts in the DAF-2 signaling pathway. *Genes Dev.* **2003**, *17*, 844–858. [CrossRef]

73. Ostrovsky, O.; Eletto, D.; Makarewich, C.; Barton, E.R.; Argon, Y. Glucose regulated protein 94 is required for muscle differentiation through its control of the autocrine production of insulin-like growth factors. *Biochim. Biophys. Acta* **2010**, *1803*, 333–341. [CrossRef] [PubMed]

74. Marzec, M.; Hawkes, C.P.; Eletto, D.; Boyle, S.; Rosenfeld, R.; Hwa, V.; Wit, J.-M.; Van Duyvenvoorde, H.A.; Oostdijk, W.; Losekoot, M.; et al. A Human Variant of Glucose-Regulated Protein 94 That Inefficiently Supports IGF Production. *Endocrinology* **2016**, *157*, 1914–1928. [CrossRef] [PubMed]

75. Liu, M.; Hodish, I.; Rhodes, C.J.; Arvan, P. Proinsulin maturation, misfolding, and proteotoxicity. *Proc. Natl. Acad. Sci. USA* **2007**, *104*, 15841–15846. [CrossRef] [PubMed]

76. Liu, M.; Hua, Q.-X.; Hu, S.-Q.; Jia, W.; Yang, Y.; Saith, S.E.; Whittaker, J.; Arvan, P.; Weiss, M.A. Deciphering the Hidden Informational Content of Protein Sequences. *J. Boil. Chem.* **2010**, *285*, 30989–31001. [CrossRef] [PubMed]

77. Hober, S.; Ljung, J.L.; Uhlén, M.; Nilsson, B. Insulin-like growth factors I and II are unable to form and maintain their native disulfides under in vivo redox conditions. *FEBS Lett.* **1999**, *443*, 271–276. [CrossRef]

78. Sohma, Y.; Hua, Q.X.; Liu, M.; Philips, N.B.; Hu, S.-Q.; Whittaker, J.; Whittaker, L.J.; Ng, A.; Roberts, C.T., Jr.; Arvan, P.; et al. Contribution of residue B5 to the folding and function of insulin and IGF-I: Constraints and fine-tuning in the evolution of a protein family. *J. Biol. Chem.* **2010**, *285*, 5040–5055. [CrossRef]

79. Eletto, D.; Maganty, A.; Eletto, D.; Dersh, D.; Makarewich, C.; Biswas, C.; Paton, J.C.; Paton, A.W.; Doroudgar, S.; Glembotski, C.C.; et al. Limitation of individual folding resources in the ER leads to outcomes distinct from the unfolded protein response. *J. Cell Sci.* **2012**, *125*, 4865–4875. [CrossRef]

80. Meunier, L.; Usherwood, Y.-K.; Chung, K.T.; Hendershot, L.M. A Subset of Chaperones and Folding Enzymes Form Multiprotein Complexes in Endoplasmic Reticulum to Bind Nascent Proteins. *Mol. Boil. Cell* **2002**, *13*, 4456–4469. [CrossRef]

81. Yakar, S.; Liu, J.-L.; Stannard, B.; Butler, A.; Accili, D.; Sauer, B.; Leroith, D. Normal growth and development in the absence of hepatic insulin-like growth factor I. *Proc. Natl. Acad. Sci. USA* **1999**, *96*, 7324–7329. [CrossRef]

82. Wu, B.X.; Hong, F.; Zhang, Y.-L.; Ansa-Addo, E.; Li, Z. GRP94/gp96 in Cancer: Biology, Structure, Immunology, and Drug Development. *Adv. Cancer Res.* **2016**, *129*, 165–190.

83. Kim, Y.K.; Lee, A.S. Transcriptional activation of the glucose-regulated protein genes and their heterologous fusion genes by beta-mercaptoethanol. *Mol. Cell. Boil.* **1987**, *7*, 2974–2976. [CrossRef] [PubMed]

84. Reddy, R.K.; Dubeau, L.; Kleiner, H.; Parr, T.; Nichols, P.; Ko, B.; Dong, D.; Ko, H.; Mao, C.; DiGiovanni, J.; et al. Cancer-inducible transgene expression by the Grp94 promoter: Spontaneous activation in tumors of various origins and cancer-associated macrophages. *Cancer Res.* **2002**, *62*, 7207–7212. [PubMed]

85. Samani, A.A.; Yakar, S.; Leroith, D.; Brodt, P. The Role of the IGF System in Cancer Growth and Metastasis: Overview and Recent Insights. *Endocr. Rev.* **2007**, *28*, 20–47. [CrossRef] [PubMed]

86. Dejeans, N.; Glorieux, C.; Guenin, S.; Beck, R.; Sid, B.; Rousseau, R.; Bisig, B.; Delvenne, P.; Calderon, P.B.; Verrax, J. Overexpression of GRP94 in breast cancer cells resistant to oxidative stress promotes high levels of cancer cell proliferation and migration: Implications for tumor recurrence. *Free. Radic. Boil. Med.* **2012**, *52*, 993–1002. [CrossRef]

87. Grimberg, A.; Cohen, P. Role of insulin-like growth factors and their binding proteins in growth control and carcinogenesis. *J. Cell. Physiol.* **2000**, *183*, 1–9. [CrossRef]

88. Yakar, S.; Rosen, C.J.; Bouxsein, M.L.; Sun, H.; Mejia, W.; Kawashima, Y.; Wu, Y.; Emerton, K.; Williams, V.; Jepsen, K.; et al. Serum complexes of insulin-like growth factor-1 modulate skeletal integrity and carbohydrate metabolism. *FASEB J.* **2008**, *23*, 709–719. [CrossRef]

89. Yakar, S.; Rosen, C.J.; Beamer, W.G.; Ackert-Bicknell, G.L.; Wu, Y.-P.; Liu, J.-L.; Ooi, G.T.; Setser, J.; Frystyk, J.; Boisclair, Y.R.; et al. Circulating levels of IGF-1 directly regulate bone growth and density. *J. Clin. Investig.* **2002**, *110*, 771–781. [CrossRef]

90. Bunn, R.C.; Fowlkes, J.L. Insulin-like growth factor binding protein proteolysis. *Trends Endocrinol. Metab.* **2003**, *14*, 176–181. [CrossRef]

Review

Implications of Insulin-Like Growth Factor-1 in Skeletal Muscle and Various Diseases

Syed Sayeed Ahmad [1,2], **Khurshid Ahmad** [1,2], **Eun Ju Lee** [1,2], **Yong-Ho Lee** [3,*] and **Inho Choi** [1,2,*]

[1] Department of Medical Biotechnology, Yeungnam University, Gyeongsan 38541, Korea; sayeedahmad4@gmail.com (S.S.A.); ahmadkhursheed2008@gmail.com (K.A.); gorapadoc0315@hanmail.net (E.J.L.)

[2] Research Institute of Cell Culture, Yeungnam University, Gyeongsan 38541, Korea

[3] Department of Biomedical Science, Daegu Catholic University, Gyeongsan 38430, Korea

* Correspondence: ylee325@cu.ac.kr (Y.-H.L.); inhochoi@ynu.ac.kr (I.C.); Fax: +82-53-810-4769

Received: 22 June 2020; Accepted: 22 July 2020; Published: 24 July 2020

Abstract: Skeletal muscle is an essential tissue that attaches to bones and facilitates body movements. Insulin-like growth factor-1 (IGF-1) is a hormone found in blood that plays an important role in skeletal myogenesis and is importantly associated with muscle mass entity, strength development, and degeneration and increases the proliferative capacity of muscle satellite cells (MSCs). IGF-1R is an IGF-1 receptor with a transmembrane location that activates PI3K/Akt signaling and possesses tyrosine kinase activity, and its expression is significant in terms of myoblast proliferation and normal muscle mass maintenance. IGF-1 synthesis is elevated in MSCs of injured muscles and stimulates MSCs proliferation and myogenic differentiation. Mechanical loading also affects skeletal muscle production by IGF-1, and low IGF-1 levels are associated with low handgrip strength and poor physical performance. IGF-1 is potentially useful in the management of Duchenne muscular dystrophy, muscle atrophy, and promotes neurite development. This review highlights the role of IGF-1 in skeletal muscle, its importance during myogenesis, and its involvement in different disease conditions.

Keywords: skeletal muscle; IGF-1; MSCs; myogenesis

---

1. Introduction

Muscles attached to the bone are referred to as skeletal muscle (SM) and account for 30–50% of body weight and are responsible for skeletal movement. In the human body, SM is one of the most plastic and dynamic tissue and utilizes up to 50–75% of all body proteins [1,2]. SM cell proliferation and differentiation are vitally required for appropriate SM development throughout embryogenesis and for postnatal SM regeneration that is essential for muscle healing after injury [3]. In multicellular organisms, cell generation in all tissues is under the control of a network of tissue-specific regulators termed growth factors (GFs). GFs are low molecular weight peptides that are active during cell proliferation and differentiation [4,5], migration, and apoptosis, and play a significant role in managing growth signal responses throughout development [6]. GFs have been reported in blood vessels and epithelial, lymphoid, neural, muscle, lymphatic, erythroid, myeloid, and hepatic systems, and few GFs and cytokines are produced in each tissue [7]. GFs also regulate cellular responses during wound healing and act as endogenous signaling molecules [8]. Wound healing is a multifaceted physiological process that involves interplay between numerous cell types, GFs, extracellular matrix (ECM) constituents, and proteinases [9].

ECM is well known to preserve SM integrity and participates throughout myogenesis. Our group has explored the contributions made by several ECM components, e.g., fibromodulin [10–12], dermatopontin [2], and matrix gla protein [13], during myogenesis. In recent decades, the number of cases of debilitating injury has increased, and the treatment of individuals suffering from different chronic injuries incurs substantial costs, especially in the United States and Europe [14–16]. At each stage of healing, specific arrangements of cytokines and GFs must cooperate with their respective receptors and ECM constituents at their target locations [17,18].

GFs play a substantial role in tissue recovery as well as in the regulation of diverse cellular processes and act as signaling molecules between cells. Because of their instabilities and soluble natures, developments are required to enable their therapeutic use [19]. GF delivery has been a theme of augmented recent research attention owing to the controlled and targeted drug delivery in addition to the development of recombinant DNA methods that have enabled GFs creation [20–22]. Heparin, a profoundly sulfated glycosaminoglycan, has been used to facilitate the local delivery of GFs from different matrices (e.g., microcapsules [23]), as it binds and potentiates the activities of GFs. Specifically, heparin has been shown to prevent the deactivation of GFs [21,24], enhance their interactions with receptors [25], increase GF loading into delivery vehicles [26], and facilitate the long-terms releases of GFs [26,27].

Components of the endocrine system, such as growth hormone (GH), insulin-like growth factor-1 (IGF-1), and androgens, are the foremost regulators of muscle metabolism. These endocrine components have substantial impacts on muscle and act as anabolic factors and significant regulators of muscle mass [28]. IGF-1 is a 70 aa polypeptide with autocrine, paracrine, and endocrine properties, and shares a ~60% similarity with IGF-2 and a 50% similarity with proinsulin structures [29]. The actions of IGF-1 and 2 are mainly facilitated by type 1 receptors. Insulin-like growth factor type 1 receptor (IGF-1R) is required for cell growth and development and to maintain the cell cycle. IGF-1 and IGF-2 are also known as mitogenic peptides that show homology with each other and with insulin [30–33]. IGF-1 is considered to play key roles in fetal development and growth up to adolescence, and in the maintenance of homeostasis in adult tissues by regulating cell proliferation, differentiation, and survival (Figure 1). It has also been reported IGF-1 has atheroprotective, neuroprotective, and insulin-like effects and that it regulates skeletal muscle metabolism and regeneration [34]. Physiological maintenance of SM requires injury or stretch stimulation, which prompts IGF-1 expression [35]. The supplementation of pro-IGF-2 could be one of the most effective therapeutic approaches for muscle injury in elderly people [36].

IGF-1 mRNA gives rise to three proforms, IGF-1Ea, IGF-1Eb, and IGF-1Ec, which yield three different C-terminal extensions called Ea, Eb, and Ec peptides [37]. IGF-1Ea and IGF-1Eb are necessary for the initiation of myogenesis in mice, but the loss of IGF-1Ea is related to greater reductions in myogenesis than IGF-1Eb [38]. Interestingly, IGF-1Ea is upregulated by a single ramp stretch of one hour but reduced by repeated cyclical stretches, whereas IGF-1Eb is upregulated by cycling stacking [39]. At the point when the typical strain and stretch are not set up, the IGF-1 signaling pathway turns into deactivated and prompts muscle atrophy, as appeared in astronauts working in the microgravity environment [40]. IGF-1 is synthesized and released from the liver along with some other tissue such as muscle, heart, adipose tissue, brain, and pancreatic β-cell [41]. IGF1 proforms can induce breast cancer cell proliferation through its receptor [42]. IGF-1 is the main regulator of growth and metabolism in mammals [31,43]. Circulating IGF-1 is controlled by members of the IGF binding protein family (IGFBP-1~6) and acid-labile subunit (ALS). GH, insulin, and nutritional status are responsible for the secretion of IGF-1 [44,45]. The maintenance of hypertrophic phenotype by IGF-1Ea involves also the activation of AMPK pathways, a factor involved in the maintenance of whole-body energy balance and an "energy sensor" controlling glucose and lipid metabolism [46]. Either IGF-1Ea or IGF-1Eb expression in muscle, activating a series of anabolic and compensatory pathways, is able to avoid muscle loss and a normal muscle-nerve interaction [47]. IGFBP belongs to a family of soluble proteins having a high affinity to bind with IGF-1 and 2. In humans, IGFBP 3 is the most abundant IGFBP and binds with a maximum amount of circulating IGF-1 [28]. The half-life from minutes to ~15 h is extended upon the incorporation of IGF-1 into the ternary complex, thus creating a stable pool of IGF-1 inside the circulation; which, further combined with the other IGFBP, can provide subtle regulation of the availability of IGF-1 to target tissues [48,49].

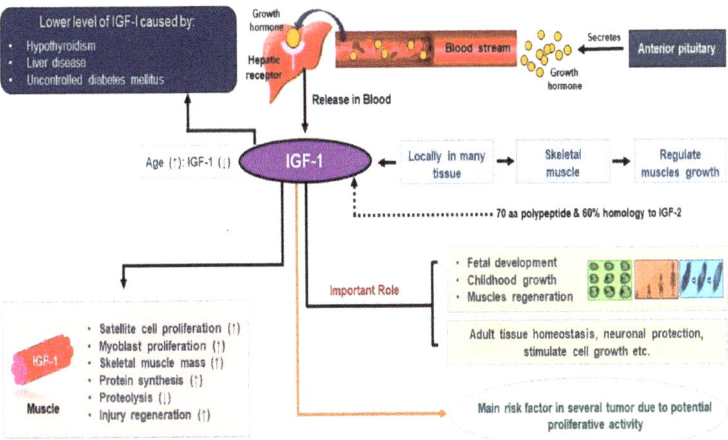

Figure 1. Role of insulin-like growth factor-1 (IGF-1) in skeletal muscle. IGF-1 is responsible for fetal development, child growth, and muscle regeneration, and elevated IGF-1 levels are required for muscle satellite cell (MSC) and myoblast proliferation, postinjury regeneration, and the increase of skeletal mass.

2. Role of IGF-1 in Skeletal Muscle

IGF-1 plays a critical role in myogenesis during embryonic development, although the mechanism responsible for IGF-1 mediated myoblast proliferation remains unclear [50]. Aging, ischemia, cancer, motor neuron degeneration, and heart failure are all associated with SM loss, for which there is no effective treatment. IGF-1 production plays an important role in muscle healing and maintenance. Preclinical experiments have shown that IGF-1 is associated with muscle mass and strength development, it reduces muscle degeneration, prevents excessive toxin-induced inflammatory expansion, and increases the proliferation capacity of muscle satellite cells (MSCs) [35]. MSCs are key players in SM regeneration [12], and IGF-1 is also produced in SM to control muscle growth in a paracrine/autocrine manner [51]. IGF-1 is also a biomarker of health and fitness; in fact, higher circulating IGF-1 concentrations are positively related to health factors associated with body structure and cardiovascular strength, and negatively related to body fat levels. Aerobic fitness and muscular stamina are positively associated with circulating IGF-1 concentrations [52]. Malnutrition, sepsis, critical sickness, high doses of exogenous glucocorticoids and inflammation, are responsible to lower the IGF-1 mRNA in muscle [51]. Like IGF-1, IGF-2 is also essential for muscle differentiation and development and acts in an autocrine manner [53]. Transforming growth factor-beta1 (TGF-β1) has been reported to diminish IGF-2 gene expression in myoblasts, decrease IGF-2 secretion, and reduce IGF-1 receptor activation [54].

3. Mechanism of IGF-1 in Skeletal Muscle

Several tissues secrete IGF-1, and the actions of IGF-1 appear to be dependent on the secretory site. Most IGF-1, also known as "somatomedin C", is secreted by the liver and transported as an endocrine hormone to other tissues [55]. The IGF-1 cascade is mediated by its interaction with IGF-1R, which has transmembrane locations and tyrosine kinase-like activity [51]. IGF-1R acts as a phosphatidylinositol 3-kinase/protein kinase B (PI3K/Akt) pathway activator and its expression is associated with myoblast proliferation and normal muscle mass maintenance [56] (Figure 2).

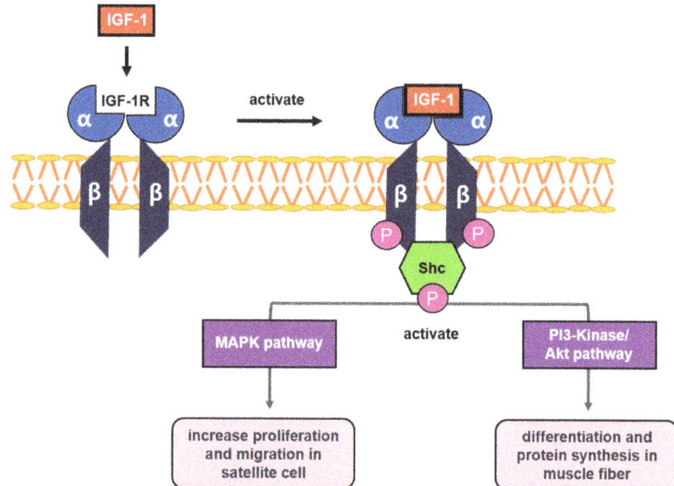

Figure 2. The molecular mechanism of IGF-1. IGF-1 interacts with its receptor (IGF-1R), and thus, activates the PI3K/Akt and mitogen-activated protein kinase (MAPK) pathways, which regulate MSC proliferation and differentiation.

It has been reported that the mitogenic activity of IGF-1 on myoblast cells is crucial and mediated by two main signaling pathways, that is, the mitogen-activated protein kinase (MAPK/ERK1/2) pathway and the PI3K/Akt pathway, which are both associated with cell cycle progression and cell survival [57]. Furthermore, the Akt-facilitated growth effect of IGF-1 in SM appears to promote protein synthesis and muscle cell development [58,59]. The PI3K-Akt cascade is the main IGF-1 signal activated in muscle. Akt1/Akt2 double-knockout mice and IGF-1R knockout mice displayed a severe growth deficiency. They both exhibited a decreased SM mass, although IGF-1R knockout mice attributed to a decrease in the number of muscle cells, whereas in the Akt1/Akt2 double-knockout mice attributed mostly to a decrease in individual cell size and suggested that IGF-1R functions during development are mostly dependent on Akt [60]. IGF-1 plays an essential role in myoblast proliferation and differentiation, and protects cells from apoptosis [61]. In the heterotetramer structure of IGF-1R, two subunits are responsible for IGF-1 binding and the other two subunits exhibit tyrosine kinase-like activity. The IGF-1 binding capability of the ligand-binding area of IGF-1R has a six-fold greater attraction for IGF-1 than IGF-2. After binding IGF-1, the intrinsic tyrosine kinase of IGF-1R autophosphorylates tyrosines that then act as docking positions for signaling proteins, which include insulin receptor substrate-1 (IRS-1). IGF-1R also phosphorylates Shc, which subsequently triggers the RAS/MAP kinase pathway to prompt mitogenesis. Muscle injury enhances IGF-1 synthesis by MSCs in rodents, which stimulates MSC proliferation and differentiation to myoblasts [35,62,63]. Mechanical loading also affects the production of IGF-1 by SM [51,64].

4. Relationship Between IGF-1 and Myostatin

IGFs and myostatin (MSTN) have contrasting roles in the regulation of SM size and growth, in particular, MSTN inhibits SM growth [65]. Circulating MSTN-attenuating mediators are being developed to treat muscle-wasting ailments, as MSTN/activin receptors are widely distributed among many nonmuscle tissues [66]. Follistatin is an inhibitor of MSTN and induces dramatic SM mass increases, upon the IGF-1 receptor/Akt/mTOR cascade [67].

IGF-1 knockout mice show muscle hypoplasia [68]. Moreover, inhibition of MSTN stimulates the Akt/mTOR/S6K pathway, which is essential for the muscle hypertrophy initiated by IGF-1 [69–71]. The regulation of IGF-1 during the muscle hypertrophy induced by MSTN inhibition is still disputed.

Elevated expressions of muscle mRNA and circulating concentrations of IGF-1 were observed following MSTN inhibition [67]. Morissette et al. reported that Akt protein levels were high in SMs of MSTN knockout mice [70].

5. Role of IGF-1 in Different Diseases

5.1. Role of IGF-1 in Duchenne Muscular Dystrophy

Duchenne muscular dystrophy (DMD) is a form of muscular dystrophy associated with X-linked recessive disorder caused by mutation of dystrophin in SM [72,73]. DMD shows a male predominance and causes muscle degeneration. Several studies have demonstrated extremely encouraging outcomes for IGF-1 treatment in DMD [74,75]. Moreover, muscle and circulating levels of IGF-1 frequently reduce in response to glucocorticoids [76]. In vitro study by Fang et al. demonstrated that glucocorticoid and IGF-1 cotreatment participate in myogenic differentiation through the Akt/GSK-3β pathway in C2C12 myoblasts. It revealed that increased phosphorylated Ser473-Akt and phosphorylated Ser9-GSK-3b as well as myogenic differentiation, provide a way for a potential alternative strategy to DMD treatment [76]. IGF-1 has been recommended for patients experiencing muscle-wasting conditions [77] and various studies have explored the functional properties of dystrophic SM after IGF-1 treatment. Lynch et al. found that four weeks of IGF-1 treatment (~2 mg/kg body mass, 50 g/h delivered subcutaneously by a miniosmotic pump) increased the mass and force-producing limit of SM from dystrophic mice. Furthermore, IGF-1 increased extensor digitorum longus (EDL) and soleus muscle masses of dystrophic mice by 20% and 29%, respectively, as compared with untreated dystrophic controls [77].

5.2. Role of IGF-1 in Muscle Atrophy

Muscle atrophy (MA) is defined as a loss of muscle mass and quality, and it is encountered in several disease conditions, for example, in malignancies, AIDS, congestive cardiovascular breakdown, chronic obstructive pulmonary disease, and renal failure and in serious burn patients [78]. Anabolic-androgenic steroids and different hormones, such as GH and IGF-1 appear to increase muscle mass in patients with MA [79]. Lama2-linked muscular dystrophy is a serious congenital muscular dystrophy produced by mutations in the LAMA2 gene, and is associated with several pathological problems such as inflammation, apoptosis, fibrosis, necrosis, severe muscle weakness, and subnominal postnatal growth. As indicated by Accorsi et al. losartan combinatorial management appeared to enhance transgenic IGF-1 overexpression, recover postnatal growth, reduce inflammation and fibrosis, increase body weights, and result in a remarkable restoration of muscle architecture and locomotory ability in DyW mice (mouse model of Lama2-related muscular dystrophy) [80].

5.3. Role of IGF-1 in Cancer

Increases in IGF-1R activity promote cancer cell proliferation, migration, and invasion and are related to tumor metastasis, treatment resistance, and reduced survival [81]. IGFBP2 has been identified as a prominent oncogene in most epithelial cancers [82]. A number of authors have proposed IGFBP2 viewed as a potential target for regulating cancer metastasis and invasion-related signaling networks, though its mode of action is keenly debated [83]. IGF-1 has been reported to upregulate angiogenesis and tumor invasion by activating matrix metalloproteinases [84], which are well known nonglycolytic proteolytic enzyme biomarkers in several cancer types [85]. Currently, therapies targeting the IGF system have attracted considerable attention in cancer research. The proliferative, antiapoptotic, and transformative impacts of IGFs are primarily activated by IGF-1R ligation [86]. Higher levels of serum IGF-1 are linked with increased risk of several common cancers comprising breast, colorectal, and prostate [87].

5.4. Role of IGF-1 in Neurodegeneration

Neurodegenerative diseases like Alzheimer's, Parkinson's diseases and prion disorders are associated with aging [88–92]. A number of promising results show that IGF-1 has a restorative impact on the brain by expanding hippocampal neurogenesis and memory accuracy in older people and potentially in individuals with neurodegenerative disorders [93]. IGF-1 has a progressively more powerful trophic impact than GH on sensory and motor neurons and on neuronal growth and recovery. IGF-1 stimulates neurite development and assumes an essential role during central and peripheral nervous system development [94,95].

It can be summarized that IGF-1 plays a crucial role in the management of various diseases and could be used in the therapeutic possibilities of several diseases, including DMD, muscle atrophy, etc. Recent IGF-1 studies are detailed in Table 1, which clearly showed the role of IGF-1 in various areas such as SM regeneration, tissue recovery, depression pathophysiology, etc.

Table 1. Recent research studies on IGF-1 in different fields.

S. No.	Role of IGF-1	Year	References
1.	IGF-1 helps in the growth and regeneration of SM and bones. Its signaling in the smooth muscle cell and in fibroblast is a critical factor of normal vascular wall growth and atheroprotection.	2020	[96,97]
2.	IGF-1 helps in the activation of IGF-1R and muscle tissue recovery. Shapiro et al. indicate that the IGFBP-3/IGF1 conjugated framework has the potential to be utilized for in-situ muscle tissue recovery.	2019	[98,99]
3.	IGF-1 have pleiotropic consequences on the skeleton during the life expectancy by prompting the bone development and resorption. Lower IGF-1 levels are related to lower handgrip strength and more terrible physical execution.	2018	[100,101]
4.	GH/IGF-1 treatment had various impacts on patients with traumatic brain injury, proving a high recuperation of neurons and clinical results.	2017	[95]
5.	IGF-1 appear in the regulation of neuronal harm, toxic insults, and a few other neurodegenerative procedures.	2016	[102]
6.	According to Kopczak et al., the signaling of IGF-1 could play a role in the pathophysiology of depression.	2015	[103]

6. Interaction Between IGF-1 and IGF-1R

Protein-protein interactions (PPIs) provide graphical illustrations of interactions between two or more proteins. PPI strategy plays an important role in the body for metabolic and signaling processes. A better understanding of the interaction between IGF-1 and IGF-1R along with several other associated proteins (Figure 3A) was obtained by SIGnaling Network Open Resource (SIGNOR; http://signor.uniroma2.it). The SIGNOR web tool can be used to predict activation/inactivation, interactions, and connections between biomolecules and signaling molecules [104]. GFs and other membrane-bound entities (e.g., ECM molecules) activate transmembrane receptors that trigger signaling responses that eventually regulate gene expressions and metabolic processes (Figure 4).

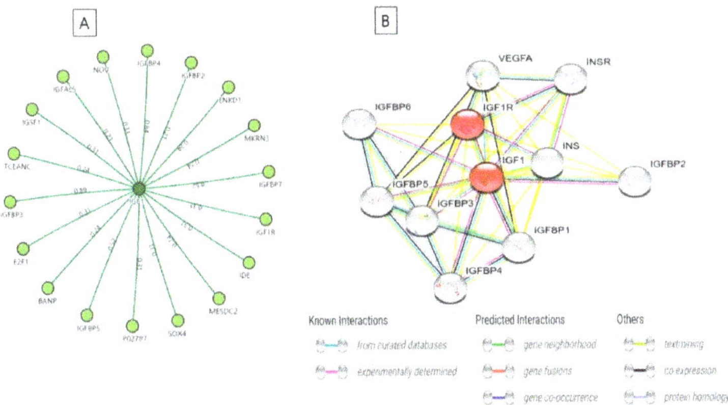

Figure 3. Protein-Protein interactions of IGF-1 with its associated proteins generated by (**A**) SIGnaling Network Open Resource, (**B**) STRING.

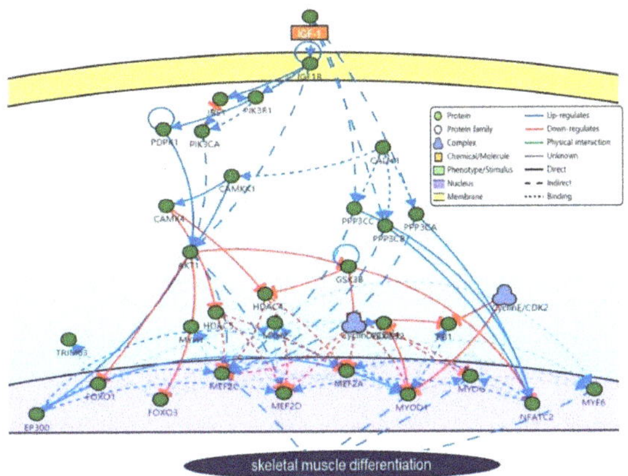

Figure 4. The mechanistic role of IGF-1 during skeletal muscle differentiation. The figure shows signaling interactions during muscle differentiation as predicted by SIGnaling Network Open Resource (SIGNOR).

The STRING database (http://string-db.org) enables critical assessments or direct (physical) and indirect (functional) PPIs. By using STRING [105], we were able to identify interacting nodes between IGF-1 and IGF-1R (Figure 3B). The interactions generated by the STRING are based on the known interactions (from the curated databases and experimentally determined), predicted interactions (e.g., gene neighborhood and gene co-occurrence) as well as few other factors viz. text mining, coexpression, etc. In this interaction, several other associated proteins such as IGFBP 1 to 6, insulin (INS), insulin to its receptor (INSR), and vascular endothelial growth factor A (VEGFA) were found to interact with each other through IGF-1 and IGF1R. Black lines represent the coexpression while the light blue line represents the protein homology. Text-mining data represents the association between proteins as shown in Figure 3B. The half-life of the IGFs are prolonged by IGFBP and helped in the growth-promoting effects of the IGFs on cell culture. INS decreases blood glucose and increases cell permeability to amino acids, monosaccharides, and fatty acids. Binding of insulin to its receptor (INSR) leads to phosphorylation of intracellular substrates, such as insulin receptor substrates (IRS1, 2, 3, 4), SHC, GAB1, and other signaling intermediates. Each of these phosphorylated proteins serve as docking proteins for other signaling proteins. VEGFA is active in angiogenesis, vasculogenesis and endothelial cell growth, it induces endothelial cell proliferation, promotes cell migration, inhibits apoptosis and induces permeabilization of blood vessels (http://string-db.org).

In Figure 4, the green circle represents the protein which binds with its receptor to direct the signaling path inside the cell. IGF-1 interaction is clearly shown in different parts of cells such as membrane to the nucleus. In this figure, the red line represents the downregulation while upregulation is represented by the blue line. The dotted line represents the binding mode between the intermediates.

The functions of proteins associated with IGF-1 are provided in Table 2, and the role played by IGF-1 in myogenesis is depicted schematically in Figure 5. The different myogenic regulatory factors such as Pax3, Pax7, MyoD, Myf5, MyoG, and Mrf4 genes are collectively expressed in the SM lineage in different tissues during development [106,107]. IGF-1 plays an important role in the activation of precursor cells and helps in the activation of the regenerative process. IGF-1 also increases the proliferation and differentiation of satellite cells and myoblast respectively. IGF-1 helps in myofiber repair. In precise, IGF-1 can favor regenerative myogenesis and support the robustness of myofibers [108]. Collectively, IGF-1 is helpful in satellite cell proliferation and differentiation. Skeletal myogenesis is an extraordinarily complex process, which is regulated at multiple levels, and transcriptional regulation naturally plays an important role during muscle formation.

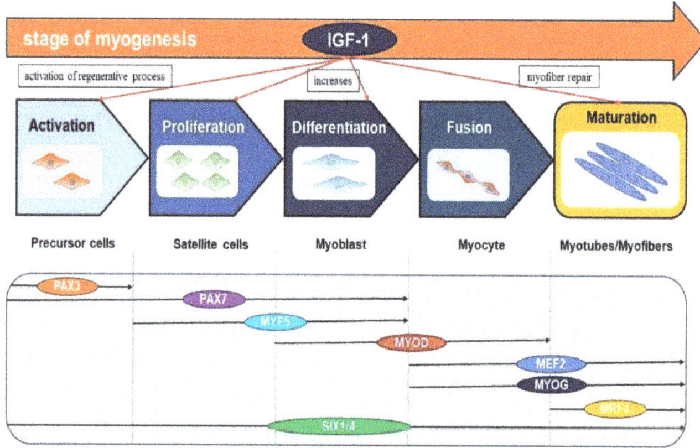

Figure 5. Role of IGF-1 in myogenesis. IGF-1 is activated during muscle regeneration and increases MSC proliferation and differentiation. In addition, IGF-1 promotes myofiber repairs.

The structure obtained by the SIGNOR network (Figure 3A) is showing the different proteins which are interlinked to IGF-1. These proteins are listed in the left part of Table 2. Now, here authors tried to elaborate in a single word about the function of these proteins as mentioned in the right part of Table 2. The IGFBP family consists of six IGFBPs, namely IGFBP1 to IGFBP6, however other proteins with low binding affinity to IGFs were known as IGFBP7, IGFBP8, IGFBP9 [109].

Overall GH is known to stimulate growth in children and adolescents with various metabolic functions [112]. Musculoskeletal injuries represent a major public health problem [113], and medications improve muscle repair and restore functions. Increasing IGF-1 levels improves SM recovery after myotoxic injury and the administration of IGF-1 has the potential for accelerating healing after trauma [114].

Table 2. Function of IGF-1 related proteins.

S. No.	Name	Function
1.	IGF-1R	Cell growth and survival control
2.	IGFBP3 and IGFBP4	Enhance the capability of IGF-1 to promote cell growth
3.	Probable E3 ubiquitin-protein ligase makorin-3 (MKRN3)	Catalyze the covalent interactions of ubiquitin moieties onto substrate proteins
4.	IGFBP-complex acid-labile subunit (IGFALS)	Regulation of the circulation of IGFs and receptor-ligand binding [110]
5.	Protein BANP	Cell cycle arrest
6.	Transcription factor E2F1	Mediate cell proliferation
7	Transcription factor SOX-4	High-affinity binding to the T-cell enhancer motif 5′-AACAAAG-3′ motif
8.	IGFBP7	Stimulates cell adhesion
9.	IGFBP5	Change the interaction of IGFs with their cell surface receptors.
10.	Immunoglobulin superfamily member 1 (IGSF1)	Essential to mediate a specific antagonistic effect of inhibin B on activin-stimulated transcription
11.	Insulin-degrading enzyme	Cellular breakdown of insulin
12.	IGFBP1, IGFBP3, IGFBP5	Stimulate IGF actions
13.	LDLR chaperone MESD (low-density lipoprotein receptors)	Help in embryonic polarity and mesoderm induction
14.	Protein NOV homolog (IGFBP9)	Binds with integrins or other membrane receptors e.g., NOTCH1 [111]

7. Concluding Remarks

IGF-1 plays an important role in the maintenance of muscle mass by acting in paracrine, autocrine, or endocrine manners. GH upregulates IGF-1 synthesis in the liver, and thereby, increases its plasma concentrations. IGF-1 is the main stimulator of SM mass since this hormone increases protein synthesis and decreases proteolysis. In addition, IGF-1 increases MSC proliferation and myoblast proliferation and differentiation during normal growth or regeneration after SM injury. Therefore, IGF-1 increases SM mass and muscle functional capacities. In addition, IGF-1 plays an important role in the prevention of muscle atrophy. The development of IGF-1 for the treatment of muscle-wasting conditions remains an important research challenge.

Author Contributions: S.S.A. and K.A. contributed to the systematic review of literature, and the design and writing of the manuscript; K.A., E.J.L., Y.-H.L. contributed to the systematic review of literature; I.C. contributed to the supervision of the manuscript editing, and the revision of major intellectual content. All authors have read and agreed to the published version of the manuscript.

Funding: This research was funded by Basic Science Research Program through the National Research Foundation of Korea (NRF) funded by the Ministry of Education (2020R1A6A1A03044512) and by the National Research Foundation of Korea (NRF) funded by the Korean government (MSIP: Grant No. NRF-2018R1A2B6001020).

Conflicts of Interest: The authors declare no conflict of interest.

Abbreviations

SM	skeletal muscle
GFs	growth factors
ECM	extracellular matrix
MSC	muscle satellite cell
IGF-1	insulin-like growth factor-1
IGF-1R	insulin-like growth factor type 1 receptor
TGF-β	transforming growth factor beta
GH	growth hormone
MSTN	myostatin
DMD	Duchenne muscular dystrophy
MA	muscle atrophy
IGFBP	IGF binding proteins

References

1. Frontera, W.R.; Ochala, J. Skeletal muscle: A brief review of structure and function. *Calcif. Tissue Int.* **2015**, *96*, 183–195. [CrossRef] [PubMed]
2. Kim, T.; Ahmad, K.; Shaikh, S.; Jan, A.T.; Seo, M.G.; Lee, E.J.; Choi, I. Dermatopontin in skeletal muscle extracellular matrix regulates myogenesis. *Cells* **2019**, *8*, 332. [CrossRef] [PubMed]
3. Langlois, S.; Cowan, K.N. Regulation of skeletal muscle myoblast differentiation and proliferation by pannexins. *Adv. Exp. Med. Biol.* **2017**, *925*, 57–73. [CrossRef] [PubMed]
4. Lee, E.J.; Pokharel, S.; Jan, A.T.; Huh, S.; Galope, R.; Lim, J.H.; Lee, D.M.; Choi, S.W.; Nahm, S.S.; Kim, Y.W.; et al. Transthyretin: A transporter protein essential for proliferation of myoblast in the myogenic program. *Int. J. Mol. Sci.* **2017**, *18*, 115. [CrossRef]
5. Halper, J. Growth factors as active participants in carcinogenesis: A perspective. *Vet. Pathol.* **2010**, *47*, 77–97. [CrossRef]
6. Vlasova-St Louis, I.; Bohjanen, P.R. Post-transcriptional regulation of cytokine and growth factor signaling in cancer. *Cytokine Growth Factor Rev.* **2017**, *33*, 83–93. [CrossRef]
7. Chen, F.M.; Zhang, M.; Wu, Z.F. Toward delivery of multiple growth factors in tissue engineering. *Biomaterials* **2010**, *31*, 6279–6308. [CrossRef]
8. Macri, L.; Clark, R.A. Tissue engineering for cutaneous wounds: Selecting the proper time and space for growth factors, cells and the extracellular matrix. *Skin Pharmacol. Physiol.* **2009**, *22*, 83–93. [CrossRef]
9. Han, G.; Ceilley, R. Chronic wound healing: A review of current management and treatments. *Adv. Ther.* **2017**, *34*, 599–610. [CrossRef]
10. Lee, E.J.; Jan, A.T.; Baig, M.H.; Ashraf, J.M.; Nahm, S.-S.; Kim, Y.-W.; Park, S.-Y.; Choi, I. Fibromodulin: A master regulator of myostatin controlling progression of satellite cells through a myogenic program. *FASEB J.* **2016**, *30*, 2708–2719. [CrossRef]
11. Lee, E.J.; Nam, J.H.; Choi, I. Fibromodulin modulates myoblast differentiation by controlling calcium channel. *Biochem. Biophys. Res. Commun.* **2018**, *503*, 580–585. [CrossRef] [PubMed]
12. Lee, E.J.; Jan, A.T.; Baig, M.H.; Ahmad, K.; Malik, A.; Rabbani, G.; Kim, T.; Lee, I.K.; Lee, Y.H.; Park, S.Y.; et al. Fibromodulin and regulation of the intricate balance between myoblast differentiation to myocytes or adipocyte-like cells. *FASEB J.* **2018**, *32*, 768–781. [CrossRef] [PubMed]
13. Ahmad, S.; Jan, A.T.; Baig, M.H.; Lee, E.J.; Choi, I. Matrix gla protein: An extracellular matrix protein regulates myostatin expression in the muscle developmental program. *Life Sci.* **2017**, *172*, 55–63. [CrossRef] [PubMed]
14. Schreml, S.; Szeimies, R.M.; Prantl, L.; Landthaler, M.; Babilas, P. Wound healing in the 21st century. *J. Am. Acad. Dermatol.* **2010**, *63*, 866–881. [CrossRef]
15. Posnett, J.; Gottrup, F.; Lundgren, H.; Saal, G. The resource impact of wounds on health-care providers in Europe. *J. Wound Care* **2009**, *18*, 154–161. [CrossRef]
16. Menke, N.B.; Ward, K.R.; Witten, T.M.; Bonchev, D.G.; Diegelmann, R.F. Impaired wound healing. *Clin. Dermatol.* **2007**, *25*, 19–25. [CrossRef]

17. Robson, M.C.; Mustoe, T.A.; Hunt, T.K. The future of recombinant growth factors in wound healing. *Am. J. Surg.* **1998**, *176*, 80S–82S. [CrossRef]

18. Park, J.W.; Hwang, S.R.; Yoon, I.S. Advanced growth factor delivery systems in wound management and skin regeneration. *Molecules* **2017**, *22*, 1259. [CrossRef]

19. Werner, S.; Grose, R. Regulation of wound healing by growth factors and cytokines. *Physiol. Rev.* **2003**, *83*, 835–870. [CrossRef]

20. Kim, S.H.; Kiick, K.L. Heparin-mimetic sulfated peptides with modulated affinities for heparin-binding peptides and growth factors. *Peptides* **2007**, *28*, 2125–2136. [CrossRef]

21. Nimni, M.E. Polypeptide growth factors: Targeted delivery systems. *Biomaterials* **1997**, *18*, 1201–1225. [CrossRef]

22. Torchilin, V.P.; Lukyanov, A.N. Peptide and protein drug delivery to and into tumors: Challenges and solutions. *Drug Discov. Today* **2003**, *8*, 259–266. [CrossRef]

23. Edelman, E.R.; Nugent, M.A.; Smith, L.T.; Karnovsky, M.J. Basic fibroblast growth factor enhances the coupling of intimal hyperplasia and proliferation of vasa vasorum in injured rat arteries. *J. Clin. Invest.* **1992**, *89*, 465–473. [CrossRef]

24. Gospodarowicz, D.; Cheng, J. Heparin protects basic and acidic FGF from inactivation. *J. Cell. Physiol.* **1986**, *128*, 475–484. [CrossRef] [PubMed]

25. Roghani, M.; Mansukhani, A.; Dell'Era, P.; Bellosta, P.; Basilico, C.; Rifkin, D.B.; Moscatelli, D. Heparin increases the affinity of basic fibroblast growth factor for its receptor but is not required for binding. *J. Biol. Chem.* **1994**, *269*, 3976–3984. [PubMed]

26. Liao, I.C.; Wan, A.C.; Yim, E.K.; Leong, K.W. Controlled release from fibers of polyelectrolyte complexes. *J. Control. Release* **2005**, *104*, 347–358. [CrossRef] [PubMed]

27. Lee, A.C.; Yu, V.M.; Lowe, J.B., 3rd; Brenner, M.J.; Hunter, D.A.; Mackinnon, S.E.; Sakiyama-Elbert, S.E. Controlled release of nerve growth factor enhances sciatic nerve regeneration. *Exp. Neurol.* **2003**, *184*, 295–303. [CrossRef]

28. Martin, A.I.; Priego, T.; Lopez-Calderon, A. Hormones and muscle atrophy. *Adv. Exp. Med. Biol.* **2018**, *1088*, 207–233. [CrossRef]

29. Le Roith, D. Seminars in medicine of the Beth Israel Deaconess Medical Center. Insulin-like growth factors. *N. Engl. J. Med.* **1997**, *336*, 633–640. [CrossRef]

30. Zhang, X.; Yee, D. Tyrosine kinase signalling in breast cancer: Insulin-like growth factors and their receptors in breast cancer. *Breast Cancer Res.* **2000**, *2*, 170–175. [CrossRef]

31. Daughaday, W.H.; Rotwein, P. Insulin-like growth factors I and II. Peptide, messenger ribonucleic acid and gene structures, serum, and tissue concentrations. *Endocr. Rev.* **1989**, *10*, 68–91. [CrossRef] [PubMed]

32. Baker, J.; Liu, J.P.; Robertson, E.J.; Efstratiadis, A. Role of insulin-like growth factors in embryonic and postnatal growth. *Cell* **1993**, *75*, 73–82. [CrossRef]

33. Sell, C.; Dumenil, G.; Deveaud, C.; Miura, M.; Coppola, D.; DeAngelis, T.; Rubin, R.; Efstratiadis, A.; Baserga, R. Effect of a null mutation of the insulin-like growth factor I receptor gene on growth and transformation of mouse embryo fibroblasts. *Mol. Cell. Biol.* **1994**, *14*, 3604–3612. [CrossRef]

34. Vitale, G.; Pellegrino, G.; Vollery, M.; Hofland, L.J. ROLE of IGF-1 system in the modulation of longevity: Controversies and new insights from a centenarians' perspective. *Front. Endocrinol.* **2019**, *10*, 27. [CrossRef] [PubMed]

35. Song, Y.H.; Song, J.L.; Delafontaine, P.; Godard, M.P. The therapeutic potential of IGF-I in skeletal muscle repair. *Trends Endocrinol. Metab.* **2013**, *24*, 310–319. [CrossRef] [PubMed]

36. Ikemoto-Uezumi, M.; Uezumi, A.; Tsuchida, K.; Fukada, S.; Yamamoto, H.; Yamamoto, N.; Shiomi, K.; Hashimoto, N. Pro-insulin-like growth factor-ii ameliorates age-related inefficient regenerative response by orchestrating self-reinforcement mechanism of muscle regeneration. *Stem Cells* **2015**, *33*, 2456–2468. [CrossRef] [PubMed]

37. Annibalini, G.; Bielli, P.; De Santi, M.; Agostini, D.; Guescini, M.; Sisti, D.; Contarelli, S.; Brandi, G.; Villarini, A.; Stocchi, V.; et al. MIR retroposon exonization promotes evolutionary variability and generates species-specific expression of IGF-1 splice variants. *Biochim. Biophys. Acta* **2016**, *1859*, 757–768. [CrossRef]

38. Matheny, R.W., Jr.; Nindl, B.C. Loss of IGF-IEa or IGF-IEb impairs myogenic differentiation. *Endocrinology* **2011**, *152*, 1923–1934. [CrossRef]

39. Cheema, U.; Brown, R.; Mudera, V.; Yang, S.Y.; McGrouther, G.; Goldspink, G. Mechanical signals and IGF-I gene splicing in vitro in relation to development of skeletal muscle. *J. Cell. Physiol.* **2005**, *202*, 67–75. [CrossRef]

40. Sandona, D.; Desaphy, J.F.; Camerino, G.M.; Bianchini, E.; Ciciliot, S.; Danieli-Betto, D.; Dobrowolny, G.; Furlan, S.; Germinario, E.; Goto, K.; et al. Adaptation of mouse skeletal muscle to long-term microgravity in the MDS mission. *PLoS ONE* **2012**, *7*, e33232. [CrossRef]

41. Kineman, R.D.; Del Rio-Moreno, M.; Sarmento-Cabral, A. 40 YEARS of IGF1: Understanding the tissue-specific roles of IGF1/IGF1R in regulating metabolism using the Cre/loxP system. *J. Mol. Endocrinol.* **2018**, *61*, T187–T198. [CrossRef] [PubMed]

42. De Santi, M.; Annibalini, G.; Barbieri, E.; Villarini, A.; Vallorani, L.; Contarelli, S.; Berrino, F.; Stocchi, V.; Brandi, G. Human IGF1 pro-forms induce breast cancer cell proliferation via the IGF1 receptor. *Cell. Oncol.* **2016**, *39*, 149–159. [CrossRef] [PubMed]

43. Ohlsson, C.; Mohan, S.; Sjogren, K.; Tivesten, A.; Isgaard, J.; Isaksson, O.; Jansson, J.O.; Svensson, J. The role of liver-derived insulin-like growth factor-I. *Endocr. Rev.* **2009**, *30*, 494–535. [CrossRef] [PubMed]

44. Wan, X.; Wang, S.; Xu, J.; Zhuang, L.; Xing, K.; Zhang, M.; Zhu, X.; Wang, L.; Gao, P.; Xi, Q.; et al. Dietary protein-induced hepatic IGF-1 secretion mediated by PPARgamma activation. *PLoS ONE* **2017**, *12*, e0173174. [CrossRef]

45. Savage, M.O. Insulin-like growth factors, nutrition and growth. *World Rev. Nutr. Diet.* **2013**, *106*, 52–59. [CrossRef] [PubMed]

46. Kjobsted, R.; Hingst, J.R.; Fentz, J.; Foretz, M.; Sanz, M.N.; Pehmoller, C.; Shum, M.; Marette, A.; Mounier, R.; Treebak, J.T.; et al. AMPK in skeletal muscle function and metabolism. *FASEB J.* **2018**, *32*, 1741–1777. [CrossRef]

47. Ascenzi, F.; Barberi, L.; Dobrowolny, G.; Villa Nova Bacurau, A.; Nicoletti, C.; Rizzuto, E.; Rosenthal, N.; Scicchitano, B.M.; Musaro, A. Effects of IGF-1 isoforms on muscle growth and sarcopenia. *Aging Cell* **2019**, *18*, e12954. [CrossRef]

48. Firth, S.M.; Baxter, R.C. Cellular actions of the insulin-like growth factor binding proteins. *Endocr. Rev.* **2002**, *23*, 824–854. [CrossRef]

49. Williams, R.M.; McDonald, A.; O'Savage, M.; Dunger, D.B. Mecasermin rinfabate: rhIGF-I/rhIGFBP-3 complex: iPLEX. *Expert Opin. Drug. Metab. Toxicol.* **2008**, *4*, 311–324. [CrossRef]

50. Yu, M.; Wang, H.; Xu, Y.; Yu, D.; Li, D.; Liu, X.; Du, W. Insulin-like growth factor-1 (IGF-1) promotes myoblast proliferation and skeletal muscle growth of embryonic chickens via the PI3K/Akt signalling pathway. *Cell Biol. Int.* **2015**, *39*, 910–922. [CrossRef]

51. Clemmons, D.R. Role of IGF-I in skeletal muscle mass maintenance. *Trends Endocrino.l Metab.* **2009**, *20*, 349–356. [CrossRef]

52. Nindl, B.C.; Santtila, M.; Vaara, J.; Hakkinen, K.; Kyrolainen, H. Circulating IGF-I is associated with fitness and health outcomes in a population of 846 young healthy men. *Growth Horm. IGF Res.* **2011**, *21*, 124–128. [CrossRef] [PubMed]

53. Alzhanov, D.T.; McInerney, S.F.; Rotwein, P. Long range interactions regulate Igf2 gene transcription during skeletal muscle differentiation. *J. Biol. Chem.* **2010**, *285*, 38969–38977. [CrossRef] [PubMed]

54. Gardner, S.; Alzhanov, D.; Knollman, P.; Kuninger, D.; Rotwein, P. TGF-beta inhibits muscle differentiation by blocking autocrine signaling pathways initiated by IGF-II. *Mol. Endocrinol.* **2011**, *25*, 128–137. [CrossRef] [PubMed]

55. Laron, Z. Insulin-like growth factor 1 (IGF-1): A growth hormone. *Mol. Pathol.* **2001**, *54*, 311–316. [CrossRef] [PubMed]

56. Hamilton, D.L.; Philp, A.; MacKenzie, M.G.; Baar, K. A limited role for PI(3,4,5)P3 regulation in controlling skeletal muscle mass in response to resistance exercise. *PLoS ONE* **2010**, *5*, e11624. [CrossRef] [PubMed]

57. Fu, S.; Yin, L.; Lin, X.; Lu, J.; Wang, X. Effects of cyclic mechanical stretch on the proliferation of L6 myoblasts and its mechanisms: PI3K/Akt and MAPK signal pathways regulated by IGF-1 receptor. *Int. J. Mol. Sci.* **2018**, *19*, 1649. [CrossRef] [PubMed]

58. Sandri, M.; Barberi, L.; Bijlsma, A.Y.; Blaauw, B.; Dyar, K.A.; Milan, G.; Mammucari, C.; Meskers, C.G.; Pallafacchina, G.; Paoli, A.; et al. Signalling pathways regulating muscle mass in ageing skeletal muscle: The role of the IGF1-Akt-mTOR-FoxO pathway. *Biogerontology* **2013**, *14*, 303–323. [CrossRef]

59. Liu, M.; Zhang, S. Amphioxus IGF-like peptide induces mouse muscle cell development via binding to IGF receptors and activating MAPK and PI3K/Akt signaling pathways. *Mol. Cell. Endocrinol.* **2011**, *343*, 45–54. [CrossRef]

60. Peng, X.D.; Xu, P.Z.; Chen, M.L.; Hahn-Windgassen, A.; Skeen, J.; Jacobs, J.; Sundararajan, D.; Chen, W.S.; Crawford, S.E.; Coleman, K.G.; et al. Dwarfism, impaired skin development, skeletal muscle atrophy, delayed bone development, and impeded adipogenesis in mice lacking Akt1 and Akt2. *Genes Dev.* **2003**, *17*, 1352–1365. [CrossRef]

61. Borselli, C.; Storrie, H.; Benesch-Lee, F.; Shvartsman, D.; Cezar, C.; Lichtman, J.W.; Vandenburgh, H.H.; Mooney, D.J. Functional muscle regeneration with combined delivery of angiogenesis and myogenesis factors. *Proc. Natl. Acad. Sci. USA* **2010**, *107*, 3287–3292. [CrossRef] [PubMed]

62. Barton-Davis, E.R.; Shoturma, D.I.; Sweeney, H.L. Contribution of satellite cells to IGF-I induced hypertrophy of skeletal muscle. *Acta Physiol. Scand.* **1999**, *167*, 301–305. [CrossRef] [PubMed]

63. Edwall, D.; Schalling, M.; Jennische, E.; Norstedt, G. Induction of insulin-like growth factor I messenger ribonucleic acid during regeneration of rat skeletal muscle. *Endocrinology* **1989**, *124*, 820–825. [CrossRef] [PubMed]

64. Yang, S.; Alnaqeeb, M.; Simpson, H.; Goldspink, G. Cloning and characterization of an IGF-1 isoform expressed in skeletal muscle subjected to stretch. *J. Muscle Res. Cell Motil.* **1996**, *17*, 487–495. [CrossRef]

65. Hennebry, A.; Oldham, J.; Shavlakadze, T.; Grounds, M.D.; Sheard, P.; Fiorotto, M.L.; Falconer, S.; Smith, H.K.; Berry, C.; Jeanplong, F.; et al. IGF1 stimulates greater muscle hypertrophy in the absence of myostatin in male mice. *J. Endocrinol.* **2017**, *234*, 187–200. [CrossRef]

66. Czaja, W.; Nakamura, Y.K.; Li, N.; Eldridge, J.A.; DeAvila, D.M.; Thompson, T.B.; Rodgers, B.D. Myostatin regulates pituitary development and hepatic IGF1. *Am. J. Physiol. Endocrinol. Metab.* **2019**, *316*, E1036–E1049. [CrossRef]

67. Barbe, C.; Kalista, S.; Loumaye, A.; Ritvos, O.; Lause, P.; Ferracin, B.; Thissen, J.P. Role of IGF-I in follistatin-induced skeletal muscle hypertrophy. *Am. J. Physiol. Endocrinol. Metab.* **2015**, *309*, E557–E567. [CrossRef]

68. Liu, J.P.; Baker, J.; Perkins, A.S.; Robertson, E.J.; Efstratiadis, A. Mice carrying null mutations of the genes encoding insulin-like growth factor I (Igf-1) and type 1 IGF receptor (Igf1r). *Cell* **1993**, *75*, 59–72. [CrossRef]

69. Lipina, C.; Kendall, H.; McPherron, A.C.; Taylor, P.M.; Hundal, H.S. Mechanisms involved in the enhancement of mammalian target of rapamycin signalling and hypertrophy in skeletal muscle of myostatin-deficient mice. *FEBS Lett.* **2010**, *584*, 2403–2408. [CrossRef]

70. Morissette, M.R.; Cook, S.A.; Buranasombati, C.; Rosenberg, M.A.; Rosenzweig, A. Myostatin inhibits IGF-I-induced myotube hypertrophy through Akt. *Am. J. Physiol. Cell Physiol.* **2009**, *297*, C1124–C1132. [CrossRef]

71. Rommel, C.; Bodine, S.C.; Clarke, B.A.; Rossman, R.; Nunez, L.; Stitt, T.N.; Yancopoulos, G.D.; Glass, D.J. Mediation of IGF-1-induced skeletal myotube hypertrophy by PI(3)K/Akt/mTOR and PI(3)K/Akt/GSK3 pathways. *Nat. Cell Biol.* **2001**, *3*, 1009–1013. [CrossRef] [PubMed]

72. Suthar, R.; Sankhyan, N. Duchenne muscular dystrophy: A practice update. *Indian J. Pediatr.* **2018**, *85*, 276–281. [CrossRef] [PubMed]

73. Ahmad, K.; Shaikh, S.; Ahmad, S.S.; Lee, E.J.; Choi, I. Cross-talk between extracellular matrix and skeletal muscle: Implications for myopathies. *Front. Pharmacol.* **2020**, *11*, 142. [CrossRef] [PubMed]

74. Noguchi, S. The biological function of insulin-like growth factor-I in myogenesis and its therapeutic effect on muscular dystrophy. *Acta Myol.* **2005**, *24*, 115–118. [PubMed]

75. Patel, K.; Macharia, R.; Amthor, H. Molecular mechanisms involving IGF-1 and myostatin to induce muscle hypertrophy as a therapeutic strategy for Duchenne muscular dystrophy. *Acta Myol.* **2005**, *24*, 230–241. [PubMed]

76. Fang, X.B.; Song, Z.B.; Xie, M.S.; Liu, Y.M.; Zhang, W.X. Synergistic effect of glucocorticoids and IGF-1 on myogenic differentiation through the Akt/GSK-3beta pathway in C2C12 myoblasts. *Int. J. Neurosci.* **2020**, *130*, 1–11. [CrossRef]

77. Lynch, G.S.; Cuffe, S.A.; Plant, D.R.; Gregorevic, P. IGF-I treatment improves the functional properties of fast- and slow-twitch skeletal muscles from dystrophic mice. *Neuromuscul. Disord.* **2001**, *11*, 260–268. [CrossRef]

78. Cao, R.Y.; Li, J.; Dai, Q.; Li, Q.; Yang, J. Muscle Atrophy: Present and Future. *Adv. Exp. Med. Biol.* **2018**, *1088*, 605–624. [CrossRef]

79. Fink, J.; Schoenfeld, B.J.; Nakazato, K. The role of hormones in muscle hypertrophy. *Phys. Sportsmed.* **2018**, *46*, 129–134. [CrossRef]

80. Accorsi, A.; Kumar, A.; Rhee, Y.; Miller, A.; Girgenrath, M. IGF-1/GH axis enhances losartan treatment in Lama2-related muscular dystrophy. *Hum. Mol. Genet.* **2016**, *25*, 4624–4634. [CrossRef]

81. Sun, Y.; Sun, X.; Shen, B. Molecular Imaging of IGF-1R in Cancer. *Mol. Imaging* **2017**, *16*, 1–7. [CrossRef]

82. Pickard, A.; McCance, D.J. IGF-Binding Protein 2—Oncogene or Tumor Suppressor? *Front. Endocrinol.* **2015**, *6*, 25. [CrossRef] [PubMed]

83. Yao, X.; Sun, S.; Zhou, X.; Guo, W.; Zhang, L. IGF-binding protein 2 is a candidate target of therapeutic potential in cancer. *Tumour Biol.* **2016**, *37*, 1451–1459. [CrossRef] [PubMed]

84. Thant, A.A.; Wu, Y.; Vadgama, J.V. Cancer invasion and angiogenesis by IGF-1-induced MMP-2 and VEGF in human Breast cancer cells. *Proc. Am. Assoc. Cancer Res.* **2005**, *46*, 695.

85. Baig, M.H.; Adil, M.; Khan, R.; Dhadi, S.; Ahmad, K.; Rabbani, G.; Bashir, T.; Imran, M.A.; Husain, F.M.; Lee, E.J.; et al. Enzyme targeting strategies for prevention and treatment of cancer: Implications for cancer therapy. *Semin. Cancer Biol.* **2019**, *56*, 1–11. [CrossRef]

86. Philippou, A.; Christopoulos, P.F.; Koutsilieris, D.M. Clinical studies in humans targeting the various components of the IGF system show lack of efficacy in the treatment of cancer. *Mutat. Res. Rev. Mutat. Res.* **2017**, *772*, 105–122. [CrossRef]

87. Major, J.M.; Laughlin, G.A.; Kritz-Silverstein, D.; Wingard, D.L.; Barrett-Connor, E. Insulin-like growth factor-I and cancer mortality in older men. *J. Clin. Endocrinol. Metab.* **2010**, *95*, 1054–1059. [CrossRef]

88. Ahmad, S.S.; Khan, H.; Danish Rizvi, S.M.; Ansari, S.A.; Ullah, R.; Rastrelli, L.; Mahmood, H.M.; Siddiqui, M.H. Computational study of natural compounds for the clearance of amyloid-betaeta: A potential therapeutic management strategy for alzheimer's disease. *Molecules* **2019**, *24*, 3233. [CrossRef]

89. Ahmad, S.S.; Kamal, M.A. Current updates on the regulation of beta-secretase movement as a potential restorative focus for management of alzheimer's disease. *Protein Pept. Lett.* **2019**, *26*, 579–587. [CrossRef]

90. Waldthaler, J.; Timmermann, L. Update on diagnostics and therapy of idiopathic Parkinson's disease. *Fortschr. Neurol. Psychiatr.* **2019**, *87*, 445–461. [CrossRef]

91. Ahmad, S.S.; Waheed, T.; Rozeen, S.; Mahmood, S.; Kamal, M.A. Therapeutic study of phytochemicals against cancer and alzheimer's disease management. *Curr. Drug Metab.* **2019**, *20*, 1006–1013. [CrossRef] [PubMed]

92. Walker, L.C.; Jucker, M. Neurodegenerative diseases: Expanding the prion concept. *Annu. Rev. Neurosci.* **2015**, *38*, 87–103. [CrossRef] [PubMed]

93. Morel, G.R.; Leon, M.L.; Uriarte, M.; Reggiani, P.C.; Goya, R.G. Therapeutic potential of IGF-I on hippocampal neurogenesis and function during aging. *Neurogenesis* **2017**, *4*, e1259709. [CrossRef]

94. Rosenbloom, A.L.; Rivkees, S.A. Off-label use of recombinant igf-I to promote growth: Is it appropriate? *J. Clin. Endocrinol. Metab.* **2010**, *95*, 505–508. [CrossRef] [PubMed]

95. Bianchi, V.E.; Locatelli, V.; Rizzi, L. Neurotrophic and neuroregenerative effects of GH/IGF1. *Int. J. Mol. Sci.* **2017**, *18*, 2441. [CrossRef]

96. Park, J.; Yan, G.; Kwon, K.C.; Liu, M.; Gonnella, P.A.; Yang, S.; Daniell, H. Oral delivery of novel human IGF-1 bioencapsulated in lettuce cells promotes musculoskeletal cell proliferation, differentiation and diabetic fracture healing. *Biomaterials* **2020**, *233*, 119591. [CrossRef]

97. Sukhanov, S.; Higashi, Y.; Shai, S.Y.; Snarski, P.; Danchuk, S.; D'Ambra, V.; Tabony, M.; Woods, T.C.; Hou, X.; Li, Z.; et al. SM22alpha (Smooth Muscle protein 22-alpha) promoter-driven IGF1R (Insulin-like Growth Factor 1 Receptor) deficiency promotes atherosclerosis. *Arterioscler. Thromb. Vasc. Biol.* **2018**, *38*, 2306–2317. [CrossRef]

98. Shapiro, L.; Elsangeedy, E.; Lee, H.; Atala, A.; Yoo, J.J.; Lee, S.J.; Ju, Y.M. In vitro evaluation of functionalized decellularized muscle scaffold for in situ skeletal muscle regeneration. *Biomed. Mater.* **2019**, *14*, 045015. [CrossRef]

99. Manzella, L.; Massimino, M.; Stella, S.; Tirro, E.; Pennisi, M.S.; Martorana, F.; Motta, G.; Vitale, S.R.; Puma, A.; Romano, C.; et al. Activation of the IGF axis in thyroid cancer: Implications for tumorigenesis and treatment. *Int. J. Mol. Sci.* **2019**, *20*, 3258. [CrossRef]

100. Maffezzoni, F.; Formenti, A.M. Acromegaly and bone. *Minerva Endocrinol.* **2018**, *43*, 168–182. [CrossRef]

101. Van Nieuwpoort, I.C.; Vlot, M.C.; Schaap, L.A.; Lips, P.; Drent, M.L. The relationship between serum IGF-1, handgrip strength, physical performance and falls in elderly men and women. *Eur. J. Endocrinol.* **2018**, *179*, 73–84. [CrossRef] [PubMed]

102. Procaccini, C.; Santopaolo, M.; Faicchia, D.; Colamatteo, A.; Formisano, L.; de Candia, P.; Galgani, M.; De Rosa, V.; Matarese, G. Role of metabolism in neurodegenerative disorders. *Metabolism* **2016**, *65*, 1376–1390. [CrossRef] [PubMed]

103. Kopczak, A.; Stalla, G.K.; Uhr, M.; Lucae, S.; Hennings, J.; Ising, M.; Holsboer, F.; Kloiber, S. IGF-I in major depression and antidepressant treatment response. *Eur. Neuropsychopharmacol.* **2015**, *25*, 864–872. [CrossRef] [PubMed]

104. Licata, L.; Lo Surdo, P.; Iannuccelli, M.; Palma, A.; Micarelli, E.; Perfetto, L.; Peluso, D.; Calderone, A.; Castagnoli, L.; Cesareni, G. SIGNOR 2.0, the SIGnaling network open resource 2.0: 2019 update. *Nucleic Acid. Res.* **2020**, *48*, D504–D510. [CrossRef]

105. Szklarczyk, D.; Gable, A.L.; Lyon, D.; Junge, A.; Wyder, S.; Huerta-Cepas, J.; Simonovic, M.; Doncheva, N.T.; Morris, J.H.; Bork, P.; et al. STRING v11: Protein-protein association networks with increased coverage, supporting functional discovery in genome-wide experimental datasets. *Nucleic Acid. Res.* **2019**, *47*, D607–D613. [CrossRef]

106. Bentzinger, C.F.; Wang, Y.X.; Rudnicki, M.A. Building muscle: Molecular regulation of myogenesis. *Cold Spring Harb. Perspect. Biol.* **2012**, *4*, a008342. [CrossRef]

107. Buckingham, M.; Relaix, F. PAX3 and PAX7 as upstream regulators of myogenesis. *Semin. Cell Dev. Biol.* **2015**, *44*, 115–125. [CrossRef]

108. Forcina, L.; Miano, C.; Scicchitano, B.M.; Musaro, A. Signals from the niche: Insights into the role of IGF-1 and IL-6 in modulating skeletal muscle fibrosis. *Cells* **2019**, *8*, 232. [CrossRef]

109. Ding, H.; Wu, T. Insulin-like growth factor binding proteins in autoimmune diseases. *Front. Endocrinol.* **2018**, *9*, 499. [CrossRef] [PubMed]

110. Boisclair, Y.R.; Rhoads, R.P.; Ueki, I.; Wang, J.; Ooi, G.T. The acid-labile subunit (ALS) of the 150 kDa IGF-binding protein complex: An important but forgotten component of the circulating IGF system. *J. Endocrinol.* **2001**, *170*, 63–70. [CrossRef]

111. Tzeng, H.E.; Chen, J.C.; Tsai, C.H.; Kuo, C.C.; Hsu, H.C.; Hwang, W.L.; Fong, Y.C.; Tang, C.H. CCN3 increases cell motility and MMP-13 expression in human chondrosarcoma through integrin-dependent pathway. *J. Cell. Physiol.* **2011**, *226*, 3181–3189. [CrossRef]

112. Bidlingmaier, M.; Strasburger, C.J. Growth hormone. *Handb. Exp. Pharmacol.* **2010**, *195*, 187–200. [CrossRef]

113. Jones, B.H.; Hauschild, V.D.; Canham-Chervak, M. Musculoskeletal training injury prevention in the U.S. Army: Evolution of the science and the public health approach. *J. Sci. Med. Sport* **2018**, *21*, 1139–1146. [CrossRef] [PubMed]

114. Martins, K.J.; Gehrig, S.M.; Naim, T.; Saenger, S.; Baum, D.; Metzger, F.; Lynch, G.S. Intramuscular administration of PEGylated IGF-I improves skeletal muscle regeneration after myotoxic injury. *Growth Horm. IGF Res.* **2013**, *23*, 128–133. [CrossRef] [PubMed]

Review

Effects of GH/IGF on the Aging Mitochondria

Sher Bahadur Poudel [1,†], **Manisha Dixit** [1,†], **Maria Neginskaya** [1,†], **Karthik Nagaraj** [2,†], **Evgeny Pavlov** [1], **Haim Werner** [2] and **Shoshana Yakar** [1,*]

[1] David B. Kriser Dental Center, Department of Molecular Pathobiology, New York University College of Dentistry New York, NY 10010–4086, USA; sbp4@nyu.edu (S.B.P.); dixitm01@nyu.edu (M.D.); mn2452@nyu.edu (M.N.); ep37@nyu.edu (E.P.)

[2] Department of Human Molecular Genetics and Biochemistry, Sackler School of Medicine, Tel Aviv University, Tel Aviv 69978, Israel; mailkartz@gmail.com (K.N.); hwerner@tauex.tau.ac.il (H.W.)

* Correspondence: sy1007@nyu.edu; Tel.: +212-998-9721

† These authors contributed equally to this work.

Received: 16 April 2020; Accepted: 26 May 2020; Published: 2 June 2020

Abstract: The mitochondria are key organelles regulating vital processes in the eukaryote cell. A decline in mitochondrial function is one of the hallmarks of aging. Growth hormone (GH) and the insulin-like growth factor-1 (IGF-1) are somatotropic hormones that regulate cellular homeostasis and play significant roles in cell differentiation, function, and survival. In mammals, these hormones peak during puberty and decline gradually during adulthood and aging. Here, we review the evidence that GH and IGF-1 regulate mitochondrial mass and function and contribute to specific processes of cellular aging. Specifically, we discuss the contribution of GH and IGF-1 to mitochondrial biogenesis, respiration and ATP production, oxidative stress, senescence, and apoptosis. Particular emphasis was placed on how these pathways intersect during aging.

Keywords: mitochondria; growth hormone; insulin-like growth factor-1; aging; oxidative stress; senescence

---

1. Overview of the growth hormone (GH)/insulin-like growth factor (IGF) Axis

Growth hormone (GH) and the insulin-like growth factor-1 (IGF-1) are part of the somatotropic hypothalamic-pituitary axis that regulates somatic growth and aging. As such, GH and IGF-1 peak during puberty to support lean and fat mass gain as well as to enhance skeletal acquisition and linear growth. During aging, GH and IGF-1 levels are significantly reduced, a state termed somatopause. Somatopause associates with many pathologies, such as osteopenia [1], sarcopenia [2], cardiovascular disorders [3,4], and more. For the sake of this review, it is important to distinguish between congenital somatopause, which results from germline mutations in components of the GH/IGF-1 axis (resulting in life-long epigenetic changes), and age-induced somatopause, which refers to the natural decline in GH and IGF-1 levels during aging.

The GH/IGF-1 axis includes the hypothalamic GH-releasing hormone (GHRH) and its receptor (GHRHR), expressed mainly in the pituitary (Figure 1) [5]. Upon activation, the GHRHR stimulates pituitary secretion of GH, which is released to the circulation and binds to its specific membrane receptor (GHR) on cells of different tissues. Among the many actions of GH in liver, its major role is stimulating the transcription of the IGF-1 gene, the IGF-binding proteins (IGFBPs) [6], and the acid labile subunit (ALS) that carry IGF-1 in circulation [7]. Other members of this axis are the proteases that release the IGF-1 from its complex with IGFBPs, pregnancy-associated plasma protein A (PAPP-A), and the PAPP-A2 [8]. The somatotropic axis is tightly regulated under normal physiological conditions. IGF-1 in serum provides a negative feedback to the pituitary to specifically inhibit GH secretion. Somatostatin (SST), secreted by the hypothalamus, inhibits several pituitary hormones including GH. In addition, SST binds to several SST-receptors (SSTR) on peripheral tissues to inhibit production of pancreatic (glucagon and insulin) and gastric (secretin and gastrin) hormones. Secretion of GH is also regulated by the gastrointenstinal hormone ghrelin. Ghrelin is one of the hormones that regulate food intake [9]. During hunger, ghrelin levels in circulation are increased and activate hypothalamic ghrelin receptor to initiate appetite. In addition, ghrelin acts as a GH secretagogue and can induce pituitary secretion of GH [10].

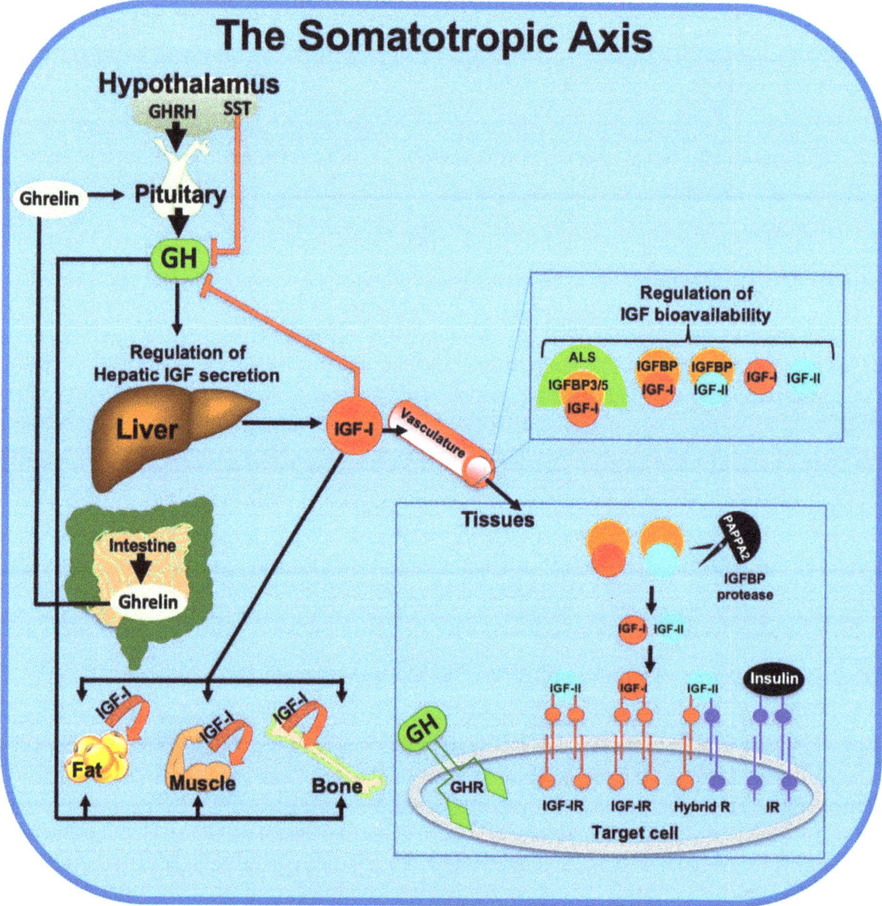

Figure 1. Schematic summary of the major molecules composing the somatotropic axis. GHRH—Growth hormone-releasing hormone, SST—Somatostatin, GH—Growth hormone, GHR—Growth hormone receptor, IGF—Insulin-like growth factor, IGFBP—Insulin-like growth factor binding protein, IGF-IR—insulin like growth factor receptor-1, IR—Insulin receptor, ALS—Acid labile subunit, PAPPA2—Pregnancy-associated plasma protein-A2.

Receptors to GH and IGF-1 are present in virtually all cells. The IGF-1 receptor (IGF-1R) is a tyrosine kinase receptor that activates multiple pathways including (but not limited to) the Phosphoinositide-3-kinase (PI3K)/Protein kinase B (PKB or AKT), Ras/Raf/Mitogen-activated protein kinase (MAPK), and Shc [11]. These pathways are implicated in cellular growth, proliferation, differentiation, survival, metabolism, gene transcription, and protein translation. Apparently, the PI3K/AKT pathway is central to regulation of cell metabolism and cell fate (apoptosis), both of which involve mitochondrial function and integrity. The GHR is a cytokine-like receptor that, upon ligand binding, activates the Janus kinase 2 (JAK2)/Signal transducer and activator of transcription 5B (STAT5b) pathway to induce transcriptional activation of genes involved in cellular growth (among them is *Igf1*) and metabolic homeostasis [11].

Among other aspects of cellular homeostasis, GH and IGF-1 have important roles in mitochondrial function. Mitochondria participate in a wide range of processes with vital roles in cellular function. A key function of mitochondria is generation of adenosine triphosphate (ATP) from metabolic processing of carbohydrates, fats, and amino acids. In addition, mitochondria regulate other functions such as calcium homeostasis, cell death programming (apoptosis), inflammation, heat production, and other tissue-specific functions. Given their critical roles in the cell, it is not surprising that mitochondria are considered central regulators of aging. This review summarizes a few aspects of mitochondrial regulation by the GH/IGF-1 axis. We could not possibly include all of the studies that reported how GH/IGF-1 control mitochondrial function, including the effects of this axis on mitochondria in cancer cells (which can be found in [12–14]). Instead, our review focuses mostly on different aspects of aging and how GH/IGF-1 affects mitochondria in that context.

2. GH/IGF-1 Effects on Mitochondrial Biogenesis

Mitochondria are dynamic organelles. The number and size of mitochondria change according to the cellular metabolic and physiologic conditions. In addition, mitochondria abundance involves fusion (merging of two originally distinct mitochondria into one), fission (division of a single mitochondrion into two distinct mitochondria), and biogenesis. Mitochondrial biogenesis requires coupling of mitochondrial genome replication and fission.

The mitochondrial genome (mtDNA) encodes 13 proteins, 22 transfer-RNAs, and two ribosomal-RNAs [15]. The 13 proteins encoded by mitochondria take part in oxidative phosphorylation (OXPHOS) including cytochrome b of complex III; ATP synthase subunits 6 and 8 of the F_0 ATP-synthase complex; cytochrome c (Cyt c) oxidase subunits 1, 2, and 3; and six subunits of the nicotinamide adenine dinucleotide (NADH) dehydrogenase of complex I (ND1 through ND6). However, there are approximately 1500 proteins encoded by the nucleus which are necessary for mitochondrial biogenesis and function [16,17]. As such, replication of the mitochondrial genome is a complex mechanism that requires activation and translation of a series of nuclear encoded proteins. These mitochondrially targeted proteins are transported to the mitochondria via channels composed of transporters located at the outer mitochondrial membrane (TOMs) [18]. One of the proteins that plays key roles in mitochondrial biogenesis is peroxisome proliferator-activated receptor gamma coactivator 1 α (PGC1α). Activation of PGC1α leads to subsequent activations of the transcriptional regulators, nuclear respiratory factors (NRF1 and 2), and peroxisome proliferator-activated receptors (PPARs), which initiate transcription of nuclear genes involved in mitochondrial biogenesis and function (Figure 2).

The possible roles of the GH/IGF-1 axis in mitochondrial biogenesis were initially recognized in the early 1970s. Following injection of radiolabelled bovine GH (^{125}I-bGH) to Sprague Dawley rats, radioactive signals were detected in the mitochondria, nuclei, and other cytoplasmic compartments of liver and kidney cells [19]. However, a later study found that, in fact, a very negligible amount of radiolabelled ^{125}I-bGH gets transported to the mitochondria compared to other subcellular compartments, including microsomes, endosomes, lysosomes, and Golgi bodies [20], suggesting that GH does not act directly on mitochondria. Instead, it was found that GH affects mitochondrial protein synthesis. Injection of human GH (hGH) to intact and hypophysectomized (hypox) rats revealed that the mitochondrial protein synthetic capacity of liver, measured by radioactive leucine incorporation in vitro as well as in vivo, significantly increased in hGH-treated rats [21]. However, a later study showed that GH is not the only player in mitochondrial biogenesis. Thus, hypox SD rats that were treated with hGH (120 µg/day), triiodothyronine (T3, 10 ng/day), or a combination of both hormones for six days showed that cellular respiration was recovered with T3 treatment but not with hGH alone [20].

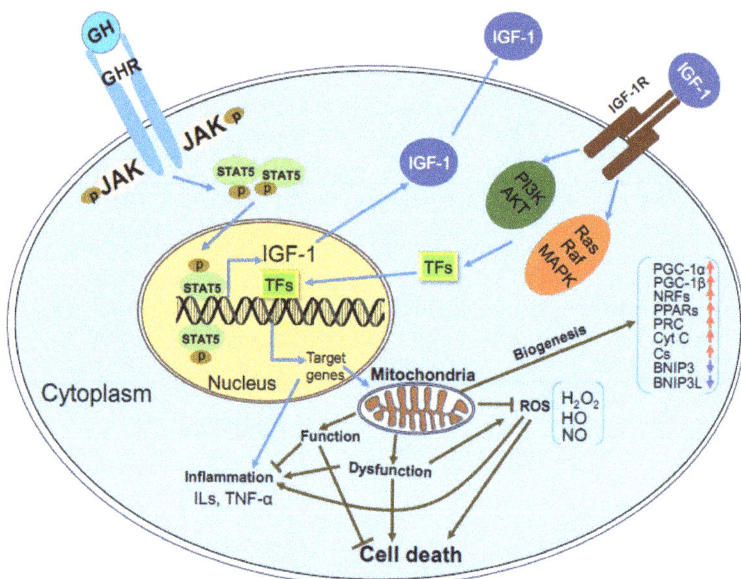

Figure 2. Schematic summary of the major effects of GH/IGF-1 on mitochondrial gene expression: Upon binding of GH to the GHR, the Janus kinase (JAK)-Signal transducer and activator of transcription 5 (STAT5) signaling pathway is activated, leading mostly to increases in IGF-1 transcription. Binding of IGF-1 to the tyrosine kinase IGF-1R stimulates several signaling pathways including the Phosphoinositide-3-kinase (PI3K)/Protein kinase B (PKB or AKT) and Ras/Raf/Mitogen-activated protein kinase (MAPK), involving phosphorylation and dephosphorylation of candidate proteins. This cascade leads to transcriptional activity of genes involved in mitochondrial biogenesis, control of Reactive oxygen species (ROS), cell survival (antiapoptotic), and genes involved in metabolism.

Identification of the tight relationship between GH and IGF-1 in control of cellular growth and replication led to numerous studies aimed at assessing the effects of IGF-1 on mitochondrial biogenesis. However, as was found with GH, IGF-1 plays indirect effects on mitochondrial biogenesis, likely in synergy with other growth factors. As such, exogenous IGF-1 increased cell volume but not mitochondria in rat sciatic Schwann cells [22]. However, addition of the mitogenic factor neuregulin (NRG) together with IGF-1 increased mitochondrial mass and DNA replication. In MCF-7 and ZR75.1 breast cancer cells, IGF-1 induced the expression of PGC-1β and PGC-1α-related coactivator (PRC), which are required for mitochondrial biogenesis (Figure 2) [13]. Increased mitochondrial mass in these cells associated with reduced expression of the mitophagy mediators BNIP3 and BNIP3L (BCL2/adenovirus E1B 19 kDa protein-interacting protein 3 and L) and impaired mitophagy suggest that IGF-1 indirectly regulates mitochondrial turnover via BNIP3. Increased levels of PGC-1α, Nrf1, Cyt c, and citrate synthase activity were also seen in gastrocnemius muscle cells of old animals treated with GH, likely mediated by IGF-1 (Figure 2) [23]. Age-induced mitochondrial dysfunction in reproductive organs causes reductions in oocyte quality leading to infertility. GH partially rescued the age-induced loss of ovulation stimulation, oocyte formation, and maturation through its effects on mitochondria [24]. Treating young (4 weeks) and aged (32 weeks) female mice with low, medium, or high doses of hGH for 8 weeks resulted in increased oocyte number as well as maturation in aged females that was similar to young females, likely via IGF-1. However, in the latter study, despite these oocyte functional improvements, there was no change in mitochondrial DNA copy number, suggesting again that mitochondrial biogenesis was not affected directly by hGH treatment.

3. GH/IGF-1 Effects on Mitochondrial Respiration and ATP Production During Aging

Mitochondrial function involves several processes, including cellular respiration, energy production via the tricarboxylic acid (TCA) cycle coupled with OXPHOS, calcium homeostasis, cellular replication, apoptosis, and generation of (and protection from) reactive oxygen species (ROS). These functions are fundamental during growth and development and play major roles during aging. Although age-induced mitochondrial dysfunction and overall reductions in the secretion and action of GH and IGF-1 coincide, a clear causality between both processes has not yet been established.

Congenital (life-long) reductions in the activity of the GH/IGF-1 axis was associated with increased longevity in mice [25], flies [26], and nematodes [27,28]. In the long-lived Ames dwarf mice, the activity and expression of several complexes of the electron transport chain (ETC) in liver and in kidney increased [29]. Interestingly, high activity of complex IV of the ETC in Ames mice was in agreement with enhanced oxygen metabolism found also in the long-lived GH receptor null (GHRKO) mice [30]. Elevated expression of key proteins of the mitochondrial respiration complexes in Ames mice coincided with reduced production of H_2O_2 in state 3 (ATP production) and state 4 (reflecting mitochondrial leak) of respiration in liver mitochondria [31]. Accordingly, the long-lived GHRKO and Ames mice exhibit improved glucose homeostasis and energy metabolism evidenced by decreased respiratory quotient (RQ) and increased oxygen consumption (VO_2). In contrast, bGH mice, with high GH/IGF-1 levels show decreased VO(2) and increased RQ [30,32,33]. Taken together, these animal models suggest that diminished GH/IGF-1 activity improves mitochondrial flexibility and increases the capacity for fat oxidation. Interestingly, a recent study found that housing the GHRKO mice at thermoneutral temperature (30 °C) resulted in decreased expression of thermogenic genes in brown adipose tissue (BAT) and elevated core body temperature [32]. However, these mice still maintained their extended longevity phenotype at 30 °C [32], suggesting an intrinsic advantage of mitochondrial function in the GHRKO mice. Together, congenital decreases in the GH/IGF signaling in genetically modified mice were associated with reduced levels of ROS, upregulated activity of the mitochondrial ETC, and overall enhanced mitochondrial function [31,34].

In humans, GH and IGF-1 levels peak during puberty, decline during aging [35], and coincide with unfavorable effects on mitochondrial metabolism in brain (cognition [36]), muscle (sarcopenia [2]), and skeletal tissue (osteopenia [1]). IGF-1 haploinsufficiency (IGF-1$^{+/-}$) in mice was associated with mitochondrial dysfunction accompanied by increased level of lipid peroxidation, protein carboxylation, and intramitochondrial ROS in hepatocytes [37]. The reductions in mitochondrial membrane potential and OXPHOS in the IGF-1$^{+/-}$ mice were resolved with IGF-1 treatment. IGF-1 administration to old Wister rats (103 weeks) for 30 days rescued mitochondrial membrane potential, oxygen consumption rate, proton leak, ATP production, cytochrome oxidase, and ATPase complexes activities in isolated mitochondria from livers [38]. Additionally, the age-associated increases in mitochondrial ROS production, reduced antioxidants activities, and increased apoptosis in isolated mitochondria were resolved by IGF-1 treatment, suggesting cytoprotective effects of IGF-1.

IGF-1 plays essential roles in the function of mitochondria in the central nervous system [39]. Hippocampal mitochondria of 18-month-old mice with adult-induced liver IGF-1 deficiency (iLID) (at 5 months of age) showed lower level of OXPHOS and increased mitochondrial uncoupling, lower level of ATP production, and increased level of oxidative damage, as compared to aged controls [39]. This compromised mitochondrial function in the iLID mice manifested as impaired spatial acquisition and reversal learning. Similarly, specific ablation of IGF-1R in astrocytes [40] caused impaired mitochondrial energy metabolism, OXPHOS, and decreased glucose and amyloid β uptake. Overall, data suggest that increases in IGF-1 signaling in astrocytes may rescue from age-related mitochondrial dysfunction and cognitive decline.

Sarcopenia, age-associated reductions in skeletal muscle-mass, is thought to result from reduced level of GH. Low doses of GH administration to 22-month-old Wistar rats for 8 weeks increased circulating IGF-1 levels, enhanced the synthesis of mitochondrial proteins and antioxidant enzyme activities (catalase, glutathione peroxidase, and glucose-6-phosphate dehydrogenase (G6PDH)), and reduced oxidative damage (measured by the levels of 8-OHdG) in the skeletal muscle [23]. GH administration induced the activation of the anabolic AKT, mTOR, p70S6K, and Myf-5 factors while inhibiting p21, p38, and muscle RING finger-1 (MuRF-1) catabolic signals. Similarly, it was reported that a decline in GH/IGF-1 signaling in muscles of old rats associated with mitochondrial dysfunction. Following treatment with antioxidants, GH and IGF-1 levels in serum increased and associated with improvement of cristae structure and clustering of muscle mitochondria [41]. Likewise, IGF-1 alleviated dysfunctional mitochondria of cardiomyocytes from obese mice [42]. Glucose uptake, ATP production, and aconitase activity increased while lipid peroxidation, ROS production, protein carbonyl content, and apoptosis decreased in transgenic mice with overexpression of cardiomyocyte-specific IGF-1 that were fed a high fat diet. Furthermore, cardiomyocyte-specific IGF-1 induced the expression of Cyt c, PGC-1α, and UCP2 as well as the essential intracellular Ca^{2+} regulatory proteins SERCA2a and the Na^+/Ca^{2+} exchanger.

Mitochondrial dysfunction was also reported for cortical bone osteocytes in GHRKO mice, which exhibited reduced mitochondrial membrane potential, decreased ATP production, and reduced maximal respiration in both young and old mice [43].

Overall, it seems that GH/IGF-1 signaling involvement in mitochondrial function is multifaceted, with important tissue-, organ-, and age-dependent features. Importantly, there are fundamental differences between the effects of age-induced reductions in GH/IGF-1 on mitochondrial function and those of life-long, congenital, ablations of members of the GH/IGF-1 axis on mitochondrial function.

4. GH/IGF-1 Effects on Oxidative Stress During Aging

According to the free radical theory of aging, ROSs generated by the ETC in the mitochondria or via nitric oxide metabolism in the cytosol have the potential to result in oxidative damage to DNA, proteins, and lipids and thus to accelerate aging (Figure 3) [44–47]. Superoxide anion (O_2^-) and H_2O_2 are produced in mitochondria as byproducts of OXPHOS. Further, H_2O_2 can be converted into a dangerous hydroxyl radical (HO) during Fenton's reaction in the presence of Fe^{2+} [48]. Antioxidant defense includes enzymatic activation of superoxide dismutases (SODs), which are metalloproteins that convert superoxide to hydrogen peroxide and molecular oxygen. There are three types of SODs: Cu/Zn-SOD, predominantly located in the cytosolic fractions; Mn-SOD, located in the mitochondria; and EC-SOD, which is found in the extracellular space [49]. Catalase is a heme protein located predominantly in peroxisomes and the inner mitochondrial membrane.

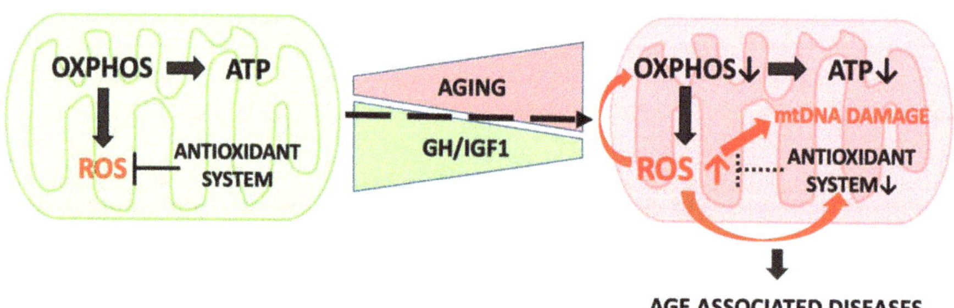

Figure 3. Mitochondria produce ATP and reactive oxygen species (ROS) as byproducts of oxidative phosphorylation (OXPHOS): In youth, ROSs are neutralized by the antioxidant system. Accumulation of proteins and enzymes damaged by escaped ROS leads to impairment of mitochondrial function during aging. Mitochondrial dysfunction is correlated with the decline in GH/IGF1 signaling and is linked to a variety of age-related diseases.

There are numerous reports showing that GH/IGF-1 signaling controls the expression and activity of antioxidant enzymes and thus regulates the level of oxidative stress. Congenital IGF-1R haploinsufficiency in mice (*Igf1r+/−*) alters sensitivity to oxidative stress. Embryonic fibroblasts from *Igf1r+/−* mice are more resistant to hydrogen peroxide-induced cell death [25]. Accordingly, we found that IGF-1 null mice that exclusively express hepatic IGF-1 transgene (KO-HIT mice) [50] show increased levels of lipid peroxidation products in serum and increased mortality rate at 18 months of age in both sexes, suggesting that elevations in serum IGF-1 are harmful. Mutations that affect pituitary development (*Prop1* and *Pit1*) and consequently lead to decreases in GH, thyroid stimulating hormone, or prolactin are associated with resistance to oxidative stress. Ames dwarf mice have very low serum IGF-1 levels and increased activities of catalase and Cu/Zn SOD [51]. These mice show reduced levels of DNA and protein oxidation in liver [52] and reduced serum and liver F2-isoprostanes, which are a stable lipid peroxidation product [53]. Similarly, the GH-deficient Snell and *lit/lit* dwarfs show resistance to oxidative stress [54,55]. Supplementation of GH to Ames dwarf mice for 7 days increased plasma IGF-1 levels and body and liver weights. However, mitochondrial glutathione S-transferase (GST) proteins (GSTK1 and GSTM4) significantly reduced with treatment [56,57]. Furthermore, glutaredoxins, which are localized in the mitochondria and sense cellular oxidative stress, significantly reduced with GH administration. In line with these data, it was reported that GHRKO mice exhibit increases in glutathione (GSH) and methionine (MET) metabolism in several tissues, making them more resistant to oxidative damage and delayed aging [58]. However, GHRKO mice, although resistant to oxidative stress, do not show improved free radical scavenging in the liver or kidney [59], though it is possible that other tissues such as muscle respond differently. Although informative, data from congenital models of the GH/IGF-1 axis cannot be directly extrapolated to normal aging in other animal models or in humans. These models exhibit altered developmental programming that affect their aging.

GHRKO mice resemble the human Laron syndrome (LS), which is caused by a deletion or an inactivating mutation of the *GHR* gene [60–63]. Similar to the GHRKO mice, LS patients have short stature, increased body adiposity, and low IGF-1 in serum [63,64]. Genome-wide microarray studies conducted on lymphocytes from LS patients identified a series of genes that are differentially expressed in various pathways, including oxidative stress, apoptosis, metabolism, Jak-STAT, and PI3K-AKT signaling. Among the overexpressed genes in LS, thioredoxin-interacting protein (*TXNIP*) was identified as a new target for IGF-1 and insulin action. TXNIP belongs to the α-arrestin family [65]. It binds to the catalytic active-center of reduced thioredoxin (TRX) and inhibits its expression and activity, highlighting the key role of TXNIP in redox regulation. Oxidative stress leads to TXNIP shuttling from the nucleus into the mitochondria. TXNIP inhibits proliferation via activation of the apoptosis signal regulating kinase 1 (ASK1) [66] and functions as a tumor suppressor, being commonly silenced in cancer cells [67–71]. Similar to LS patients, GHRKO mice show reduced tumor incidence in experimental models of cancer [72]. In accordance with its enhanced expression in LS-derived cells, qPCR revealed that *TXNIP* expression increased approximately two-fold in livers of GHRKO mice, while it decreased in HIT mice overexpressing IGF-1 in the liver.

LS lymphocytes were shown to display protection from oxidative stress [73]. Accordingly, induction of oxidative stress in lymphocytes from LS patient (Figure 4A) leads to upregulation of *TXNIP*. The capacity of IGF-1 to downregulate the oxidative stress-induced TXNIP upregulation (Figure 4B) indicates that IGF-1 could rescue the cells by downregulating TXNIP. TXNIP acts as an oxidative stress mediator by inhibiting TRX activity or by limiting its bioactivity [74]. The redox-related protein complex TRX/TXNIP, or "redoxisome," is a critical regulator of ROS signalling and is involved in the pathogenesis of various diseases, including autoimmune and degenerative conditions [68]. The finding that TXNIP levels are increased in response to oxidation in LS patient-derived but not control lymphoblastoid cells is of major translational relevance.

Apart from oxidative stress, TXNIP was reported to function as a strong glucose sensor as its expression increased upon high-glucose stress [75]. *TXNIP* knockout mice (*TXNIP* KO) show impaired metabolic homeostasis, including adipogenesis and reduced gluconeogenesis [76], and decreased glucose uptake [77]. Similarly, a recently described human mutation demonstrated that diminished TXNIP function is linked to inefficient utilization of glucose [78]. Accordingly, glucose stress (hyperglycemia) upregulated TXNIP levels in 3T3LY adipocytes and was downregulated by IGF-1 and insulin (Figure 5A) [79]. Overall, these studies show that oxidative and glucose stresses induced TXNIP levels at both the transcriptional and translational levels (Figure 5B) and that IGF-1 indirectly regulates cellular TXNIP, protecting the cells from apoptosis (Figure 5C) [79]. The potential involvement of epigenetic mechanisms, particularly DNA methylation and histone acetylation, in inhibitory regulation of *TXNIP* gene expression by IGF-1 is currently unknown.

Notably, there are numerous studies indicating that administration of GH/IGF-1 in vivo or in vitro protects from oxidative stress, specifically during aging. Old Wistar rats treated with GH showed increased circulating IGF-1 levels and reductions in age-associated oxidative stress in skeletal muscle [23]. This was accompanied by increased levels of the antioxidant enzymes catalase, glutathione peroxidase, and G6PDH. Similarly, supplementation of IGF-1 to aged rats associated with reduced oxidative damage and restored levels of SOD, glutathione peroxidase, and catalase in hippocampus [34]. Hepatic tissue level of catalase was also restored with IGF-1 treatment in old rats, suggesting antioxidant properties of IGF-1 in brain and liver [34].

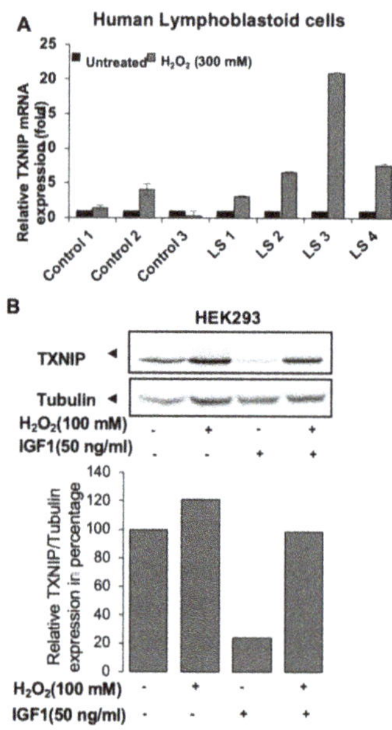

Figure 4. Effect of oxidative stress on thioredoxin-interacting protein (TXNIP) levels [79]: (**A**) Effect of oxidative stress on TXNIP levels in Laron syndrome (LS)-derived and control lymphoblastoids. Four individual LS and three control lymphoblastoid cell lines were treated with 300 mM of H_2O_2 for 2 h, and levels of TXNIP mRNA were measured by RT-QPCR. A value of 1 was given to TXNIP mRNA levels in untreated cells (solid bars). (**B**) Serum-starved HEK293 cells were treated with H_2O_2 (100 mM) or IGF1 (50 ng/mL) or both for 2 h. TXNIP and tubulin were detected by Western blotting.

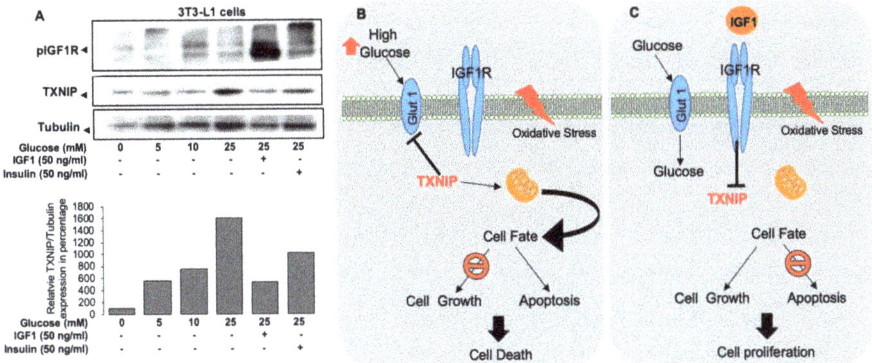

Figure 5. Effect of glucose stress on TXNIP levels [79]: (**A**) Serum-starved 3T3-L1 cells were maintained in medium with different concentrations of glucose in the presence or absence of IGF1 or insulin for 6 h. TXNIP, phospho-IGF-1R, and tubulin were detected by Western blotting. Schematic representation of the regulation of TXNIP by IGF1 signaling: (**B**) Under normal serum-free conditions, TXNIP is upregulated upon oxidative and glucose stresses. The activated TXNIP inhibits glucose uptake and is capable of mediating mitochondrial mediated apoptosis. (**C**) Upon IGF1 stimulation, TXNIP is downregulated even under oxidative and glucose stresses. Suppression of TXNIP leads to inhibition of apoptosis with ensuing increase in cell proliferation.

IGF-1 is reported to be protective against oxidative stress in cardiac and skeletal muscle. Pathologic left ventricular remodeling and functional loss following myocardial infarction were more severe in dwarf rats (*dw/dw*) with significantly reduced GH/IGF-1 [80]. Using an ex vivo murine model of myocardial ischemia/reperfusion injury, it was found that IGF-1 protects ischemic myocardium from further reperfusion injury, likely via maintenance of mitochondrial to nuclear DNA ratio within heart tissue [81] and by activation of the PI3K-Akt and/or Erk 1/2 kinase cascades [82]. Likewise, hearts of IGF-1 transgenic mice were protected from ischemia and reperfusion [83], and streptozotocin-induced diabetic cardiomyopathy in mice resulted in accumulation of nitrotyrosine (a reactive oxygen product) in vivo and the formation of H_2O_2 in myocytes in vitro that were rescued in IGF-1 transgenic mice [84]. Additionally, IGF-1 protected from 2:4 dinitrophenol-induced oxidative stress in rat muscle in vitro [85] and myoblasts protected from H_2O_2 stress-induced apoptosis [86].

The protective effects of the GH/IGF-1 axis from oxidative stress was demonstrated for many cell types. Endothelial cells from Ames dwarf mice show elevated levels of H_2O_2, increased mitochondrial ROS, and decreased antioxidative enzymes such as SODs and glutathione peroxidase. These observations were in accordance with in vitro findings using cultured aortic segments and human coronary arterial endothelial cells. Treatment of these cultures with GH and IGF-1 lead to elevations in the level of antioxidants and reduced prooxidant levels [87]. In astrocytes, downregulation of the IGF-1R increased mitochondrial ROS production and reduced resistance to external oxidative damage [40]. IGF-1 promoted the survival of rat primary cerebellar neurons and of immortalized hypothalamic rat GT1-7 cells following H_2O_2-induced oxidative stress [88]. There is a long list of studies utilizing numerous cell types, showing the protective effects of GH/IGF-1 from oxidative stress that, although important, were not included in our review.

Antioxidant administration to old animals can elevate circulating GH and IGF-1 levels. Thus, treatment of Wistar rats or senescence-accelerated OXYS rats with antioxidants and, in particular, with mitochondrial antioxidants prevented the age-associated decrease in serum levels of GH and IGF-1 [41,89]. These were accompanied by improvements in pathologies such as retinopathy and cataract, learning ability and memory, and immune system decline [89,90]. Treatments of aged rats with mitochondrial antioxidants also led to improvement of mitochondrial structure-disorganization developed with age in muscle tissues. Disorders of muscles tissue mitochondrial apparatus in old rats could be driven by a decline in GH/IGF-1 signaling, as improvement of cristae structure and clustering of muscle mitochondria was correlated with an increase of GH and IGF-1 levels in serum after treatment with mitochondrial antioxidants [41].

In summary, insufficient levels of GH/IGF-1 are associated with organ-specific impairment of free radical scavenging systems. GH/IGF-1 plays significant roles in regulating oxidative stress, which is clearly only one of many mechanisms affecting mitochondrial function during aging. Finally, there are fundamental differences in handling oxidative stress between age-associated decline in GH/IGF-1 (somatopause) and congenital impairments in the GH/IGF-1 axis.

5. GH/IGF-1 Effects on Cellular Senescence

Cellular senescence is characterized by an irreversible block of the cell cycle. This mechanism was initially thought to function as protection against cancer [91], but later studies have found that senescence is linked to aging and age-related diseases. Senescent cells undergo numerous phenotypic and metabolic modifications. They show increased cells size [92], dysfunctional mitochondria and telomeres, impaired DNA damage response, increased secretory functions, formation of heterochromatic foci, and a senescence-associated secretory phenotype (SASP). In fact, the SASP is a main feature of senescent cells, which secrete pro-inflammatory cytokines and chemokines, growth factors, and proteases, forming a toxic and harmful microenvironment to non-senescent cells [93–97].

Mitochondria undergo drastic changes in morphology and function in senescent cells. Given the complex functions of mitochondria, it is hard to dissect the specific mitochondria-mediated mechanisms leading to cell senescence. Several potential mechanisms, mediated by mitochondria, that may lead to cell senescence include increased ROS production and elevations in damage-associated molecular patterns (DAMPs) leading to DNA damage response (DDR), which locks the cells in a senescence mode, leading to SASP. Senescent cells exhibit increased mitochondrial mass but, at the same time, show reduced OXPHOS and rely mainly on ATP from glycolysis. Increased mitochondrial mass is driven by upregulation of PGC-1β [98]. Accordingly, deletion of PGC-1β in mice delayed several aspects of senescence [98]. In line with these data, it was shown that senescent pancreatic beta cells exhibit increased mitochondrial biogenesis [99].

Pituitary adenomas expressing and secreting GH exhibit a senescent phenotype. Autocrine/paracrine GH acts in pituitary cells as an apoptosis switch for p53-mediated senescence, likely preventing the pituitary adenoma cells from progression to malignancy [100]. Skin fibroblasts from acromegalic patients (with excess GH and IGF-1) exhibit shortened telomeres and cellular senescence [101]. Similarly, senescence-associated gene expression of p16 and IL-6 increased in white adipose tissue of 10-month-old female bGH transgenic mice as compared to controls. In addition, β-galactosidase (β-gal)-positive cells (senescence-associated) were elevated in GH-injected 19-month-old female mice as compared to age-matched saline-injected controls [102]. Accordingly, mice with a specific deletion of the IGF-1R in cardiomyocytes show delayed development of aging-associated myocardial pathologies [103]. Cultured cardiomyocytes treated with IGF-1 exhibited increased senescence, while inhibition of phosphoinositide 3-kinase prevented the IGF-1-mediated increase in interleukin (IL)-1α, IL-1β, receptor activator of nuclear factor-κB ligand, and p21 protein levels [103]. Finally, prolonged IGF-1 treatment of MCF7 cells inhibited SIRT1 deacetylase activity, associated with increased p53 acetylation and activation, and lead to premature cellular senescence [104]. Similarly, prolonged exposure of primary human mesangial cells to glycated albumin (GA) was associated with IGF-1 release, activation of the IGF-1R, and enhanced cellular senescence. GA-induced IGF-1R activation associated with a reduction in the catalase content likely through activation of the Ras and Erk1/2 pathway. Downregulation of the IGF-1R via overexpression of klotho lowered p53 and reversed the senescence phenotype [105]. In vitro, cell senescence is often observed once cells reach confluency. IGF-1 promoted senescence in mouse embryonic fibroblasts and human fibroblasts as well as in rat vascular smooth muscle cells (VSMCs) after attaining confluency. The IGF-1-induced senescence in these cells was associated with elevated cellular ROS, p53, and p21 protein levels and the DNA damage marker γH2AX [106].

On the other hand, there are studies showing that GH/IGF-1 protects from senescence. GH treatment (0.4 mg/d for 7 days) of endothelial progenitor cells from patients with atherosclerosis reversed age-related dysfunction and attenuated senescence (indicated by increased telomerase activity) [107]. These findings were in line with another study in which GH was incorporated into reconstituted high-density lipoprotein (rHDL) and delivered to zebrafish. In this study, GH enhanced anti-atherosclerotic activity and antisenescence activity with inhibition of fructose-mediated glycation [108]. Likewise, IGF-1 gene transfer to CCl_4-treated rats (with liver injury and fibrosis) relieved hepatocyte oxidative stress and premature senescence, likely mediated by the p53/progerin pathway [109]. In vitro, IGF-1 treatment of H_2O_2-exposed hepatocytes reversed oxidative stress-induced premature senescence via enhancing cytoplasmic AKT1–p53 interaction and by subsequently inhibiting nuclear p53–progerin interaction [109].

6. GH/IGF-1 Effects on Mitochondria-Mediated Apoptosis During Aging

Similar to senescence, programmed cell death, or apoptosis, is believed to be a consequence of cellular stress and mitochondrial dysfunction. However, the roles of apoptosis during aging are uncertain. In general, the determinants of whether a cell activates senescence or apoptosis pathways are cell-type, stress-type, and stress-intensity dependent. It is important to note that apoptosis occurs mainly through activation of a mitochondrial intrinsic pathway, which depends on caspase 9 and apoptotic protease activating factor 1 (Apaf1) or via an extrinsic pathway mediated by the Fas ligand and tumor necrosis factor receptor (TNFR) and by activation of caspase 8. Herein, we will describe shortly the roles of the GH/IGF-1 axis in mitochondrial (intrinsic)-mediated apoptosis.

A key step in mitochondria-mediated apoptosis is the release of Cyt c from the inner mitochondrial space (Figure 6). Several pathways are known to trigger Cyt c release including opening of the mitochondrial permeability transition pore (PTP). Opening of the PTP leads to mitochondrial swelling, depolarization of the membrane potential, subsequent rupture of the outer membrane, and nonselective release of intermembrane space proteins. Once released, Cyt c binds to Apaf1 and activates a group of cysteine proteases called caspases that cleave an array of substrates and proteins that are vital to cellular function. The Apaf1-cyt c-caspase 9 form apoptosome complexes. Apoptosome, central to the apoptotic pathway, binds to other pro-enzymes and cleaves them to their active forms.

Cyt c release from mitochondria is a primary signal for B-cell lymphoma 2 (Bcl-2)-regulated inhibition of apoptosis [110,111]. Bcl-2 is localized in the outer mitochondrial membrane (Figure 6). Bcl-2 belongs to a large family of proteins with antiapoptotic properties (i.e., Bcl-XL, Bcl-w, Mcl-1, A1, Bcl-Rambo, Bcl-L10, and Bcl-G) and proapoptotic properties (i.e., Bax, Bak, and Bok) [112]. There are several other proteins that block the antiapoptotic activity of Bcl-2, which are termed proapoptotic BH3-only proteins (i.e., Puma, Noxa, Bid, Bad, Bim, Bik, Hrk, and Bmf) [113]. However, it appears that Bax/Bak are the key regulatory targets where many intracellular signals converge and determine a cell's fate [114]. It was widely accepted that the role of the antiapoptotic BCL-2-like proteins is to inhibit their proapoptotic counteracting partners, such that the balance between anti- and proapoptotic BCL-2 family proteins determines cell fate. However, this model is oversimplified in view of later discoveries, including the occasional interconversion of anti- and proapoptotic activities of these proteins, and the findings that BCL-2 family members exhibit nonapoptotic functions [115,116].

The insulin and IGF-1 signaling pathway (IIS) are the most evolutionarily conserved pathway of aging. IGF-1 has been recognized as a survival factor of numerous cell types. Activation of the insulin or IGF-1 receptors elicits activation of several pathways; among them are the mitogen-activated protein kinase (MAPK) pathway and the PI3K-AKT pathway. Activation of AKT leads to phosphorylation and inactivation of a group of forkhead box transcription factors of the class O (FOXO) factors by retaining them in the cytoplasm. Phosphorylation of FOXOs and their binding to the regulatory proteins 14-3-3 sequester them from the nucleus, leading to suppression of FOXO-dependent transcription (mostly the BH3-only proteins) [117–119]. Thus, the survival factor IGF-1 attenuates apoptosis via activation of the PI3K-AKT pathway and inhibition of FOXO proteins.

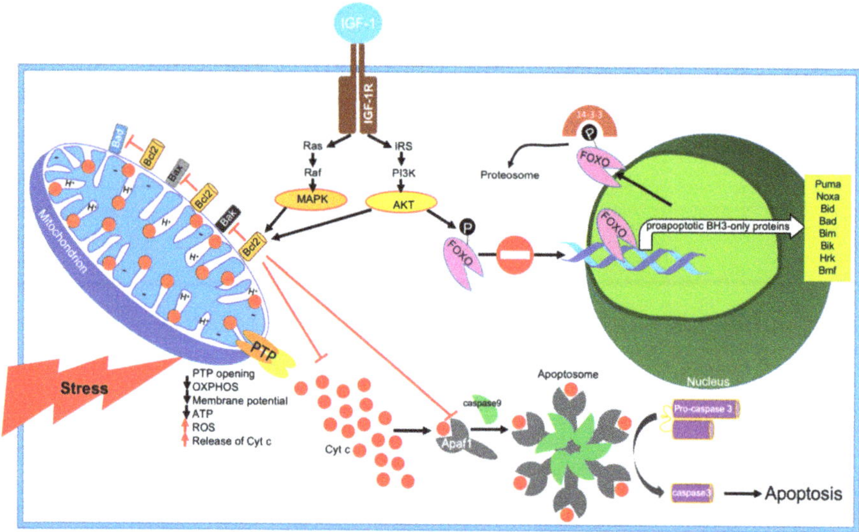

Figure 6. The effects of IGF-1 on mitochondria-mediated apoptosis: Cellular stress leads to permeability transition pore (PTP) opening, reduction in OXPHOS function, membrane potential, and ATP production, while ROS production is increased, altogether leading to cyt c release. Cytosolic Cyt c activates the apoptosome, leading to cell death. IGF-1R-mediated inhibition of apoptosis occurs via activation of two major signaling pathways that trigger phosphorylation cascade of cytosolic and nuclear proteins regulating transcription and activation of proteins involved in protection from apoptosis. The AKT pathway leads to phosphorylation of Forkhead box transcription factors of the class O (FOXO) proteins and subsequent inhibition of their transcriptional activity. The MAPK pathway, stimulated by IGF-1 binding, activates the Bcl antiapoptotic family of proteins.

FOXOs are the most important transcriptional effectors of the IIS, and as such, they are tightly regulated post-translationally [120]. Mammals have four FOXO genes: FOXO1, FOXO3, FOXO4, and FOXO6. FOXO1-null mice are embryonically lethal, while FOXO3, FOXO4, and FOXO6 show mild phenotypes [121]. Tissue-specific approaches to delete FOXO proteins 1, 3, and 4 have established the importance of these factors in regulation of cell fate. For example, ablation of Foxo1, 3, and 4 proteins in osteoblasts associated with increases in oxidative stress and osteoblast apoptosis [122], while osteoblast-specific overexpression of Foxo3 decreased oxidative stress and apoptosis.

In summary, the GH/IGF-1 axis regulates mitochondrial (intrinsic)-mediated apoptosis mostly via activation of the PI3K-AKT/FOXO pathway and possibly via cell-specific transcriptional regulation of antiapoptotic genes. However, while during development and adult stages apoptosis is required for the normal cell turnover in specific tissues (such as endothelium, intestinal epithelium, etc.), the role of apoptosis during aging is not clear. In fact, apoptosis was shown to decrease in adipose mesenchymal stem cells from healthy young, middle-aged, and aged volunteers [123]; in bone marrow mesenchymal stem cells from old mice [124]; in livers of old rats challenged with DNA-damaging agents as compared to their young controls [125]; and even in response to radiation of peripheral blood lymphocytes from old mice [126]. Accordingly, reductions in markers of apoptosis during normal aging was also found in human serum [127]. Perhaps during aging, mitochondrial dysfunction leads mostly to cell senescence as opposed to apoptosis.

7. GH/IGF-1 Effects on Mitochondrial Function During Inflammation

Inflammation is a defense mechanism against harmful stimuli that damage cellular homeostasis [128]. Inappropriate response to cellular damage may lead to chronic inflammation, including autoimmune diseases, diabetes, allergies, cardiovascular diseases, asthma, cancer, arthritis, and aging [129,130]. Response to various immunological challenges entails particular metabolic configurations for energy generation required for biosynthesis of molecules [131]. Dysfunctional mitochondria with excessive ROS, abnormal calcium and potassium ion mobilization, and unusual ATP or NAD$^+$ levels can trigger immune response [132]. On the other hand, inflammatory cytokines acting upon cells can trigger intracellular signaling cascades altering mitochondrial dynamics and mitophagy and eventually resulting in cell death [133,134].

The GHR and IGF-1R are expressed on immune cells and participate in thymic development and differentiation of immune cells such as T-cell, B- cell, and natural killer cells as well as in antigen presentation, antibody production, etc. [135,136]. Additionally, there are studies showing local production of GH and IGF-1 by lymphocytes, suggesting an autocrine/paracrine regulation of immune cells [137,138]. Further, human lymphocytes treated with super physiological levels (abuse) of IGF-1 can undergo cytoskeletal reorganization and overproduction of cytokines, augmenting the inflammatory response [139]. On the other hand, children with chronic inflammatory disorders exhibit excessive pro-inflammatory cytokines production and show growth failure and pubertal abnormalities [140]. Similarly, the elderly, with reduced GH or IGF-1 signals show poor response to immune stimuli [141].

The effects of GH/IGF-1 on mitochondria of inflammatory cells are underexplored. Aging is often associated with metabolic abnormalities accompanied by increased chronic inflammation in many tissues, including the brain. In fact, GH signaling was found to positively modulate brain inflammation in aged mice [142]. Accordingly, GHRKO mice showed reduced inflammation in the hypothalamus. GHRKO and Ames mice displayed lower glial fibrillary acidic protein (GFAP) and tumor necrosis factor (TNF)-α levels in the hypothalamus, indicative of reduced inflammation. Consequently, administration of GH to GHRKO and Ames mice led to increased GFAP and TNF-α and increased brain inflammation [142]. As mentioned above, the GHRKO is a mouse model with congenital ablation of the GH/IGF-1 axis. It is unclear whether this axis plays a role in systemic or central inflammation, particularly during natural aging where a gradual fall in GH/IGF-1 signals happens.

8. Summary

GH and IGF-1 are pleotropic hormones affecting multiple cellular functions, including cell proliferation, differentiation, metabolism, and cell survival. Both hormones activate many signaling cascades implicated in regulation of mitochondrial proteins expression and function. Most evidence indicates that the effects of GH on mitochondrial mass and function are indirect and mostly mediated by IGF-1. IGF-1 affects mitochondrial mass via increased transcriptional activities of key factors involved in mitochondrial biogenesis such as PGC-1α. Additionally, it appears that the effects of IGF-1 on mitochondrial respiration are indirect and coincide with enhanced synthesis of mitochondrial proteins such as Cyt c and UCP. With respect to oxidative stress, the literature is divided and studies showing either positive or negative effects of GH or IGF-1 have been published. While congenital mouse models with life-long decreases in GH/IGF-1 axis signaling indicate mostly protection from oxidative stress, models of age-induced decreases in GH/IGF-1 signals, as seen in humans, associate with increased oxidative stress.

Finally, it is widely accepted that GH/IGF-1 are involved in cell senescence and apoptosis. The molecular mechanisms involved in GH/IGF-1-mediated cellular senescence are still poorly understood. Both hormones exert a dual function and promote, on one hand, cell proliferation and, on the other hand, cellular senescence. Therefore, it is conceivable that the dose and duration of GH/IGF-1 exposure might regulate senescence, and that the effects of GH/IGF-1 on senescence are tissue- and cell type-specific. On the other hand, the protective roles of IGF-1 from mitochondrial-mediated apoptosis have been better defined. Studies of numerous cell types and animal models have shown that IGF-1-mediated activation of the PI3K-AKT/FOXO pathway upregulates transcription of antiapoptotic genes.

Funding: Financial support was received from National Institutes of Health Grant R01AG056397 to S.Y. and from the United States-Israel Binational Science Foundation grant (2013282) to S.Y. and H.W.

Conflicts of Interest: The authors declare no conflict of interest.

Abbreviations

iLID	Adult-induced liver IGF-1 deficiency
ALS	Acid labile subunit
ATP	Adenosine triphosphate
ASK1	Apoptosis signal regulating kinase 1
Bcl-2	B-cell lymphoma 2
β-gal	β-galactosidase
bGH	Bovine growth hormone
BNIP3 and BNIP3L	BCL2/adenovirus E1B 19 kDa protein-interacting protein 3 and L
BAT	Brown adipose tissue
CS	Cockayne syndrome protein
Cyt c	Cytochrome c
DAMPs	Damage-associated molecular patterns
DDR	DNA damage response
ETC	Electron transport chain
FOXO	Forkhead box transcription factors of the class O
GH	Growth hormone
GHR	Growth hormone receptor
GHRKO	GH receptor null
GHRH	GH-releasing hormone
GHRHR	GH-releasing hormone receptor
G6PDH	Glucose-6-phosphate dehydrogenase
GST	Glutathione S-transferase
GSH	Glutathione
GA	Glycated albumin
HO	Heme oxygenase
HIT	Hepatic IGF-1 transgene
hGH	Human GH
H_2O_2	Hydrogen peroxide
hypx	Hypophysectomized
IGF-1	Insulin-like growth factor-1
IGFBPs	IGF-binding proteins
IGF-1R	Insulin-like growth factor-1 receptor
IIS	Insulin and IGF-1 signaling pathway
IL	Interleukin
JAK2	Janus kinase 2
LS	Laron syndrome
MET	Methionine
mtDNA	Mitochondrial genome
MuRF-1	Muscle RING finger-1
ND1 through ND6	NADH dehydrogenase of complex I
TNF	Necrosis factor
NRG	Neuregulin
NO	Nitric oxide
NRF1 and 2	Nuclear respiratory factors
OXPHOS	Oxidative phosphorylation
PTP	Permeability transition pore

PGC1α	Peroxisome proliferator-activated receptor gamma coactivator 1 α
PPARs	Peroxisome proliferator-activated receptors
PRC	PGC-1α-related coactivator
PI3K/AKT	Phosphoinositide-3-kinase (PI3K)/Protein kinase B (PKB, or Akt)
MAPK	Rat sarcoma GTPase protein (Ras)/Rapidly accelerated fibrosarcoma protein kinase (Raf)/Mitogen-activated protein kinase
ROS	Reactive oxygen species
RQ	Respiratory quotient
rHDL	Reconstituted high-density lipoprotein
SST	Somatostatin
SSTR	SST-receptors
Shc	Src homology 2 domain containing transforming protein
STAT5b	Signal transducer and activator of transcription 5B
SERCA2a	Sarco/endoplasmic reticulum Ca^{2+}-ATPase-2a
SASP	Senescence-associated secretory phenotype
TF	Transcription factors
TXNIP	Thioredoxin-interacting protein
TRX	Thioredoxin
T3	Triiodothyronine
TOMs	Transporters located at the outer mitochondrial membrane
TCA	Tricarboxylic acid
TNFa	Tumor necrosis factor-alpha
VSMCs	Vascular smooth muscle cells

References

1. Niu, T.; Rosen, C.J. The insulin-like growth factor-I gene and osteoporosis: A critical appraisal. *Gene* **2005**, *361*, 38–56. [CrossRef] [PubMed]
2. Barclay, R.D.; Burd, N.A.; Tyler, C.; Tillin, N.A.; Mackenzie, R.W. The Role of the IGF-1 Signaling Cascade in Muscle Protein Synthesis and Anabolic Resistance in Aging Skeletal Muscle. *Front. Nutr.* **2019**, *6*, 146. [CrossRef] [PubMed]
3. Harada, K.; Hanayama, Y.; Obika, M.; Itoshima, K.; Okada, K.; Otsuka, F. Clinical relevance of insulin-like growth factor-1 to cardiovascular risk markers. *Aging Male* **2019**, 1–9. [CrossRef] [PubMed]
4. Graham, M.R.; Evans, P.; Thomas, N.E.; Davies, B.; Baker, J.S. Changes in endothelial dysfunction and associated cardiovascular disease morbidity markers in GH-IGF axis pathology. *Am. J. Cardiovasc. Drugs* **2009**, *9*, 371–381. [CrossRef] [PubMed]
5. Milman, S.; Huffman, D.M.; Barzilai, N. The Somatotropic Axis in Human Aging: Framework for the Current State of Knowledge and Future Research. *Cell Metab.* **2016**, *23*, 980–989. [CrossRef]
6. Haywood, N.J.; Slater, T.A.; Matthews, C.J.; Wheatcroft, S.B. The insulin like growth factor and binding protein family: Novel therapeutic targets in obesity & diabetes. *Mol. Metab.* **2019**, *19*, 86–96.
7. Domene, H.M.; Hwa, V.; Jasper, H.G.; Rosenfeld, R.G. Acid-labile subunit (ALS) deficiency. *Best Pract. Res. Clin. Endocrinol. Metab.* **2011**, *25*, 101–113. [CrossRef]
8. Banaszak-Ziemska, M.; Niedziela, M. PAPP-A2 a new key regulator of growth. *Endokrynol. Pol.* **2017**, *68*, 682–691. [CrossRef]
9. Broglio, F.; Arvat, E.; Benso, A.; Papotti, M.; Muccioli, G.; Deghenghi, R.; Ghigo, E. Ghrelin: Endocrine and non-endocrine actions. *J. Pediatr. Endocrinol. Metab.* **2002**, *15*, 1219–1227. [CrossRef]
10. Arvat, E.; Broglio, F.; Aimaretti, G.; Benso, A.; Giordano, R.; Deghenghi, R.; Ghigo, E. Ghrelin and synthetic GH secretagogues. *Best Pract. Res. Clin. Endocrinol. Metab.* **2002**, *16*, 505–517. [CrossRef]
11. Hakuno, F.; Takahashi, S.I. IGF1 receptor signaling pathways. *J. Mol. Endocrinol.* **2018**, *61*, T69–T86. [CrossRef] [PubMed]
12. Riis, S.; Murray, J.B.; O'Connor, R. IGF-1 Signalling Regulates Mitochondria Dynamics and Turnover through a Conserved GSK-3beta-Nrf2-BNIP3 Pathway. *Cells* **2020**, *9*, 147. [CrossRef] [PubMed]

13. Lyons, A.; Coleman, M.; Riis, S.; Favre, C.; O'Flanagan, C.H.; Zhdanov, A.V.; Papkovsky, D.B.; Hursting, S.D.; O'Connor, R. Insulin-like growth factor 1 signaling is essential for mitochondrial biogenesis and mitophagy in cancer cells. *J. Biol. Chem.* **2017**, *292*, 16983–16998. [CrossRef] [PubMed]

14. Kurmasheva, R.T.; Houghton, P.J. IGF-I mediated survival pathways in normal and malignant cells. *Biochim. Biophys. Acta* **2006**, *1766*, 1–22. [CrossRef]

15. Craigen, W.J. Mitochondrial DNA mutations: an overview of clinical and molecular aspects. *Methods Mol. Biol.* **2012**, *837*, 3–15. [PubMed]

16. Murphy, M.P.; Hartley, R.C. Mitochondria as a therapeutic target for common pathologies. *Nat. Rev. Drug Discov.* **2018**, *17*, 865–886. [CrossRef]

17. El-Hattab, A.W.; Suleiman, J.; Almannai, M.; Scaglia, F. Mitochondrial dynamics: Biological roles, molecular machinery, and related diseases. *Mol. Genet. Metab.* **2018**, *125*, 315–321. [CrossRef]

18. Phu, L.; Rose, C.M.; Tea, J.S.; Wall, C.E.; Verschueren, E.; Cheung, T.K.; Kirkpatrick, D.S.; Bingol, B. Dynamic Regulation of Mitochondrial Import by the Ubiquitin System. *Mol. Cell* **2020**, *77*, e1107–e1123. [CrossRef]

19. Groves, W.E.; Houts, G.E.; Bayse, G.S. Subcellular distribution of 125 I-labeled bovine growth hormone in rat liver and kidney. *Biochim. Biophys. Acta* **1972**, *264*, 472–480. [CrossRef]

20. Mutvei, A.; Husman, B.; Andersson, G.; Nelson, B.D. Thyroid hormone and not growth hormone is the principle regulator of mammalian mitochondrial biogenesis. *Acta Endocrinol. (Copenh)* **1989**, *121*, 223–228. [CrossRef]

21. Maddaiah, V.T.; Sharma, R.K.; Balachandar, V.; Rezvani, I.; Collipp, P.J.; Chen, S.Y. Effect of growth hormone on mitochondrial protein synthesis. *J. Biol. Chem.* **1973**, *248*, 4263–4268. [PubMed]

22. Echave, P.; Machado-da-Silva, G.; Arkell, R.S.; Duchen, M.R.; Jacobson, J.; Mitter, R.; Lloyd, A.C. Extracellular growth factors and mitogens cooperate to drive mitochondrial biogenesis. *J. Cell Sci.* **2009**, *122*, 4516–4525. [CrossRef] [PubMed]

23. Brioche, T.; Kireev, R.A.; Cuesta, S.; Gratas-Delamarche, A.; Tresguerres, J.A.; Gomez-Cabrera, M.C.; Vina, J. Growth hormone replacement therapy prevents sarcopenia by a dual mechanism: Improvement of protein balance and of antioxidant defenses. *J. Gerontol. A Biol. Sci. Med. Sci.* **2014**, *69*, 1186–1198. [CrossRef] [PubMed]

24. Hou, H.Y.; Wang, X.; Yu, Q.; Li, H.Y.; Li, S.J.; Tang, R.Y.; Guo, Z.X.; Chen, Y.Q.; Hu, C.X.; Yang, Z.J.; et al. Evidence that growth hormone can improve mitochondrial function in oocytes from aged mice. *Reproduction* **2018**, *157*, 345–358. [CrossRef] [PubMed]

25. Holzenberger, M.; Dupont, J.; Ducos, B.; Leneuve, P.; Geloen, A.; Even, P.C.; Cervera, P.; Le Bouc, Y. IGF-1 receptor regulates lifespan and resistance to oxidative stress in mice. *Nature* **2003**, *421*, 182–187. [CrossRef] [PubMed]

26. Tatar, M.; Kopelman, A.; Epstein, D.; Tu, M.P.; Yin, C.M.; Garofalo, R.S. A mutant Drosophila insulin receptor homolog that extends life-span and impairs neuroendocrine function. *Science* **2001**, *292*, 107–110. [CrossRef] [PubMed]

27. Kimura, K.D.; Tissenbaum, H.A.; Liu, Y.; Ruvkun, G. daf-2, an insulin receptor-like gene that regulates longevity and diapause in Caenorhabditis elegans. *Science* **1997**, *277*, 942–946. [CrossRef]

28. Kenyon, C.; Chang, J.; Gensch, E.; Rudner, A.; Tabtiang, R. A C. elegans mutant that lives twice as long as wild type. *Nature* **1993**, *366*, 461–464. [CrossRef]

29. Brown-Borg, H.M.; Johnson, W.T.; Rakoczy, S.G. Expression of oxidative phosphorylation components in mitochondria of long-living Ames dwarf mice. *Age (Dordr)* **2012**, *34*, 43–57. [CrossRef]

30. Westbrook, R.; Bonkowski, M.S.; Strader, A.D.; Bartke, A. Alterations in oxygen consumption, respiratory quotient, and heat production in long-lived GHRKO and Ames dwarf mice, and short-lived bGH transgenic mice. *J. Gerontol. A Biol. Sci. Med. Sci.* **2009**, *64*, 443–451. [CrossRef]

31. Brown-Borg, H.; Johnson, W.T.; Rakoczy, S.; Romanick, M. Mitochondrial oxidant generation and oxidative damage in Ames dwarf and GH transgenic mice. *J. Am. Aging Assoc.* **2001**, *24*, 85–96. [CrossRef] [PubMed]

32. Fang, Y.; McFadden, S.; Darcy, J.; Hascup, E.R.; Hascup, K.N.; Bartke, A. Lifespan of long-lived growth hormone receptor knockout mice was not normalized by housing at 30 degrees C since weaning. *Aging Cell* **2020**, e13123. [CrossRef]

33. Darcy, J.; McFadden, S.; Fang, Y.; Berryman, D.E.; List, E.O.; Milcik, N.; Bartke, A. Increased environmental temperature normalizes energy metabolism outputs between normal and Ames dwarf mice. *Aging (Albany NY)* **2018**, *10*, 2709–2722. [CrossRef] [PubMed]

34. Garcia-Fernandez, M.; Delgado, G.; Puche, J.E.; Gonzalez-Baron, S.; Castilla Cortazar, I. Low doses of insulin-like growth factor I improve insulin resistance, lipid metabolism, and oxidative damage in aging rats. *Endocrinology* **2008**, *149*, 2433–2442. [CrossRef] [PubMed]

35. Zadik, Z.; Chalew, S.A.; McCarter, R.J.Jr.; Meistas, M.; Kowarski, A.A. The influence of age on the 24-hour integrated concentration of growth hormone in normal individuals. *J. Clin. Endocrinol. Metab.* **1985**, *60*, 513–516. [CrossRef]

36. Colon, G.; Saccon, T.; Schneider, A.; Cavalcante, M.B.; Huffman, D.M.; Berryman, D.; List, E.; Ikeno, Y.; Musi, N.; Bartke, A.; et al. The enigmatic role of growth hormone in age-related diseases, cognition, and longevity. *Geroscience* **2019**, *41*, 759–774. [CrossRef]

37. Olleros Santos-Ruiz, M.; Sadaba, M.C.; Martin-Estal, I.; Munoz, U.; Sebal Neira, C.; Castilla-Cortazar, I. The single IGF-1 partial deficiency is responsible for mitochondrial dysfunction and is restored by IGF-1 replacement therapy. *Growth Horm. IGF Res.* **2017**, *35*, 21–32. [CrossRef]

38. Puche, J.E.; Garcia-Fernandez, M.; Muntane, J.; Rioja, J.; Gonzalez-Baron, S.; Castilla Cortazar, I. Low doses of insulin-like growth factor-I induce mitochondrial protection in aging rats. *Endocrinology* **2008**, *149*, 2620–2627. [CrossRef]

39. Pharaoh, G.; Owen, D.; Yeganeh, A.; Premkumar, P.; Farley, J.; Bhaskaran, S.; Ashpole, N.; Kinter, M.; Van Remmen, H.; Logan, S. Disparate Central and Peripheral Effects of Circulating IGF-1 Deficiency on Tissue Mitochondrial Function. *Mol. Neurobiol.* **2020**, *57*, 1317–1331. [CrossRef]

40. Logan, S.; Pharaoh, G.A.; Marlin, M.C.; Masser, D.R.; Matsuzaki, S.; Wronowski, B.; Yeganeh, A.; Parks, E.E.; Premkumar, P.; Farley, J.A.; et al. Insulin-like growth factor receptor signaling regulates working memory, mitochondrial metabolism, and amyloid-beta uptake in astrocytes. *Mol. Metab.* **2018**, *9*, 141–155. [CrossRef]

41. Vays, V.B.; Eldarov, C.M.; Vangely, I.M.; Kolosova, N.G.; Bakeeva, L.E.; Skulachev, V.P. Antioxidant SkQ1 delays sarcopenia-associated damage of mitochondrial ultrastructure. *Aging (Albany NY)* **2014**, *6*, 140–148. [CrossRef] [PubMed]

42. Zhang, Y.; Yuan, M.; Bradley, K.M.; Dong, F.; Anversa, P.; Ren, J. Insulin-like growth factor 1 alleviates high-fat diet-induced myocardial contractile dysfunction: Role of insulin signaling and mitochondrial function. *Hypertension* **2012**, *59*, 680–693. [CrossRef] [PubMed]

43. Liu, Z.; Solesio, M.E.; Schaffler, M.B.; Frikha-Benayed, D.; Rosen, C.J.; Werner, H.; Kopchick, J.J.; Pavlov, E.V.; Abramov, A.Y.; Yakar, S. Mitochondrial Function Is Compromised in Cortical Bone Osteocytes of Long-Lived Growth Hormone Receptor Null Mice. *J. Bone Miner. Res.* **2019**, *34*, 106–122. [CrossRef] [PubMed]

44. Sohal, R.S.; Orr, W.C. The redox stress hypothesis of aging. *Free Radic. Biol. Med.* **2012**, *52*, 539–555. [CrossRef]

45. Finkel, T.; Holbrook, N.J. Oxidants, oxidative stress and the biology of ageing. *Nature* **2000**, *408*, 239–247. [CrossRef]

46. Cadenas, E.; Davies, K.J. Mitochondrial free radical generation, oxidative stress, and aging. *Free Radic. Biol. Med.* **2000**, *29*, 222–230. [CrossRef]

47. Harman, D. The biologic clock: the mitochondria? *J. Am. Geriatr. Soc.* **1972**, *20*, 145–147. [CrossRef]

48. Winterbourn, C.C. Toxicity of iron and hydrogen peroxide: The Fenton reaction. *Toxicol. Lett.* **1995**, *82–83*, 969–974. [CrossRef]

49. Fridovich, I. Biological effects of the superoxide radical. *Arch. Biochem. Biophys.* **1986**, *247*, 1–11. [CrossRef]

50. Elis, S.; Wu, Y.; Courtland, H.W.; Sun, H.; Rosen, C.J.; Adamo, M.L.; Yakar, S. Increased serum IGF-1 levels protect the musculoskeletal system but are associated with elevated oxidative stress markers and increased mortality independent of tissue igf1 gene expression. *Aging Cell* **2011**, *10*, 547–550. [CrossRef]

51. Hauck, S.J.; Bartke, A. Effects of growth hormone on hypothalamic catalase and Cu/Zn superoxide dismutase. *Free Radic. Biol. Med.* **2000**, *28*, 970–978. [CrossRef]

52. Yamamoto, M.; Clark, J.D.; Pastor, J.V.; Gurnani, P.; Nandi, A.; Kurosu, H.; Miyoshi, M.; Ogawa, Y.; Castrillon, D.H.; Rosenblatt, K.P.; et al. Regulation of oxidative stress by the anti-aging hormone klotho. *J. Biol. Chem.* **2005**, *280*, 38029–38034. [CrossRef] [PubMed]

53. Choksi, K.B.; Roberts, L.J., 2nd; DeFord, J.H.; Rabek, J.P.; Papaconstantinou, J. Lower levels of F2-isoprostanes in serum and livers of long-lived Ames dwarf mice. *Biochem. Biophys. Res. Commun.* **2007**, *364*, 761–764. [CrossRef] [PubMed]

54. Salmon, A.B.; Murakami, S.; Bartke, A.; Kopchick, J.; Yasumura, K.; Miller, R.A. Fibroblast cell lines from young adult mice of long-lived mutant strains are resistant to multiple forms of stress. *Am. J. Physiol. Endocrinol. Metab.* **2005**, *289*, E23–E29. [CrossRef]

55. Bartke, A.; Brown-Borg, H. Life extension in the dwarf mouse. *Curr. Top. Dev. Biol.* **2004**, *63*, 189–225.

56. Rojanathammanee, L.; Rakoczy, S.; Brown-Borg, H.M. Growth hormone alters the glutathione S-transferase and mitochondrial thioredoxin systems in long-living Ames dwarf mice. *J. Gerontol. A Biol. Sci. Med. Sci.* **2014**, *69*, 1199–1211. [CrossRef]

57. Brown-Borg, H.M.; Bode, A.M.; Bartke, A. Antioxidative mechanisms and plasma growth hormone levels: Potential relationship in the aging process. *Endocrine* **1999**, *11*, 41–48. [CrossRef]

58. Brown-Borg, H.M.; Rakoczy, S.G.; Sharma, S.; Bartke, A. Long-living growth hormone receptor knockout mice: Potential mechanisms of altered stress resistance. *Exp. Gerontol.* **2009**, *44*, 10–19. [CrossRef]

59. Hauck, S.J.; Aaron, J.M.; Wright, C.; Kopchick, J.J.; Bartke, A. Antioxidant enzymes, free-radical damage, and response to paraquat in liver and kidney of long-living growth hormone receptor/binding protein gene-disrupted mice. *Horm. Metab. Res.* **2002**, *34*, 481–486. [CrossRef]

60. Laron, Z. Laron syndrome (primary growth hormone resistance or insensitivity): The personal experience 1958–2003. *J. Clin. Endocrinol. Metab.* **2004**, *89*, 1031–1044. [CrossRef]

61. Amselem, S.; Duquesnoy, P.; Attree, O.; Novelli, G.; Bousnina, S.; Postel-Vinay, M.C.; Goossens, M. Laron dwarfism and mutations of the growth hormone-receptor gene. *N. Engl. J. Med.* **1989**, *321*, 989–995. [CrossRef] [PubMed]

62. Godowski, P.J.; Leung, D.W.; Meacham, L.R.; Galgani, J.P.; Hellmiss, R.; Keret, R.; Rotwein, P.S.; Parks, J.S.; Laron, Z.; Wood, W.I. Characterization of the human growth hormone receptor gene and demonstration of a partial gene deletion in two patients with Laron-type dwarfism. *Proc. Natl. Acad. Sci. USA* **1989**, *86*, 8083–8087. [CrossRef] [PubMed]

63. Laron, Z.; Pertzelan, A.; Mannheimer, S. Genetic pituitary dwarfism with high serum concentration of growth hormone—a new inborn error of metabolism? *Isr. J. Med. Sci.* **1966**, *2*, 152–155. [PubMed]

64. Laron, Z.; Pertzelan, A.; Karp, M.; Kowadlo-Silbergeld, A.; Daughaday, W.H. Administration of growth hormone to patients with familial dwarfism with high plasma immunoreactive growth hormone: measurement of sulfation factor, metabolic and linear growth responses. *J. Clin. Endocrinol. Metab.* **1971**, *33*, 332–342. [CrossRef] [PubMed]

65. Shalev, A. Minireview: Thioredoxin-interacting protein: regulation and function in the pancreatic beta-cell. *Mol. Endocrinol.* **2014**, *28*, 1211–1220. [CrossRef] [PubMed]

66. Saxena, G.; Chen, J.; Shalev, A. Intracellular shuttling and mitochondrial function of thioredoxin-interacting protein. *J. Biol. Chem.* **2010**, *285*, 3997–4005. [CrossRef]

67. Baldan, F.; Mio, C.; Lavarone, E.; Di Loreto, C.; Puglisi, F.; Damante, G.; Puppin, C. Epigenetic bivalent marking is permissive to the synergy of HDAC and PARP inhibitors on TXNIP expression in breast cancer cells. *Oncol. Rep.* **2015**, *33*, 2199–2206. [CrossRef]

68. Yoshihara, E.; Masaki, S.; Matsuo, Y.; Chen, Z.; Tian, H.; Yodoi, J. Thioredoxin/Txnip: Redoxisome, as a redox switch for the pathogenesis of diseases. *Front. Immunol.* **2014**, *4*, 514. [CrossRef]

69. Zhou, J.; Chng, W.J. Roles of thioredoxin binding protein (TXNIP) in oxidative stress, apoptosis and cancer. *Mitochondrion* **2013**, *13*, 163–169. [CrossRef]

70. Minn, A.H.; Pise-Masison, C.A.; Radonovich, M.; Brady, J.N.; Wang, P.; Kendziorski, C.; Shalev, A. Gene expression profiling in INS-1 cells overexpressing thioredoxin-interacting protein. *Biochem. Biophys. Res. Commun.* **2005**, *336*, 770–778. [CrossRef]

71. Han, S.H.; Jeon, J.H.; Ju, H.R.; Jung, U.; Kim, K.Y.; Yoo, H.S.; Lee, Y.H.; Song, K.S.; Hwang, H.M.; Na, Y.S. VDUP1 upregulated by TGF-beta1 and 1,25-dihydorxyvitamin D3 inhibits tumor cell growth by blocking cell-cycle progression. *Oncogene* **2003**, *22*, 4035–4046. [CrossRef] [PubMed]

72. Wang, Z.; Prins, G.S.; Coschigano, K.T.; Kopchick, J.J.; Green, J.E.; Ray, V.H.; Hedayat, S.; Christov, K.T.; Unterman, T.G.; Swanson, S.M. Disruption of growth hormone signaling retards early stages of prostate carcinogenesis in the C3(1)/T antigen mouse. *Endocrinology* **2005**, *146*, 5188–5196. [CrossRef] [PubMed]

73. Lapkina-Gendler, L.; Rotem, I.; Pasmanik-Chor, M.; Gurwitz, D.; Sarfstein, R.; Laron, Z.; Werner, H. Identification of signaling pathways associated with cancer protection in Laron syndrome. *Endocr. Relat. Cancer* **2016**, *23*, 399–410. [CrossRef] [PubMed]

74. Patwari, P.; Higgins, L.J.; Chutkow, W.A.; Yoshioka, J.; Lee, R.T. The interaction of thioredoxin with Txnip. Evidence for formation of a mixed disulfide by disulfide exchange. *J. Biol. Chem.* **2006**, *281*, 21884–21891. [CrossRef] [PubMed]

75. Hong, S.Y.; Hagen, T. 2-Deoxyglucose induces the expression of thioredoxin interacting protein (TXNIP) by increasing O-GlcNAcylation—Implications for targeting the Warburg effect in cancer cells. *Biochem. Biophys. Res. Commun.* **2015**, *465*, 838–844. [CrossRef]

76. Chutkow, W.A.; Patwari, P.; Yoshioka, J.; Lee, R.T. Thioredoxin-interacting protein (Txnip) is a critical regulator of hepatic glucose production. *J. Biol. Chem.* **2008**, *283*, 2397–2406. [CrossRef]

77. Waldhart, A.N.; Dykstra, H.; Peck, A.S.; Boguslawski, E.A.; Madaj, Z.B.; Wen, J.; Veldkamp, K.; Hollowell, M.; Zheng, B.; Cantley, L.C.; et al. Phosphorylation of TXNIP by AKT Mediates Acute Influx of Glucose in Response to Insulin. *Cell Rep.* **2017**, *19*, 2005–2013. [CrossRef]

78. Katsu-Jimenez, Y.; Vazquez-Calvo, C.; Maffezzini, C.; Halldin, M.; Peng, X.; Freyer, C.; Wredenberg, A.; Gimenez-Cassina, A.; Wedell, A.; Arner, E.S.J. Absence of TXNIP in Humans Leads to Lactic Acidosis and Low Serum Methionine Linked to Deficient Respiration on Pyruvate. *Diabetes* **2019**, *68*, 709–723. [CrossRef]

79. Nagaraj, K.; Lapkina-Gendler, L.; Sarfstein, R.; Gurwitz, D.; Pasmanik-Chor, M.; Laron, Z.; Yakar, S.; Werner, H. Identification of thioredoxin-interacting protein (TXNIP) as a downstream target for IGF1 action. *Proc. Natl. Acad. Sci. USA* **2018**, *115*, 1045–1050. [CrossRef]

80. Cittadini, A.; Grossman, J.D.; Stromer, H.; Katz, S.E.; Morgan, J.P.; Douglas, P.S. Importance of an intact growth hormone/insulin-like growth factor 1 axis for normal post-infarction healing: Studies in dwarf rats. *Endocrinology* **2001**, *142*, 332–338. [CrossRef]

81. Davani, E.Y.; Brumme, Z.; Singhera, G.K.; Cote, H.C.; Harrigan, P.R.; Dorscheid, D.R. Insulin-like growth factor-1 protects ischemic murine myocardium from ischemia/reperfusion associated injury. *Crit. Care* **2003**, *7*, R176–R183. [CrossRef] [PubMed]

82. Hausenloy, D.J.; Yellon, D.M. New directions for protecting the heart against ischaemia-reperfusion injury: Targeting the Reperfusion Injury Salvage Kinase (RISK)-pathway. *Cardiovasc. Res.* **2004**, *61*, 448–460. [CrossRef] [PubMed]

83. Yamashita, K.; Kajstura, J.; Discher, D.J.; Wasserlauf, B.J.; Bishopric, N.H.; Anversa, P.; Webster, K.A. Reperfusion-activated Akt kinase prevents apoptosis in transgenic mouse hearts overexpressing insulin-like growth factor-1. *Circ. Res.* **2001**, *88*, 609–614. [CrossRef] [PubMed]

84. Kajstura, J.; Fiordaliso, F.; Andreoli, A.M.; Li, B.; Chimenti, S.; Medow, M.S.; Limana, F.; Nadal-Ginard, B.; Leri, A.; Anversa, P. IGF-1 overexpression inhibits the development of diabetic cardiomyopathy and angiotensin II-mediated oxidative stress. *Diabetes* **2001**, *50*, 1414–1424. [CrossRef] [PubMed]

85. Lian, J.D.; al-Jumah, M.; Cwik, V.; Brooke, M.H. Neurotrophic factors decrease the release of creatine kinase and prostaglandin E2 from metabolically stressed muscle. *Neuromuscul. Disord.* **1998**, *8*, 7–13. [CrossRef]

86. Matheny, R.W.Jr.; Adamo, M.L. PI3K p110 alpha and p110 beta have differential effects on Akt activation and protection against oxidative stress-induced apoptosis in myoblasts. *Cell Death Differ.* **2010**, *17*, 677–688. [CrossRef] [PubMed]

87. Csiszar, A.; Labinskyy, N.; Perez, V.; Recchia, F.A.; Podlutsky, A.; Mukhopadhyay, P.; Losonczy, G.; Pacher, P.; Austad, S.N.; Bartke, A.; et al. Endothelial function and vascular oxidative stress in long-lived GH/IGF-deficient Ames dwarf mice. *Am. J. Physiol. Heart Circ. Physiol.* **2008**, *295*, H1882–H1894. [CrossRef]

88. Heck, S.; Lezoualc'h, F.; Engert, S.; Behl, C. Insulin-like growth factor-1-mediated neuroprotection against oxidative stress is associated with activation of nuclear factor kappaB. *J. Biol. Chem.* **1999**, *274*, 9828–9835. [CrossRef]

89. Kolosova, N.G.; Stefanova, N.A.; Muraleva, N.A.; Skulachev, V.P. The mitochondria-targeted antioxidant SkQ1 but not N-acetylcysteine reverses aging-related biomarkers in rats. *Aging (Albany NY)* **2012**, *4*, 686–694. [CrossRef]

90. Farr, S.A.; Poon, H.F.; Dogrukol-Ak, D.; Drake, J.; Banks, W.A.; Eyerman, E.; Butterfield, D.A.; Morley, J.E. The antioxidants alpha-lipoic acid and N-acetylcysteine reverse memory impairment and brain oxidative stress in aged SAMP8 mice. *J. Neurochem.* **2003**, *84*, 1173–1183. [CrossRef]

91. Campisi, J.; d'Adda di Fagagna, F. Cellular senescence: When bad things happen to good cells. *Nat. Rev. Mol. Cell Biol.* **2007**, *8*, 729–740. [CrossRef] [PubMed]

92. Sharpless, N.E.; Sherr, C.J. Forging a signature of in vivo senescence. *Nat. Rev. Cancer* **2015**, *15*, 397–408. [CrossRef] [PubMed]

93. Acosta, J.C.; Banito, A.; Wuestefeld, T.; Georgilis, A.; Janich, P.; Morton, J.P.; Athineos, D.; Kang, T.W.; Lasitschka, F.; Andrulis, M.; et al. A complex secretory program orchestrated by the inflammasome controls paracrine senescence. *Nat. Cell Biol.* **2013**, *15*, 978–990. [CrossRef] [PubMed]

94. Coppe, J.P.; Rodier, F.; Patil, C.K.; Freund, A.; Desprez, P.Y.; Campisi, J. Tumor suppressor and aging biomarker p16(INK4a) induces cellular senescence without the associated inflammatory secretory phenotype. *J. Biol. Chem.* **2011**, *286*, 36396–36403. [CrossRef]

95. Rodier, F.; Coppe, J.P.; Patil, C.K.; Hoeijmakers, W.A.; Munoz, D.P.; Raza, S.R.; Freund, A.; Campeau, E.; Davalos, A.R.; Campisi, J. Persistent DNA damage signalling triggers senescence-associated inflammatory cytokine secretion. *Nat. Cell Biol.* **2009**, *11*, 973–979. [CrossRef]

96. Coppe, J.P.; Patil, C.K.; Rodier, F.; Sun, Y.; Munoz, D.P.; Goldstein, J.; Nelson, P.S.; Desprez, P.Y.; Campisi, J. Senescence-associated secretory phenotypes reveal cell-nonautonomous functions of oncogenic RAS and the p53 tumor suppressor. *PLoS Biol.* **2008**, *6*, 2853–2868. [CrossRef]

97. Kuilman, T.; Michaloglou, C.; Vredeveld, L.C.; Douma, S.; van Doorn, R.; Desmet, C.J.; Aarden, L.A.; Mooi, W.J.; Peeper, D.S. Oncogene-induced senescence relayed by an interleukin-dependent inflammatory network. *Cell* **2008**, *133*, 1019–1031. [CrossRef]

98. Correia-Melo, C.; Marques, F.D.; Anderson, R.; Hewitt, G.; Hewitt, R.; Cole, J.; Carroll, B.M.; Miwa, S.; Birch, J.; Merz, A. Mitochondria are required for pro-ageing features of the senescent phenotype. *EMBO J.* **2016**, *35*, 724–742. [CrossRef]

99. Helman, A.; Klochendler, A.; Azazmeh, N.; Gabai, Y.; Horwitz, E.; Anzi, S.; Swisa, A.; Condiotti, R.; Granit, R.Z.; Nevo, Y.; et al. p16(Ink4a)-induced senescence of pancreatic beta cells enhances insulin secretion. *Nat. Med.* **2016**, *22*, 412–420. [CrossRef]

100. Chesnokova, V.; Zhou, C.; Ben-Shlomo, A.; Zonis, S.; Tani, Y.; Ren, S.G.; Melmed, S. Growth hormone is a cellular senescence target in pituitary and nonpituitary cells. *Proc. Natl. Acad. Sci. USA* **2013**, *110*, E3331–E3339. [CrossRef]

101. Matsumoto, R.; Fukuoka, H.; Iguchi, G.; Odake, Y.; Yoshida, K.; Bando, H.; Suda, K.; Nishizawa, H.; Takahashi, M.; Yamada, S.; et al. Accelerated Telomere Shortening in Acromegaly; IGF-I Induces Telomere Shortening and Cellular Senescence. *PLoS ONE* **2015**, *10*, e0140189. [CrossRef] [PubMed]

102. Stout, M.B.; Tchkonia, T.; Pirtskhalava, T.; Palmer, A.K.; List, E.O.; Berryman, D.E.; Lubbers, E.R.; Escande, C.; Spong, A.; Masternak, M.M.; et al. Growth hormone action predicts age-related white adipose tissue dysfunction and senescent cell burden in mice. *Aging (Albany NY)* **2014**, *6*, 575–586. [CrossRef] [PubMed]

103. Ock, S.; Lee, W.S.; Ahn, J.; Kim, H.M.; Kang, H.; Kim, H.S.; Jo, D.; Abel, E.D.; Lee, T.J.; Kim, J. Deletion of IGF-1 Receptors in Cardiomyocytes Attenuates Cardiac Aging in Male Mice. *Endocrinology* **2016**, *157*, 336–345. [CrossRef] [PubMed]

104. Tran, D.; Bergholz, J.; Zhang, H.; He, H.; Wang, Y.; Zhang, Y.; Li, Q.; Kirkland, J.L.; Xiao, Z.X. Insulin-like growth factor-1 regulates the SIRT1-p53 pathway in cellular senescence. *Aging Cell* **2014**, *13*, 669–678. [CrossRef] [PubMed]

105. Del Nogal-Avila, M.; Troyano-Suarez, N.; Roman-Garcia, P.; Cannata-Andia, J.B.; Rodriguez-Puyol, M.; Rodriguez-Puyol, D.; Kuro, O.M.; Ruiz-Torres, M.P. Amadori products promote cellular senescence activating insulin-like growth factor-1 receptor and down-regulating the antioxidant enzyme catalase. *Int. J. Biochem. Cell Biol.* **2013**, *45*, 1255–1264. [CrossRef] [PubMed]

106. Handayaningsih, A.E.; Takahashi, M.; Fukuoka, H.; Iguchi, G.; Nishizawa, H.; Yamamoto, M.; Suda, K.; Takahashi, Y. IGF-I enhances cellular senescence via the reactive oxygen species-p53 pathway. *Biochem. Biophys. Res. Commun.* **2012**, *425*, 478–484. [CrossRef]

107. Thum, T.; Hoeber, S.; Froese, S.; Klink, I.; Stichtenoth, D.O.; Galuppo, P.; Jakob, M.; Tsikas, D.; Anker, S.D.; Poole-Wilson, P.A.; et al. Age-dependent impairment of endothelial progenitor cells is corrected by growth-hormone-mediated increase of insulin-like growth-factor-1. *Circ. Res.* **2007**, *100*, 434–443. [CrossRef]

108. Kim, S.H.; Lee, E.Y.; Cho, K.H. Incorporation of human growth hormone-2 into proteoliposome enhances tissue regeneration with anti-oxidant and anti-senescence activities. *Rejuvenation Res.* **2015**, *18*, 20–29. [CrossRef]

109. Luo, X.; Jiang, X.; Li, J.; Bai, Y.; Li, Z.; Wei, P.; Sun, S.; Liang, Y.; Han, S.; Li, X.; et al. Insulin-like growth factor-1 attenuates oxidative stress-induced hepatocyte premature senescence in liver fibrogenesis via regulating nuclear p53-progerin interaction. *Cell Death Dis.* **2019**, *10*, 451. [CrossRef]

110. Kluck, R.M.; Bossy-Wetzel, E.; Green, D.R.; Newmeyer, D.D. The release of cytochrome c from mitochondria: A primary site for Bcl-2 regulation of apoptosis. *Science* **1997**, *275*, 1132–1136. [CrossRef]

111. Yang, J.; Liu, X.; Bhalla, K.; Kim, C.N.; Ibrado, A.M.; Cai, J.; Peng, T.I.; Jones, D.P.; Wang, X. Prevention of apoptosis by Bcl-2: release of cytochrome c from mitochondria blocked. *Science* **1997**, *275*, 1129–1132. [CrossRef] [PubMed]

112. Cory, S.; Adams, J.M. The Bcl2 family: Regulators of the cellular life-or-death switch. *Nat. Rev. Cancer* **2002**, *2*, 647–656. [CrossRef] [PubMed]

113. Youle, R.J.; Strasser, A. The BCL-2 protein family: Opposing activities that mediate cell death. *Nat. Rev. Mol. Cell Biol.* **2008**, *9*, 47–59. [CrossRef] [PubMed]

114. Zong, W.X.; Lindsten, T.; Ross, A.J.; MacGregor, G.R.; Thompson, C.B. BH3-only proteins that bind pro-survival Bcl-2 family members fail to induce apoptosis in the absence of Bax and Bak. *Genes Dev.* **2001**, *15*, 1481–1486. [CrossRef]

115. Gonzalez, J.M.; Esteban, M. A poxvirus Bcl-2-like gene family involved in regulation of host immune response: sequence similarity and evolutionary history. *Virol. J.* **2010**, *7*, 59. [CrossRef]

116. Peterson, J.S.; Bass, B.P.; Jue, D.; Rodriguez, A.; Abrams, J.M.; McCall, K. Noncanonical cell death pathways act during Drosophila oogenesis. *Genesis* **2007**, *45*, 396–404. [CrossRef]

117. Fu, Z.; Tindall, D.J. FOXOs, cancer and regulation of apoptosis. *Oncogene* **2008**, *27*, 2312–2319. [CrossRef]

118. Murphy, C.T.; McCarroll, S.A.; Bargmann, C.I.; Fraser, A.; Kamath, R.S.; Ahringer, J.; Li, H.; Kenyon, C. Genes that act downstream of DAF-16 to influence the lifespan of Caenorhabditis elegans. *Nature* **2003**, *424*, 277–283. [CrossRef]

119. Guo, S.; Rena, G.; Cichy, S.; He, X.; Cohen, P.; Unterman, T. Phosphorylation of serine 256 by protein kinase B disrupts transactivation by FKHR and mediates effects of insulin on insulin-like growth factor-binding protein-1 promoter activity through a conserved insulin response sequence. *J. Biol. Chem.* **1999**, *274*, 17184–17192. [CrossRef]

120. Martins, R.; Lithgow, G.J.; Link, W. Long live FOXO: Unraveling the role of FOXO proteins in aging and longevity. *Aging Cell* **2016**, *15*, 196–207. [CrossRef]

121. Paik, J.H.; Kollipara, R.; Chu, G.; Ji, H.; Xiao, Y.; Ding, Z.; Miao, L.; Tothova, Z.; Horner, J.W.; Carrasco, D.R.; et al. FoxOs are lineage-restricted redundant tumor suppressors and regulate endothelial cell homeostasis. *Cell* **2007**, *128*, 309–323. [CrossRef]

122. Ambrogini, E.; Almeida, M.; Martin-Millan, M.; Paik, J.H.; Depinho, R.A.; Han, L.; Goellner, J.; Weinstein, R.S.; Jilka, R.L.; O'Brien, C.A.; et al. FoxO-mediated defense against oxidative stress in osteoblasts is indispensable for skeletal homeostasis in mice. *Cell Metab.* **2010**, *11*, 136–146. [CrossRef] [PubMed]

123. Alt, E.U.; Senst, C.; Murthy, S.N.; Slakey, D.P.; Dupin, C.L.; Chaffin, A.E.; Kadowitz, P.J.; Izadpanah, R. Aging alters tissue resident mesenchymal stem cell properties. *Stem Cell Res.* **2012**, *8*, 215–225. [CrossRef] [PubMed]

124. Wilson, A.; Shehadeh, L.A.; Yu, H.; Webster, K.A. Age-related molecular genetic changes of murine bone marrow mesenchymal stem cells. *BMC Genomics* **2010**, *11*, 229. [CrossRef] [PubMed]

125. Suh, Y.; Lee, K.A.; Kim, W.H.; Han, B.G.; Vijg, J.; Park, S.C. Aging alters the apoptotic response to genotoxic stress. *Nat. Med.* **2002**, *8*, 3–4. [CrossRef]

126. Polyak, K.; Wu, T.T.; Hamilton, S.R.; Kinzler, K.W.; Vogelstein, B. Less death in the dying. *Cell Death Differ.* **1997**, *4*, 242–246. [CrossRef]

127. Kavathia, N.; Jain, A.; Walston, J.; Beamer, B.A.; Fedarko, N.S. Serum markers of apoptosis decrease with age and cancer stage. *Aging (Albany NY)* **2009**, *1*, 652–663. [CrossRef]

128. Medzhitov, R. Inflammation 2010: New adventures of an old flame. *Cell* **2010**, *140*, 771–776. [CrossRef]

129. Bennett, J.M.; Reeves, G.; Billman, G.E.; Sturmberg, J.P. Inflammation-Nature's Way to Efficiently Respond to All Types of Challenges: Implications for Understanding and Managing "the Epidemic" of Chronic Diseases. *Front. Med. (Lausanne)* **2018**, *5*, 316. [CrossRef]

130. Chen, L.; Deng, H.; Cui, H.; Fang, J.; Zuo, Z.; Deng, J.; Li, Y.; Wang, X.; Zhao, L. Inflammatory responses and inflammation-associated diseases in organs. *Oncotarget* **2018**, *9*, 7204–7218. [CrossRef]

131. O'Brien, K.L.; Finlay, D.K. Immunometabolism and natural killer cell responses. *Nat. Rev. Immunol.* **2019**, *19*, 282–290. [CrossRef] [PubMed]

132. Breda, C.N.S.; Davanzo, G.G.; Basso, P.J.; Saraiva Camara, N.O.; Moraes-Vieira, P.M.M. Mitochondria as central hub of the immune system. *Redox Biol.* **2019**, *26*, 101255. [CrossRef] [PubMed]

133. Strickland, M.; Yacoubi-Loueslati, B.; Bouhaouala-Zahar, B.; Pender, S.L.F.; Larbi, A. Relationships Between Ion Channels, Mitochondrial Functions and Inflammation in Human Aging. *Front. Physiol.* **2019**, *10*, 158. [CrossRef] [PubMed]

134. Van Horssen, J.; van Schaik, P.; Witte, M. Inflammation and mitochondrial dysfunction: A vicious circle in neurodegenerative disorders? *Neurosci. Lett.* **2019**, *710*, 132931. [CrossRef]

135. Kermani, H.; Goffinet, L.; Mottet, M.; Bodart, G.; Morrhaye, G.; Dardenne, O.; Renard, C.; Overbergh, L.; Baron, F.; Beguin, Y.; et al. Expression of the growth hormone/insulin-like growth factor axis during Balb/c thymus ontogeny and effects of growth hormone upon ex vivo T cell differentiation. *Neuroimmunomodulation* **2012**, *19*, 137–147. [CrossRef]

136. Smith, T.J. Insulin-like growth factor-I regulation of immune function: A potential therapeutic target in autoimmune diseases? *Pharmacol. Rev.* **2010**, *62*, 199–236. [CrossRef]

137. Weigent, D.A.; Blalock, J.E. Expression of growth hormone by lymphocytes. *Int. Rev. Immunol.* **1989**, *4*, 193–211. [CrossRef]

138. Rappolee, D.A.; Mark, D.; Banda, M.J.; Werb, Z. Wound macrophages express TGF-alpha and other growth factors in vivo: Analysis by mRNA phenotyping. *Science* **1988**, *241*, 708–712. [CrossRef]

139. Spaziani, S.; Imperlini, E.; Mancini, A.; Caterino, M.; Buono, P.; Orru, S. Insulin-like growth factor 1 receptor signaling induced by supraphysiological doses of IGF-1 in human peripheral blood lymphocytes. *Proteomics* **2014**, *14*, 1623–1629. [CrossRef]

140. Wong, S.C.; Dobie, R.; Altowati, M.A.; Werther, G.A.; Farquharson, C.; Ahmed, S.F. Growth and the Growth Hormone-Insulin Like Growth Factor 1 Axis in Children With Chronic Inflammation: Current Evidence, Gaps in Knowledge, and Future Directions. *Endocr. Rev.* **2016**, *37*, 62–110. [CrossRef]

141. Masternak, M.M.; Bartke, A. Growth hormone, inflammation and aging. *Pathobiol. Aging Age Relat. Dis.* **2012**, *2*. [CrossRef] [PubMed]

142. Sadagurski, M.; Landeryou, T.; Cady, G.; Kopchick, J.J.; List, E.O.; Berryman, D.E.; Bartke, A.; Miller, R.A. Growth hormone modulates hypothalamic inflammation in long-lived pituitary dwarf mice. *Aging Cell* **2015**, *14*, 1045–1054. [CrossRef] [PubMed]

Review

IGFBP-3/IGFBP-3 Receptor System as an Anti-Tumor and Anti-Metastatic Signaling in Cancer

Qing Cai [1], Mikhail Dozmorov [1,2] and Youngman Oh [1,*]

[1] Department of Pathology, Medical College of Virginia Campus, Virginia Commonwealth University, Richmond, VA 23298, USA; qing.cai@vcuhealth.org (Q.C.); mikhail.dozmorov@vcuhealth.org (M.D.)

[2] Department of Biostatistics, Massey Cancer Center, Virginia Commonwealth University, Richmond, VA 23298, USA

[*] Correspondence: youngman.oh@vcuhealth.org; Tel.: +1-804-827-1324

Received: 14 April 2020; Accepted: 18 May 2020; Published: 20 May 2020

Abstract: Insulin-like growth factor binding protein-3 (IGFBP-3) is a p53 tumor suppressor-regulated protein and a major carrier for IGFs in circulation. Among six high-affinity IGFBPs, which are IGFBP-1 through 6, IGFBP-3 is the most extensively investigated IGFBP species with respect to its IGF/IGF-I receptor (IGF-IR)-independent biological actions beyond its endocrine/paracrine/autocrine role in modulating IGF action in cancer. Disruption of IGFBP-3 at transcriptional and post-translational levels has been implicated in the pathophysiology of many different types of cancer including breast, prostate, and lung cancer. Over the past two decades, a wealth of evidence has revealed both tumor suppressing and tumor promoting effects of IGF/IGF-IR-independent actions of IGFBP-3 depending upon cell types, post-translational modifications, and assay methods. However, IGFBP-3's anti-tumor function has been well accepted due to identification of functional IGFBP-3-interacting proteins, putative receptors, or crosstalk with other signaling cascades. This review mainly focuses on transmembrane protein 219 (TMEM219), which represents a novel IGFBP-3 receptor mediating antitumor effect of IGFBP-3. Furthermore, this review delineates the potential underlying mechanisms involved and the subsequent biological significance, emphasizing the clinical significance of the IGFBP-3/TMEM219 axis in assessing both the diagnosis and the prognosis of cancer as well as the therapeutic potential of TMEM219 agonists for cancer treatment.

Keywords: IGF system; IGFBP-3; IGFBP-3R; TMEM219; anti-tumor; anti-metastatic; agonists; mAb therapy

1. Introduction

The insulin-like growth factor (IGF) system comprises of ligands IGF-I, IGF-II, its corresponding cell-membrane receptors IGF-I receptor (IGF-IR), IGF-II receptor (IGF-IIR), IGF-binding proteins (IGFBPs), and IGFBP degrading enzymes known as proteases. The IGF system plays a critical role in somatic growth in an endocrine fashion as well as cell proliferation, survival, and differentiation of normal and malignant cells in a paracrine/autocrine fashion. Dysregulation of the IGF system attributes to pathophysiology of a variety of human diseases such as cancer, diabetes, chronic inflammatory disease, and malnutrition. In particular, IGF/IGF-IR-independent actions of IGFBP-3 have been extensively investigated and their involvement in initiation and progression of various cancers has been recognized.

2. IGFBP-3

2.1. Structure-Function Analysis

Human IGFBP-3 is comprised of 264 amino acids, of which the molecular mass is 28.7 kDa without any post-translational modifications [1]. The primary structures of human IGFBP-3 consist of three distinct domains: a highly conserved cysteine-rich *N*- and *C*-terminal domains and a nonconserved central domain. Each domain contains various functional motifs/sequences that confer IGFBP-3's diverse IGF/IGF-IR-dependent and IGF/IGF-IR-independent actions (Figure 1) [2–6]. These distinctive functional motifs/sequences include a caveolin scaffolding docking domain, a metal binding domain, heparin binding motifs, a retinoic acid binding motif, and a nuclear localization sequence.

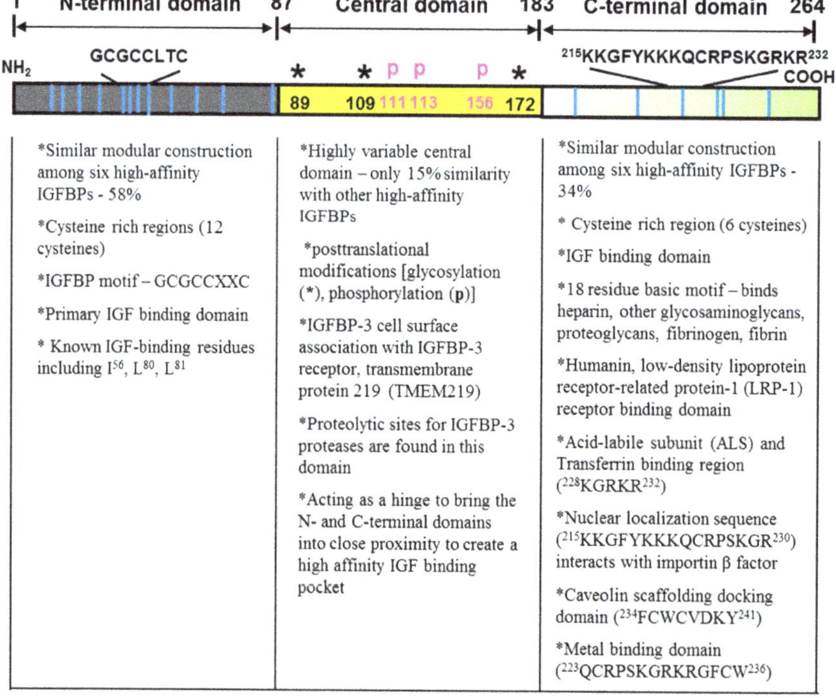

*Similar modular construction among six high-affinity IGFBPs - 58% *Cysteine rich regions (12 cysteines) *IGFBP motif – GCGCCXXC *Primary IGF binding domain * Known IGF-binding residues including I56, L80, L81	*Highly variable central domain – only 15% similarity with other high-affinity IGFBPs *posttranslational modifications [glycosylation (*), phosphorylation (p)] *IGFBP-3 cell surface association with IGFBP-3 receptor, transmembrane protein 219 (TMEM219) *Proteolytic sites for IGFBP-3 proteases are found in this domain *Acting as a hinge to bring the N- and C-terminal domains into close proximity to create a high affinity IGF binding pocket	*Similar modular construction among six high-affinity IGFBPs - 34% * Cysteine rich region (6 cysteines) *IGF binding domain *18 residue basic motif – binds heparin, other glycosaminoglycans, proteoglycans, fibrinogen, fibrin *Humanin, low-density lipoprotein receptor-related protein-1 (LRP-1) receptor binding domain *Acid-labile subunit (ALS) and Transferrin binding region (228KGRKR232) *Nuclear localization sequence (215KKGFYKKKQCRPSKGR230) interacts with importin β factor *Caveolin scaffolding docking domain (234FCWCVDKY241) *Metal binding domain (223QCRPSKGRKRGFCW236)

Figure 1. Structure of the mature human IGFBP-3. This figure depicts the three distinct domains of the IGFBP-3 and lists the important functions and motifs/residues within each domain [3]. The vertical blue lines represent 18 cysteine residues in highly conserved *N*-terminal and *C*-terminal domains.

2.1.1. The Conserved *N*-Terminal Domain

In the mature IGFBP-3 peptide, amino acid residues 1–87 comprise the conserved *N*-terminal domain, which shares approximately 58% similarity with other high-affinity IGFBPs. A well conserved IGFBP motif (GCGCCXXC) present in all IGFBP species is located in this domain. Ten to 12 of the 16–20 cysteines are located in the *N*-terminal domain of high-affinity IGFBPs. Among a total of 18 cysteines in IGFBP-3, 12 cysteines reside in the *N*-terminal domain, which results in the formation of six disulfide bonds within the domain and providing a highly organized tertiary structure. Thus, this conserved *N*-terminal domain shares not only amino acid similarity but also conformational similarities among high-affinity IGFBPs. Important IGF-binding residues including I56, L80, and L81 are also located within this domain [2,3,7].

2.1.2. The Variable Central Domain

The central domain contains 95 amino acids and spans residues 88–183. This domain separates the *N*-terminal domain from the *C*-terminal domain and shares less than 15% similarity with other high-affinity IGFBPs [2]. However, it appears that this domain structurally acts as a hinge between the *N*- and *C*-terminal domains and bring two domains together into close proximity to create a high affinity IGF binding pocket. Post-translational modifications such as glycosylation, phosphorylation, and proteolysis of IGFBP-3 have been found in this domain [8–12]. The functional significance of those post-translational modifications has been reported that glycosylation can affect cell interactions, that phosphorylation can affect IGF-binding affinity and susceptibility to proteases, and that proteolysis can affect both IGF-dependent and IGF-independent actions [4,11–13]. Three *N*-linked glycosylation sites at asparagine 89, 109, and 172, and phosphorylation sites at serine 111, 113, 156, 165, and at threonine 170, as well as proteolytic sites for metalloproteases (MMPs) and serine proteases exist in this domain [8–10,12,13]. The central domain is responsible for the interaction with the IGFBP-3 specific receptor known as transmembrane protein 219 (TMEM219) [14,15].

2.1.3. The Conserved *C*-Terminal Domain

This domain spans residues 184–264, containing six cysteines with three disulfide bonds. This domain is also important in IGF binding [16–19]. Since the IGFBP-3 fragment that contains only *N*- or *C*-terminal domains has significantly reduced affinity for IGFs, it requires an IGF-binding pocket involving both domains for high affinity binding to IGFs. Several functionally important sequences/motifs are present in this domain such as binding for heparin, glycosaminoglycans, proteoglycans, fibronectin, fibrin, transferrin, plasminogen, acid-labile subunit (ALS), and metals such as iron, zinc, and nickel [3–5,20–24]. Both IGFBP-3 and IGFBP-3-IGF complexes bind fibrinogen, fibrin, and plasminogen. Furthermore, a nuclear localization sequence (NLS) [25] and a caveolin-scaffolding domain consensus sequence [26] also reside in this domain.

2.2. IGF/IGF-IR Dependent Actions of IGFBP-3

The principal action of IGFBP-3 is to transport IGF-I and IGF-II in circulation, and, thereby, prolong the half-life of IGFs. IGFBP-3 has a higher affinity for IGFs (Kd approximately 10^{-10} M) than their respective receptors. In serum, most of the IGFs circulate as a 150 kDa complex, consisting of 7·5 kDa IGF-I or IGF-II and 45 kDa glycosylated IGFBP-3 and 90 kDa ALS [3–5,27–29]. The biological activity of circulating IGFs in the tissues is determined by the transition of IGF from 150 kDa complex to the 55 kDa IGF-IGFBP-3 complex and subsequent proteolysis of the complex to release IGF in the circulation or in the local body fluid. In addition to functioning as an IGF transporter, IGFBP-3 also functions as modulators of IGF availability and activity at the cellular levels in an autocrine or paracrine manner [25,28–32].

IGFBP-3 can inhibit or enhance IGF actions, depending on cell types, the cellular environment, IGFBP-3 concentration, and post-translational modifications such as glycosylation, proteolysis, and phosphorylation [4,11–13]. IGFBP-3 has shown to inhibit IGF activity by competitively binding IGFs and preventing its binding to IGFRs [29,30]. On the other hand, IGFBP-3 can enhance IGF activity by increasing IGF concentration in the extracellular microenvironment by binding to heparin and proteoglycans, and, thereby, acting as a reservoir of IGFs [20–22].

2.3. IGF/IGF-IR Independent Actions of IGFBP-3

The IGF/IGF-IR-independent actions of IGFBP-3 have been shown to contribute to the pathophysiology of various human diseases such as cancer, diabetes, obesity, fatty liver disease, ischemia, and Alzheimer's disease [15,33–47]. In an early era of IGFBP-3 research in cancer, many studies demonstrated that IGFBP-3 is upregulated by different types of cell growth inhibitors at the transcriptional level in a variety of human cancer cells. These include anti-estrogens (Tamoxifen,

ICI-182780), TGF-β, retinoic acid, TNF-α, vitamin D, histone deacetylase inhibitor sodium butyrate, and anti-cancer dietary components including silibinin, apigenin, lycopene, resveratrol, curcumin, and quercetin [48–57]. In particular, the tumor suppressor gene p53 has been shown to upregulate IGFBP-3 at the transcriptional level [58,59]. Two p53 binding sites, Box A and Box B, were identified in the first and second introns of the IGFBP-3 gene, based on homology to the p53 binding consensus sequence [58]. Further studies using p53 mutants have revealed a link between p53's activation of IGFBP-3 transcription and its induction of apoptosis by showing that the mutants that lost the ability to activate IGFBP-3 could not induce apoptosis [60]. Further research also demonstrated that the transfection of doxycycline-inducible p53 plasmids resulted in increased expression of p53 and IGFBP-3 and, subsequently, induced apoptosis in p53-negative PC-3 prostate cancer cells [40]. This p53-depedent induction of apoptosis was inhibited by treating with IGF-I, IGFBP-3 blocking antibodies, and IGFBP-3 antisense oligonucleotides, which demonstrated p53-dependent IGFBP-3's proapoptotic function. In light of p53 dependency of IGFBP-3 expression, ΔNp63α, an isoform of tumor suppressor p63 with both dominant negative (ΔN) activities and a potent repressor of p53-mediated transactivation has been demonstrated to suppress expression of IGFBP-3 [61]. It appears that ΔNp63α binds the p53 binding sites, Box A and Box B, in the IGFBP-3 gene, and, thereby, inhibits p53-dependent IGFBP-3 expression and presumably suppresses IGFBP-3-induced apoptosis. However, evidence also supports that IGFBP-3 can be induced in a p53-independet manner [40]. Treatment with genotoxic drugs such as etoposide and Adriamycin resulted in increased IGFBP-3 expression in p53-negative PC-3 prostate cancer cells.

Moreover, several studies demonstrated that the loss of IGFBP-3 expression by DNA methylation is linked to tumorigenesis and cancer progression as well as intrinsic and/or acquired resistance to radiotherapy and chemo-drugs such as cisplatin in many different types of cancer including lung, colon, and ovarian cancers [62–69]. These findings strongly suggested that IGFBP-3 may exert anti-proliferative and anti-tumor functions beyond its ability to modulate IGF functions (IGF/IGF-IR dependent actions), but the underlying mechanisms involved remain largely unknown. Since then, there has been an intensive investigation toward characterizing the molecular and cellular mechanisms for IGF/IGF-IR-independent antitumor effects of IGFBP-3 in human cancer in vitro and in vivo. It is clear that IGFBP-3 exerts its IGF/IGF-IR-independent biological actions through interactions with a variety of binding partners on cell surfaces and within cells.

2.3.1. IGFBP-3 Binding Partners on the Cell Surface

The very first evidence for the IGF/IGF-IR-independent actions of IGFBP-3 was the identification of specific cell surface binding between IGFBP-3 and cell surface proteins and subsequent cell growth inhibition in Hs578T human triple negative breast cancer (TNBC) cells [33,70]. These initial findings demonstrated that only IGFBP-3 specifically binds to the cell surface among IGFBPs and the central domain of IGFBP-3 is necessary for the binding. IGFs attenuated the cell surface binding and the subsequent growth inhibitory effects of IGFBP-3 by forming IGF-IGFBP-3 complexes. The existence of high-affinity binding sites for IGFBP-3, which is typical of receptor-ligand interactions, were found. The binding sites further demonstrated 20- and 28-kDa cell surface proteins as putative receptors. Based on biochemical and functional characteristics, these proteins are later proven to be an IGFBP-3 receptor, TMEM219, which was identified by a yeast two-hybrid screening using the central domain of IGFBP-3 from the same Hs578T human TNBC cell line [15]. This IGFBP-3 receptor will be further discussed in Section 3.

At present, a few proteins have been identified as IGFBP-3 cell surface binding partners such as the low-density lipoprotein receptor-related protein-1 (LRP-1)/α2M receptor [71], autocrine motility factor (AMF)/phosphoglucose isomerase (PGI) [72], latent TGF-β binding protein-1 (LTBP-1), caveolin, and transferrin/transferrin receptor [26,73]. The LRP-1/α2M receptor, also known as TGF-β type V receptor, is shown to mediate IGFBP-3-induced cell growth inhibition independent of IGF [74,75]. In addition, it plays a crucial role for cellular internalization of IGFBP-3 since LRP knock-out cells exhibited significant reduction of IGFBP-3 internalization when compared with LRP-expressing mouse embryonic fibroblasts [76]. While AMF/PGI, which is a tumor-secreted cytokine, is endocytosed and regulates cell migration, proliferation, and survival, IGFBP-3 has been shown to inhibit AMF/PGI-induced cell migration in T47D and MCF-7 breast cancer cells [72]. LTBP-1, which is a component of the latent TGF-β complex and a part of structural component of the ECM, is involved in sequestration of latent TGF-β in the ECM and delivery of TGF-β to the plasma membrane [77]. Although the functional significance of IGFBP-3 binding to LTBP-1 as well as the large latent complex has not been fully elucidated, it may be a potential mechanism whereby IGFBP-3 can interact with the TGF-β system [78]. Since a substantial amount of LTBP-1 can be secreted by cells without bound TGF-β, IGFBP-3 may also involve TGF-β-independent functions of LTBP-1 [79].

IGFBP-3 interaction with caveolin-1 through a caveolin-scaffolding sequence induced IGFBP-3 internalization [26]. Furthermore, recent research indicated that caveolin-1 is an oncogenic membrane protein and is associated with endocytosis, extracellular matrix organization, cholesterol distribution, cell migration, and signaling. This strongly suggests the potential regulation of IGFBP-3 on these caveolin-1-induced functions [80]. IGFBP-3 also binds to transferrin and forms an IGFBP-3-tranferrin-transferrin receptor complex, providing another mechanism for IGFBP-3 internalization and signaling [26,73]. IGFBP-3 internalization was inhibited by co-incubation and extracellular sequestration with IGF-I, and was dependent on the transferrin-binding *C*-terminal peptide region of IGFBP-3 [26]. By the same token, blocking transferrin receptor-mediated endocytosis suppressed IGFBP-3 internalization and IGFBP-3-induced apoptosis [26]. At present, it remains unclear whether TMEM219 is a sole IGFBP-3 receptor mediating IGF/IGF-IR independent antitumor actions of IGFBP-3 or whether the previously mentioned cell surface binding partners are also partly involved in IGFBP-3 internalization and subsequent IGF/IGF-IR independent actions in cytoplasmic and nuclear compartments.

2.3.2. IGFBP-3 Binding Partners within Cells

Although IGFBP-3 can be internalized to the cytoplasmic compartment and translocated to the nucleus through the NLS in the conserved *C*-terminal domain, limited knowledge is available on whether nuclear targeting of IGFBP-3 occurs in all types of cells or requires specific cellular conditions. Nevertheless, IGFBP-3 has been shown to interact with cytoplasmic/nuclear proteins. These include humanin [37], RNA polymerase II binding subunit 3 (Rpb3) [81], GalNAc-T14 [82], glucose-regulated protein 78 (GRP78) [83,84], nuclear retinoid X receptor (RXR) [85], retinoic acid receptor (RAR) [86], Nur77 [87], vitamin D receptor (VDR) [88], and peroxisome proliferator-activated receptor-γ (PPARγ) [89].

Humanin is a mitochondrial-derived peptide that inhibits neuronal cell death induced by mutant genes in Alzheimer's disease [37]. Humanin has been shown to bind to IGFBP-3 and inhibit nuclear translocation and induction of apoptosis of IGFBP-3 in human lung cancer cells by suppressing the IGFBP-3 interaction with importin-β [90]. Rpb3, which is an essential component of the mRNA transcription apparatus, aids the recruitment of the polymerase complex to specific transcription factors. Rpb3 has been shown to interact with the NLS motif of IGFBP-3 and might lead to IGFBP-3's role in modulating gene transcription [81]. GalNAc-T14, a large subfamily of glycosyltransferases residing in the Golgi complex, catalyze the first step in the *O*-glycosylation of mammalian proteins by transferring *N*-acetyl-D-galactosamine (GalNAc) to peptide substrates [91]. Since GalNAc-T14 has been shown to be associated with poor recurrence-free survival and promote cell migration and invasion as well as metastasis through the Wnt signaling in lung cancer [92], IGFBP-3 may interfere with pro-tumorigenic and pro-metastatic GalNAc-T14 signaling by complexing with GalNAc-T14 in certain types of cancer including lung cancer. On the other hand, GalNAc-T14 has been also shown to inhibit IGFBP-3-induced cell proliferation and colony formation in glioblastoma cells. Although overexpression of IGFBP-3 induced expression of Cyclin E, CDK2, and p-ERK1/2, and overexpression of

GalNAc-T14 inhibited these IGFBP-3 effects in glioblastoma cells, no evidence was presented whether direct binding of these two proteins is involved in observed biological outcomes [93].

GRP78, which is also known as immunoglobulin heavy-chain binding immunoglobulin protein (BiP), plays a critical role for endoplasmic reticulum integrity and stress-induced autophagy in mammalian cells [94]. When unfolded or misfolded proteins accumulate in the ER (called ER stress), an unfolded protein response (UPR) is activated through the induction of GRP78 as the first defense response, which, thereby, restores normal function of the ER by attenuating global translation and increasing the folding capacity of the ER [95]. It has been shown that the elevated expression of GRP78 is correlated with cancer malignancy, metastasis, and drug resistance in a variety of cancers, including breast cancer, prostate cancer, lung cancer, and glioma [96]. In line with these findings, GRP78 has been further shown to possess pro-survival and anti-apoptotic properties [97]. Interaction of IGFBP-3 and GRP78 has been identified in human breast cancer cells using a yeast two-hybrid screening [83]. In this study, overexpression of IGFBP-3 showed that IGFBP-3 binding to GRP78 results in the disruption of the GRP78-caspase-7 complex, which, thereby, activates caspase-7, and, subsequently, induces apoptosis in anti-estrogen-resistant breast cancer cells. These findings strongly suggest that IGFBP-3 could sensitize anti-estrogen-resistant breast cancer cells to anti-estrogen such as ICI 182,780 by preventing the anti-apoptotic function of GRP78. On the contrary, IGFBP-3 has been shown to enhance the survival of cells subjected to glucose starvation and hypoxia by inducing autophagy in a GRP78-dependent manner in human breast cancer cells, which suggests that IGFBP-3 may play a key role in mediating an autophagic survival response [84]. Although the biological outcomes of IGFBP-3 are much different depending on the cellular environment, it is clear that the specific interaction of GRP78 and IGFBP-3 is attributed to the observed IGFBP-3 effects.

IGFBP-3 has also been shown to inhibit cell growth and induce apoptosis through an interaction with nuclear proteins such as retinoid X receptor (RXR)-α, retinoic acid receptor (RAR), and Nur77 [98]. RXR is involved in physiological functions of thyroid hormone, steroid hormones, embryonic development, apoptosis, and homeostasis [99–101]. RXR heterodimerizes with Nur77, a nuclear receptor transcription factor, and, thereby, enhances its DNA binding ability and regulates apoptosis in various cancers [102]. IGFBP-3 binds RXR-α and RAR and, subsequently, modulates RAR/RXR and RXR/Nur77 signaling, which, thereby, induces apoptosis [81,82]. It has been further shown that Nur77 translocates to the nucleus and initiates apoptosis in the presence of IGFBP-3 [98]. However, recent studies also showed that IGFBP-3 mutants that failed to translocate to the nucleus and lost binding ability to RXR-α, still induced apoptosis in breast cancer cells [103,104]. This suggests that IGFBP-3 may either utilize multiple mechanisms for its anti-tumor actions depending upon the cellular environment or the observed IGFBP-3 interaction with cytoplasmic/nuclear partners may not represent major IGFBP-3 anti-tumor signaling. It is clear that IGFBP-3 exerts a pro-apoptotic and anti-proliferative IGF/IGF-IR independent actions through multiple mechanisms such as an interaction with the IGFBP-3 receptor and other binding partners on the cell surface and within cells as well as nuclear association. The remainder of this review will focus on the IGFBP-3 receptor TMEM219 in human cancer by mainly providing the evidence to date regarding the IGFBP-3/TMEM219 system as an anti-tumor and anti-metastatic signaling in human cancer.

3. TMEM219 as an IGFBP-3 Specific Receptor

Early studies in the IGF/IGF-IR-independent actions of IGFBP-3 have been shown that IGFBP-3 binding to a cell surface protein is required for its anti-proliferative action in human breast cancer cells and that the central domain of IGFBP-3, which is the least conserved region among IGFBPs 1–6, is responsible for cell surface binding [33,70,105]. Furthermore, IGFBP-3 has been shown to induce apoptosis by activating caspase-8 cleavage, but not cytochrome c release or caspase-9 cleavage involved in the death receptor-mediated apoptotic pathways in MCF-7 breast cancer cells [106]. These findings strongly suggested the existence of an IGFBP-3-specific receptor mediating the direct anti-proliferative and pro-apoptotic effects of IGFBP-3 in a variety of cancer cells.

As an effort to identify a novel cell death receptor specific for IGFBP-3, yeast two-hybrid screening was employed using a cDNA construct encoding amino acid residues 88–148 of the variable central domain of IGFBP-3 as bait against an Hs578T human TNBC cell cDNA library. As a result, a functionally unknown transmembrane protein TMEM219 has been identified as an IGFBP-3 specific interacting protein and later designated as an IGFBP-3 receptor (IGFBP-3R) [15]. TMEM219 consists of four exons comprising the 915-base pair cDNA sequence on chromosome 16q13 and represents a 240-amino acid polypeptide. Further analysis of the deduced amino acid sequence indicated that the 202-residue mature human IGFBP-3R consists of an extracellular domain, a putative single-span transmembrane domain, and a short C-terminal cytoplasmic domain (Figure 2). The extracellular domain contains three potential *N*-glycosylation sites and three phosphorylation sites. The transmembrane domain contains a leucine zipper-like heptad repeat pattern of amino acids that appear to involve dimerization/oligomerization of the membrane proteins. This very unique leucine zipper sequence is also present in the single-span transmembrane domain of the erythropoietin receptor and the discoidin domain receptor [106,107]. Additionally, IGFBP-3R activates caspase-8-induced apoptosis in unconventional ways: (1) IGFBP-3R and inactive procaspase-8 is pre-complexed at the resting stage, and IGFBP-3 binding to IGFBP-3R releases procaspase-8, and, thereby, activates caspase-8-dependent apoptosis, and (2) IGFBP-3R complexes with procasepase-8 without involvement of a typical death domain (DD) sequence. The DD sequence in the intracellular portion of the receptor is required to form a death-inducing signaling complex (DISC) by recruiting adaptor proteins (FADD) and procaspase-8 after receptor activation in various death receptors such as the TNF-α receptor, TNF-related apoptosis-inducing ligand receptor 1 (TRAIL-R1/DR4), TRAIL-R2 (APO-2/DR5), and CD95 (Fas, APO-1). However, similar to IGFBP-3R, a few other proteins have been shown to interact with caspase-8 and induce apoptosis despite the lack of a DD sequence [108,109]. IGFBP-3R is located in both the plasma membrane and cytoplasm, but not in the nucleus of the cancer cells. This cell surface IGFBP-3R interacts specifically with IGFBP-3 but not with other high-affinity IGFBPs, activates procaspase-8, and mediates IGFBP-3-induced apoptosis in many different types of cancer cells and tumor suppression in both prostate and breast cancer xenograft mouse models. Further knockdown of IGFBP-3R attenuates IGFBP-3-induced caspase activities and apoptosis, whereas its overexpression elicited the opposite effects [15,65,110,111]. These findings clearly indicate that IGFBP-3R (TMEM219) is a bona fide IGFBP-3 receptor and mediates anti-tumor activities of IGFBP-3.

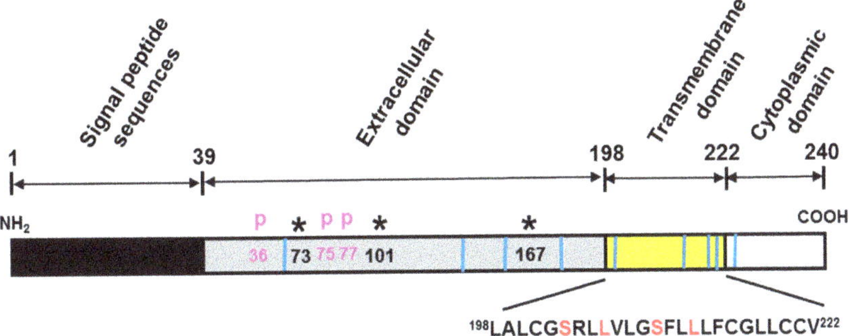

Figure 2. Structure of human IGFBP-3R (TMEM219). The 202-residue mature IGFBP-3R, omitting the 38-residue signal peptide, is comprised of three domains: extracellular, transmembrane, and cytoplasmic domains. Extracellular domain contains three potential *N*-glycosylation sites (residues 73, 101, 167) and three potential phosphorylation sites (S36, T75, T77). The single-span transmembrane domain contains a leucine zipper-like heptad repeat pattern characteristic of leucine zipper interaction domains. The letters in red correspond to a-type and d-type interfacial residues in leucine zipper interaction domains [15]. The vertical blue lines represent nine cysteine residues.

In addition, IGFBP-3 has been shown to suppress tumor-induced NF-κB activity via activation of caspase-8 and caspase-3/7 in an IGF/IGF-IR-dependent manner in prostate cancer cells [111]. IGFBP-3 suppresses NF-κB activity in a unique way. It exerts caspase-induced degradation of IκBα and NF-κB, but not other components such as IKK. IGFBP-3 also inhibited the expression of NF-κB-regulated factors such as VEGF, IL-8, ICAM-1, and VCAM-1. This inhibitory action of IGFBP-3 was IGF/IGF-IR-independent since the IGFBP-3 mutant devoid of IGF binding affinity had a similar inhibitory effect. Furthermore, IGFBP-3R has been shown to be responsible for IGFBP-3-induced suppression of NF-κB activity in cancer cells. These findings indicate that IGFBP-3 in addition to inducing apoptosis, also suppresses tumor-induced NF-κB activity, and, thereby, enhances the inhibition of tumor growth, angiogenesis, invasion, metastasis, and chemoresistance [111].

Recent reports further explored the therapeutic potential of the IGFBP-3/IGFBP-3R axis in cancer by developing an IGFBP-3R agonistic monoclonal antibody (mAb) [112]. It has been shown that activation of IGFBP-3R by IGFBP-3 and IGFBP-3R agonistic mAb inhibits cell growth by inducing apoptosis and by tumor-induced NF-κB activity specifically in cancer cells, but not in normal cells. At present, for the cancer cell, specific pro-apoptotic properties of IGFBP-3 and IGFBP-3R agonistic mAb are not fully elucidated. However, a few potential mechanisms can be speculated based on the findings in tumor-specific targeting of the death receptor (DR)-4 and DR-5 agonistic mAb therapy despite the presence of DR-4/DR-5 in normal cells [113–115]. These include: (1) decreased level of the cell surface DR-4/DR-5 in normal cells compared to cancer cells, (2) differential expression of unknown intracellular inhibitor(s) of apoptosis downstream of caspase-8, and (3) changes in apoptotic potency due to different glycosylation patterns of DRs.

In addition, IGFBP-3R agonistic mAb lost anti-proliferative effects in IGFBP-3R knockout cells. These in vitro data indicate that IGFBP-3R is indispensable for anti-tumor functions of IGFBP-3 and IGFBP-3R agonistic mAb in a variety of cancer cells. Further anti-tumor and anti-metastatic effects of IGFBP-3R agonistic mAb have been shown in vivo using MDA231 TNBC and patient-derived TNBC xenograft models [112]. Taken together, these findings provide evidence that IGFBP-3R (TMEM219) is a bona fide IGFBP-3 receptor and a potential target for cancer therapy.

4. Clinical Insights of IGFBP-3/IGFBP-3R (TMEM219) System in Cancer

Although IGFBP-3 may utilize multiple mechanisms for its anti-tumor actions, current findings suggest that the IGFBP-3/IGFBP-3R axis may constitute a novel anti-tumor/anti-metastatic signaling pathway and a novel potential therapeutic target in cancer. However, since limited knowledge is available on clinicopathologic significance and prognostic value of the IGFBP-3/IGFBP-3R system, the remainder of this review will focus on its clinical significance using data mining and analyses of publicly available databases including The Cell Index (CELLX) database (http://cellx.sourceforge.net) and The Cancer Genome Atlas (TCGA) database (https://portal.gdc.cancer.gov). Log2-transformed RSEM (RNA-Seq by Expectation-Maximization) [116] gene expression values were obtained.

4.1. IGFBP-3 and TMEM219 Gene Expression in Tumor and Normal Samples

Analysis of IGFBP-3 expression identified highly variable expression levels among different types of cancer as well as normal tissues (Figure 3). Further differential expression of IGFBP-3 was analyzed in cancers and the counterpart normal tissues where at least 12 normal samples were available (Table 1).

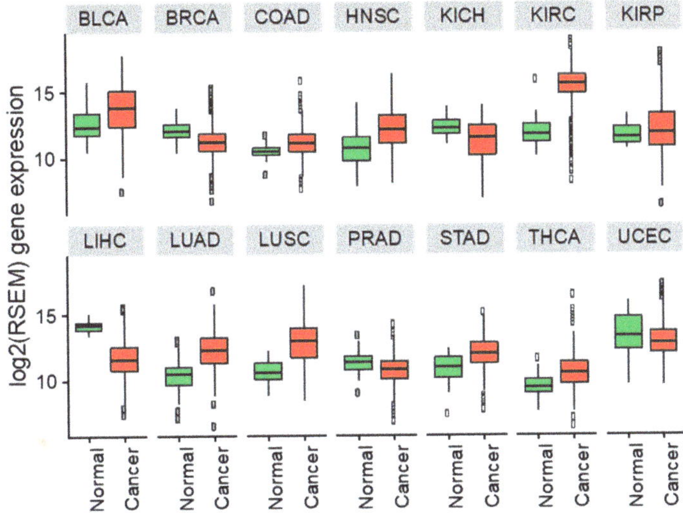

Figure 3. Cancer vs. normal IGFBP-3 log2 RSEM gene expression boxplots (all cancer). Description of abbreviations presented in Table 1.

Table 1. Differential expression of IGFBP-3. "Mean log2(RSEM)"—average gene expression in normal and cancer tissues, respectively. *t*-test *p*-values are shown.

TCGA ID	Description	log2 Fold Change	*p*-Value	Mean log2(RSEM) Normal	Mean log2(RSEM) Cancer
	Upregulated in Tumor				
KIRC	Kidney renal clear cell carcinoma	3.44	$3.11\ 10^{-53}$	11.86	15.30
LUSC	Lung squamous cell carcinoma	2.26	$1.22\ 10^{-28}$	10.61	12.86
LUAD	Lung adenocarcinoma	1.94	$9.28\ 10^{-21}$	10.31	12.26
HNSC	Head and Neck squamous cell carcinoma	1.32	$2.29\ 10^{-7}$	10.82	12.14
STAD	Stomach adenocarcinoma	1.18	$8.67\ 10^{-7}$	10.82	12.00
THCA	Thyroid carcinoma	1.11	$4.85\ 10^{-16}$	9.46	10.57
BLCA	Bladder urothelial carcinoma	1.05	$3.57\ 10^{-3}$	12.56	13.61
COAD	Colon adenocarcinoma	0.65	$4.01\ 10^{-9}$	10.45	11.10
KIRP	Kidney renal papillary cell carcinoma	0.34	$5.19\ 10^{-2}$	11.69	12.03
	Downregulated in Tumor				
LIHC	Liver hepatocellular carcinoma	−2.53	$4.17\ 10^{-81}$	14.18	11.65
KICH	Kidney Chromophobe	−1.07	$6.80\ 10^{-5}$	12.32	11.25
BRCA	Breast invasive carcinoma	−0.82	$9.74\ 10^{-24}$	12.04	11.23
PRAD	Prostate adenocarcinoma	−0.56	$1.47\ 10^{-5}$	11.27	10.72
UCEC	Uterine Corpus Endometrial Carcinoma	−0.43	$1.53\ 10^{-1}$	13.29	12.86

Increased IGFBP-3 expression was observed in kidney renal clear cell carcinoma (log2 fold change +3.44), lung squamous cell carcinoma (log2 fold change +2.26), lung adenocarcinoma (log2 fold change +1.94), head and neck squamous cell carcinoma (log2 fold change +1.32), stomach adenocarcinoma (log2 fold change +1.18), thyroid carcinoma (log2 fold change +1.11), bladder urothelial carcinoma (log2 fold change +1.05), colon adenocarcinoma (log2 fold change +0.65), and kidney renal papillary cell carcinoma (log2 fold change +0.34). On the other hand, decreased expression of IGFBP-3 was observed in liver hepatocellular carcinoma (log2 fold change −2.53), kidney chromophobe (log2 fold change −1.07), breast invasive carcinoma (log2 fold change −0.82), prostate adenocarcinoma (log2 fold change −0.56), and uterine corpus endometrial carcinoma (log2 fold change −0.43).

In the same datasets, TMEM219 expression also showed variable expression patterns, but had less variation when compared to IGFBP-3 among different types of cancer as well as normal tissues (Figure 4). Analysis of differential expression of TMEM219 revealed significant increased TMEM219 expression in 6 out of 14 tumors (Table 2). These include kidney renal papillary cell carcinoma (log2 fold change +0.53), thyroid carcinoma (log2 fold change +0.40), breast invasive carcinoma (log2 fold change +0.39), kidney renal clear cell carcinoma (log2 fold change +0.29), bladder urothelial carcinoma (log2 fold change +0.27), and uterine corpus endometrial carcinoma (log2 fold change +0.22). On the contrary, decreased TMEM219 expression was observed in lung squamous cell carcinoma (log2 fold change −0.74), stomach adenocarcinoma (log2 fold change −0.42), colon adenocarcinoma (log2 fold change −0.36), head and neck squamous cell carcinoma (log2 fold change −0.32), lung adenocarcinoma (log2 fold change −0.12), and kidney chromophobe (log2 fold change −0.10). In summary, these results suggest IGFBP-3 as a diagnostic biomarker and TMEM219 as a therapeutic target in certain types of tumors. Importantly, TMEM219 agonists may represent a novel therapy for tumors with significantly lower expression of IGFBP-3 but not TMEM219 compared to the counterpart normal tissues, such as breast invasive carcinoma, uterine corpus endometrial carcinoma, liver hepatocellular carcinoma, and prostate adenocarcinoma.

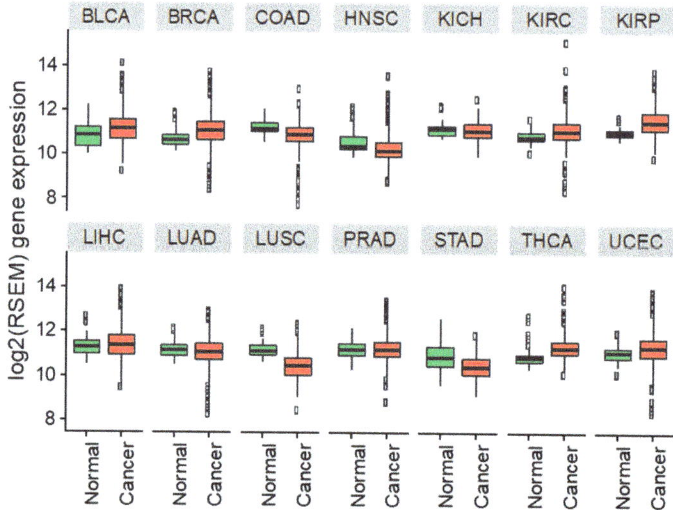

Figure 4. Cancer vs. normal TMEM219 log2 RSEM gene expression boxplots (all cancer). Description of abbreviations presented in Table 2.

Table 2. Differential expression of TMEM219. "Mean log2 (RSEM)"—average gene expression in normal and cancer tissues, respectively. *t*-test *p*-values are shown.

TCGA ID	Description	log2 Fold Change	*p*-Value	Mean log2(RSEM) Normal	Mean log2(RSEM) Cancer
	Upregulated in Tumor				
KIRP	Kidney renal papillary cell carcinoma	0.53	3.16×10^{-14}	11.07	11.60
THCA	Thyroid carcinoma	0.40	2.77×10^{-8}	10.94	11.34
BRCA	Breast invasive carcinoma	0.39	1.00×10^{-19}	10.71	11.10
KIRC	Kidney renal clear cell carcinoma	0.29	4.30×10^{-11}	10.88	11.18
BLCA	Bladder urothelial carcinoma	0.27	1.08×10^{-1}	10.92	11.19
UCEC	Uterine Corpus Endometrial Carcinoma	0.22	4.47×10^{-3}	11.10	11.32
LIHC	Liver hepatocellular carcinoma	0.09	2.11×10^{-1}	11.30	11.39
PRAD	Prostate adenocarcinoma	0.08	2.52×10^{-1}	11.18	11.26
	Downregulated in Tumor				
LUSC	Lung squamous cell carcinoma	−0.74	6.56×10^{-23}	11.18	10.45
STAD	Stomach adenocarcinoma	−0.42	5.93×10^{-4}	10.86	10.44
COAD	Colon adenocarcinoma	−0.36	1.47×10^{-7}	11.25	10.89
HNSC	Head and neck squamous cell carcinoma	−0.32	9.00×10^{-4}	10.63	10.30
LUAD	Lung adenocarcinoma	−0.12	3.97×10^{-2}	11.18	11.06
KICH	Kidney Chromophobe	−0.10	3.20×10^{-1}	11.24	11.14

4.2. Pan-Cancer Survival Effect of IGFBP-3 and TMEM219

To investigate the effect of IGFBP-3 and TMEM219 expression on survival in clinical settings, the RNA-seq data from TCGA was analyzed (Figure 5). Gene expression data summarized as RSEM values were obtained using the TCGA2STAT R package v.1.2, along with the corresponding clinical annotations. Data for each of the 34 cancers were obtained separately. The data were log2-transformed and analyzed using Kaplan-Meier curves and the Cox proportional hazard model. Each gene of interest was analyzed for its effect on survival by separating patients into high/low expression subgroups. The scanning approach KaplanScan, used on the R2 Genomics web portal [117], was used to estimate the best gene expression cutoff that separates high/low expression subgroups with differential survival (R code modified from Reference [118]). In addition to survival analysis across all cancers, further survival analysis was performed within clinical subgroups of specific cancers, e.g., in "race-black or African-American" subgroup. *p*-values were corrected for multiple testing using the Benjamin-Hochberg (FDR) method [119] and reported throughout unless noted otherwise. Only subgroups with >40 patients were considered. This approach allowed us to understand the effect of IGFBP3 and TMEM219 expression on the level of individual cancers and in specific population subgroups.

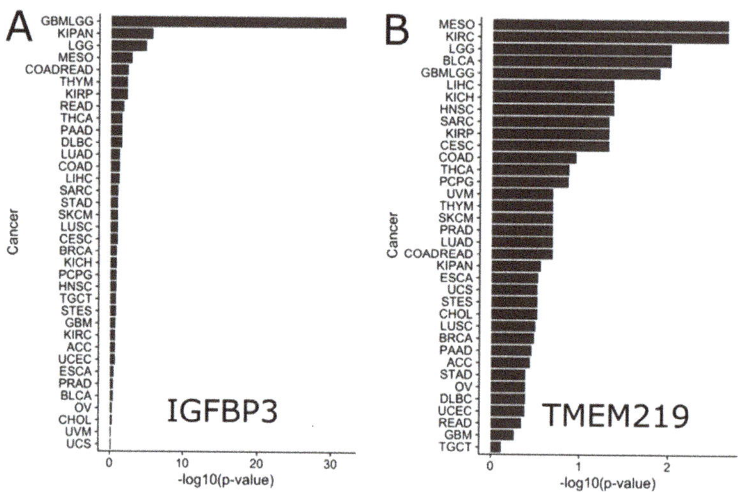

Figure 5. Survival effect of IGFBP-3 (**A**) and TMEM219 (**B**) in various cancers. Larger size of each bar corresponds to the more significant effect on survival.

Analysis of the survival effect of IGFBP-3 expression in all cancers identified its highly significant effect on survival in glioma (FDR = $1.51 \cdot 10^{-32}$ (Hazard Ratio, HR = 4.39)) (Figures 5A and 6A). Survival in pan-kidney cohort (KICH+KIRC+KIRP), lower grade glioma, mesothelioma, colorectal adenocarcinoma was similarly affected by IGFBP-3 to a lesser extent (FDR = $1.74 \cdot 10^{-6}$ (HR = 2.73), $1.25 \cdot 10^{-5}$ (HR = 2.36), $1.22 \cdot 10^{-3}$ (HR = 2.98), $3.87 \cdot 10^{-3}$ (HR = 2.20), respectively) (Figure 6B,C). On the other hand, higher IGFBP-3 was suggestive of better survival outcome in lymphoid neoplasm diffuse large B-cell lymphoma (FDR = $2.58 \cdot 10^{-2}$ (HR = 0.14)), breast cancer ($1.20 \cdot 10^{-1}$ (HR = 0.74)), prostate adenocarcinoma ($3.43 \cdot 10^{-1}$ (HR = 0.51)), cholangiocarcinoma ($4.84 \cdot 10^{-1}$ (HR = 0.57)), bladder urothelial carcinoma ($3.63 \cdot 10^{-1}$ (HR = 0.84)), and uterine carcinosarcoma ($6.65 \cdot 10^{-1}$ (HR = 0.860)) (Figure 6D–F).

In summary, these results suggest IGFBP-3 as a prognostic biomarker in glioma, mesothelioma, kidney, and colorectal cancers with lower expression suggestive of better survival outcome, whereas diffuse large B-cell lymphoma, cholangiocarcinoma, bladder urothelial carcinoma, uterine carcinosarcoma, breast, and prostate cancer with higher expression suggestive of better survival outcome. Of note, the observed dichotomy of IGFBP-3 expression and patients' survival in various cancers may be attributed to other factors such as IGF-1/IGF-2 expression, IGFBP-3 polymorphism status, tumor suppressor p53 family status, tumor metabolic characteristics, and others. In addition, functional IGFBP-3 protein levels in circulation or in tumor and ratio of IGF-I and IGFBP-3 in circulation should be further factored to interpret the TCGA data.

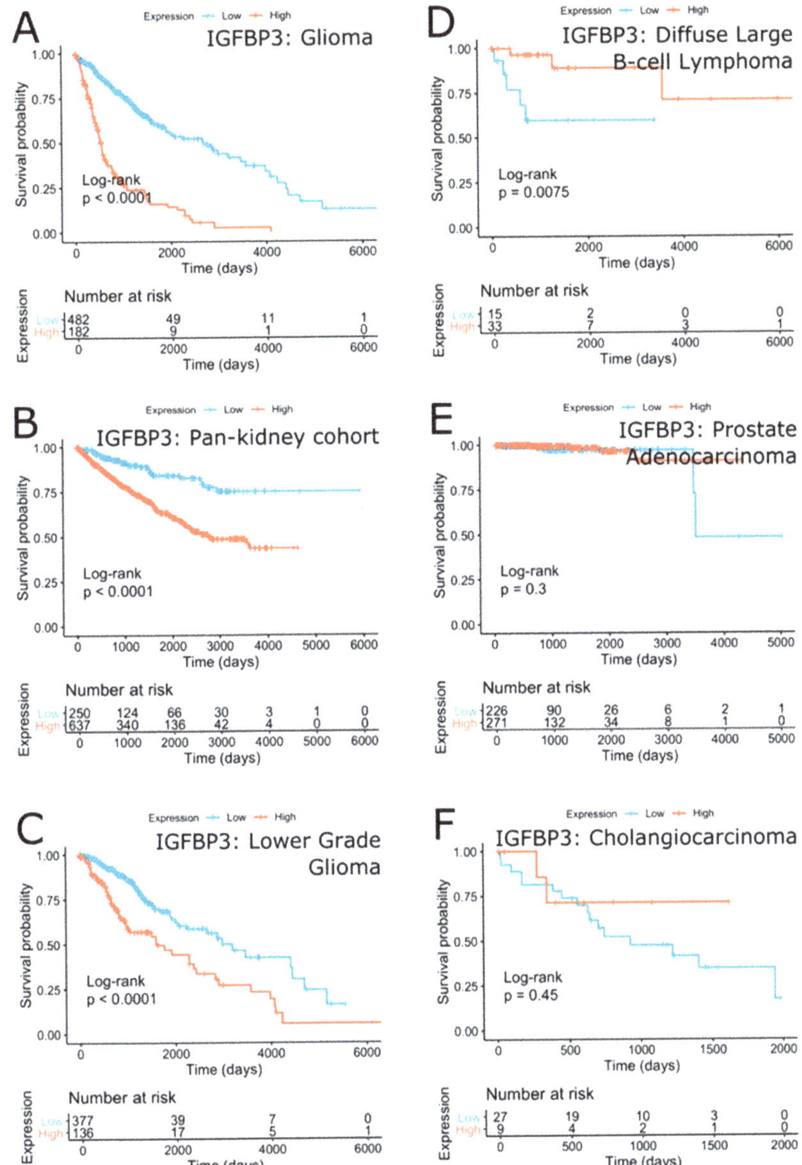

Figure 6. Kaplan-Meier plots of top cancers with survival outcomes separated by the expression of IGFBP-3 with a negative correlation (**A–C**) and a positive correlation (**D–F**). Unadjusted *p*-values are shown.

The TMEM219 survival effect was less significant than that of IGFBP-3 (Figure 5B). Nevertheless, the lower expression of the TMEM219 gene was associated with survival in kidney renal clear cell carcinoma (FDR = $2.14 \cdot 10^{-3}$ (HR = 1.85)), glioma (FDR = $1.27 \cdot 10^{-2}$ (HR = 1.56)), lymphoid neoplasm diffuse large B-cell lymphoma (FDR = 4.110.54)), head and neck squamous cell carcinoma (FDR = 4.17E-2 (HR = 1.43)) and pan-kidney cohort (KICH+KIRC+KIRP) (FDR = 2.76E-1 (HR = 1.2)) (Figure 7A–C). On the contrary, the higher expression of TMEM219 was better for survival in mesothelioma (FDR =

2.14E-3 (HR = 0.39)), lower grade glioma (FDR = 9.54E-3 (HR =0.54)), prostate adenocarcinoma (FDR = $2.02 \cdot 10^{-1}$ (HR = 0.18)), thyroid carcinoma (FDR = $1.332 \cdot 10^{-1}$ (HR = 0.39)) and bladder urothelial carcinoma (FDR = $9.54 \cdot 10^{-3}$ (HR = 0.61)) (Figure 7D–F).

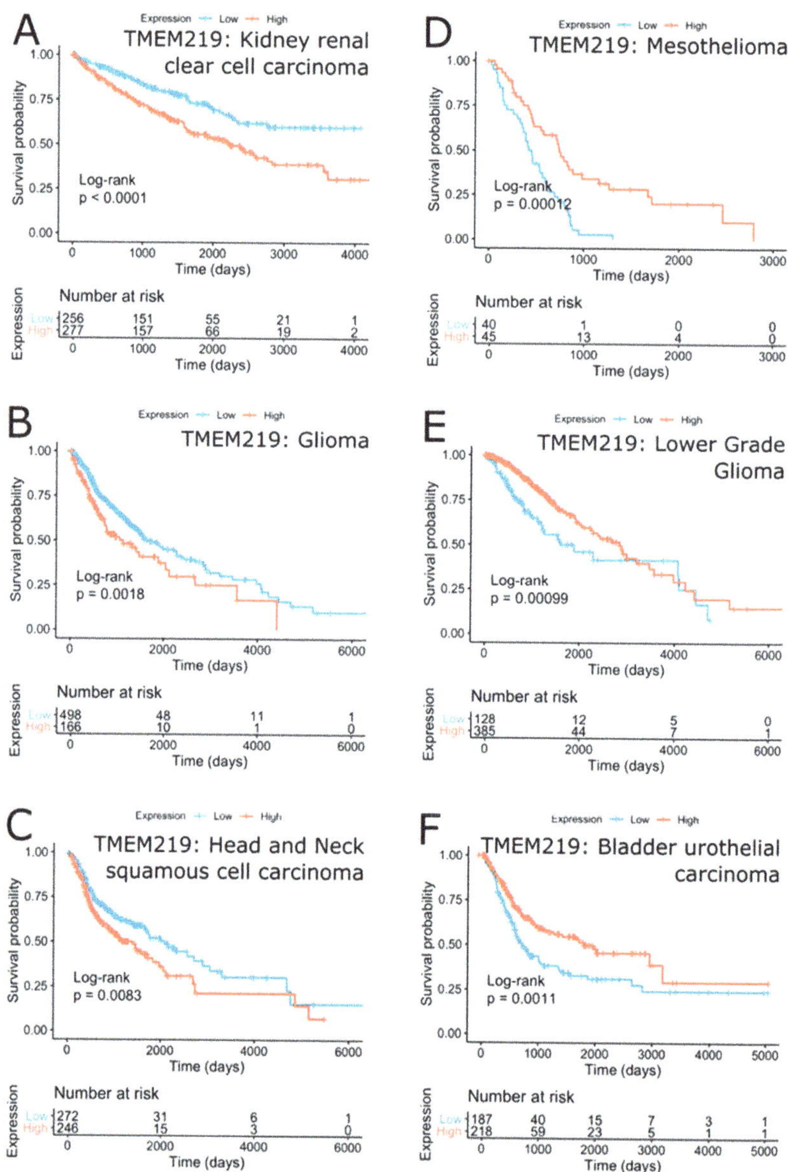

Figure 7. Kaplan-Meier plots of top cancers with survival outcomes separated by the expression of TMEM219 with a negative correlation (**A–C**) and a positive correlation (**D–F**). Unadjusted *p*-values are shown.

In addition to the cancer types expected to be affected by the IGFBP-3/TMEM219 system, bladder urothelial carcinoma and head and neck squamous cell carcinoma appear to be significantly associated with TMEM219 but not with IGFBP-3 expression. Higher TMEM219 expression was associated with better survival in bladder urothelial carcinoma (FDR = $9.54 \cdot 10^{-3}$ (HR = 0.61)), while the reverse was true for head and neck squamous cell carcinoma (FDR = $4.17 \cdot 10^{-2}$ (HR = 1.43)). On the other hand, the expression of TMEM219 was not significantly associated with survival in breast cancer, while IGFBP-3 expression was positively associated with survival outcome (Figure 8). These results indicate that the effect of TMEM219 expression on survival is less pronounced and highly cancer-specific.

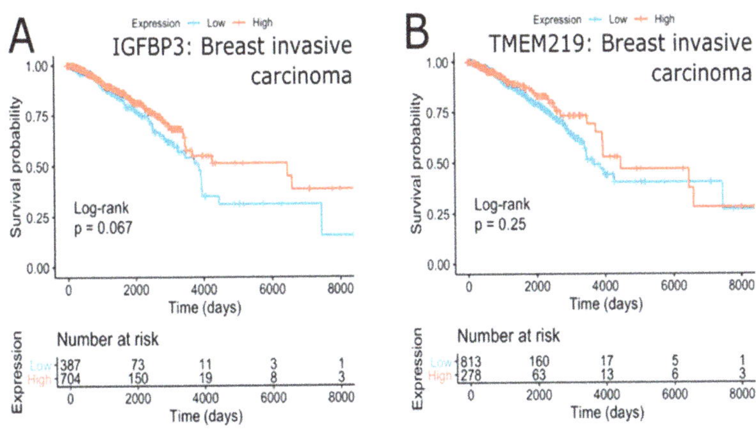

Figure 8. Kaplan-Meier plots of breast cancer cohort with survival outcomes separated by the expression of IGFBP-3 (**A**) and TMEM219 (**B**). Unadjusted *p*-values are shown.

4.3. Survival Effect of IGFBP-3 and TMEM219 in Clinical Subcategories

By taking advantage of the availability of clinical annotations, survival analysis of the effect of IGFBP-3 and TMEM219 expression in clinical subgroups, e.g., "race-black or African-American" was further performed. Similar to all cancer analyses, *p*-values were corrected for multiple testing across all tested subgroups in a given cancer. The advantage of such analyses is that they provide detailed insights into the effect of IGFBP-3 and TMEM219 in different subgroups of patients. The disadvantage is that some subgroups have an insufficient number of patients, e.g., in the "race-black or African-American" subgroup, which limits the cross-cancer comparisons.

Given the high significance of IGFBP-3 gene expression on survival outcome in glioma (FDR = $1.51 \cdot 10^{-32}$ (HR = 4.39)), it was unsurprising that IGFBP-3 expression affected survival in nearly all glioma subgroups (FDR < $2.90 \cdot 10^{-2}$), with lower expression being associated with better survival outcome. Similarly, all subgroups in lower grade glioma were significantly associated with IGFBP-3 expression, with lower expression indicative of better survival outcome (FDR < $5.82 \cdot 10^{-2}$). Similar results were observed for clinical subgroups in mesothelioma, the pan-kidney cohort, rectum adenocarcinoma, colorectal adenocarcinoma, and colon adenocarcinoma cancers, where low expression of IGFBP-3 was similarly associated with better survival outcome. These results confirm previous observations that the expression of IGFBP-3 may affect survival in glioma, mesothelioma, kidney, and colorectal cancers.

Further analyses of the effect of IGFBP-3 and TMEM219 expression in specific clinical subgroups revealed that kidney renal papillary cell carcinoma is the only cancer where the expression of both IGFBP-3 and TMEM219 is marginally associated with survival in "race-black or the African-American" subgroup. Lower IGFBP-3 expression was beneficial for survival (FDR = $5.84 \cdot 10^{-2}$ (HR = 5.12)), while higher TMEM219 expression was associated with better survival in the "race-black or African-American" subgroup (FDR = $6.28 \cdot 10^{-2}$ (HR = 0.16)), Figure 9. These results suggest the importance of the IGFBP-3/TMEM219 system in kidney renal papillary cell carcinoma in the "race-black or African-American" subgroup.

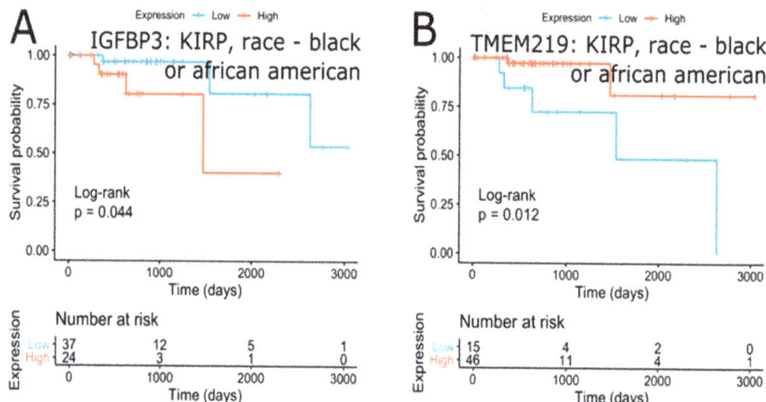

Figure 9. Survival effect of IGFBP3 (**A**) and TMEM219 (**B**) in "race-black or African-American." Kidney renal papillary cell carcinoma. Unadjusted p-values are shown.

4.4. Survival Effect of IGFBP-3 and TMEM219 in Breast Cancer

Since the expression of IGFBP-3 and TMEM219 was not significantly associated with survival outcome in breast cancer, it is possible that heterogeneity of the disease may prevent the detection of significant associations. Consequent analysis of the survival effect of IGFBP-3 and TMEM219 in clinical subgroups of the breast cancer cohort revealed that clinical subgroup annotated as "histological type-Infiltrating Lobular Carcinoma" show marginally significant association of IGFBP-3 (FDR = $2.01 \cdot 10^{-1}$ (HR = 0.36)) and TMEM219 (($3.96 \cdot 10^{-2}$ (HR = 0.25)) with the survival outcome (Figure 10B,C). For both genes, high expression was associated with a better prognosis. These results suggest that targeting the IGFBP-3/TMEM219 system in patients diagnosed with infiltrating lobular carcinoma, which is the second most common type of breast cancer, may be beneficial. Other clinical subgroups of breast cancer patients included "breast_carcinoma_surgical_procedure_name-Modified Radical Mastectomy" (IGFBP-3, FDR = $1.26 \cdot 10^{-1}$ (HR = 0.44), Figure 10A), "lab_proc_her2_neu_immunohistochemistry_receptor_status-Equivocal" (TMEM219, FDR = $3.27 \cdot 10^{-1}$ (HR = 3.29)). Of note were race-specific survival effects with high expression of IGFBP-3 being beneficial in the "race-black or African-American" subgroup (FDR = $2.01 \cdot 10^{-1}$ (HR = 0.42), Figure 10E) and TMEM219 high expression being beneficial in the "race-Asian" subgroup (FDR = $1.16 \cdot 10^{-1}$ (HR = 0.00), Figure 10D). Confirming our previous observations, the survival benefits of IGFBP-3 expression in breast cancer were consistently associated with high IGFBP-3 expression, while the effect of TMEM219 was more diverse and subgroup-specific.

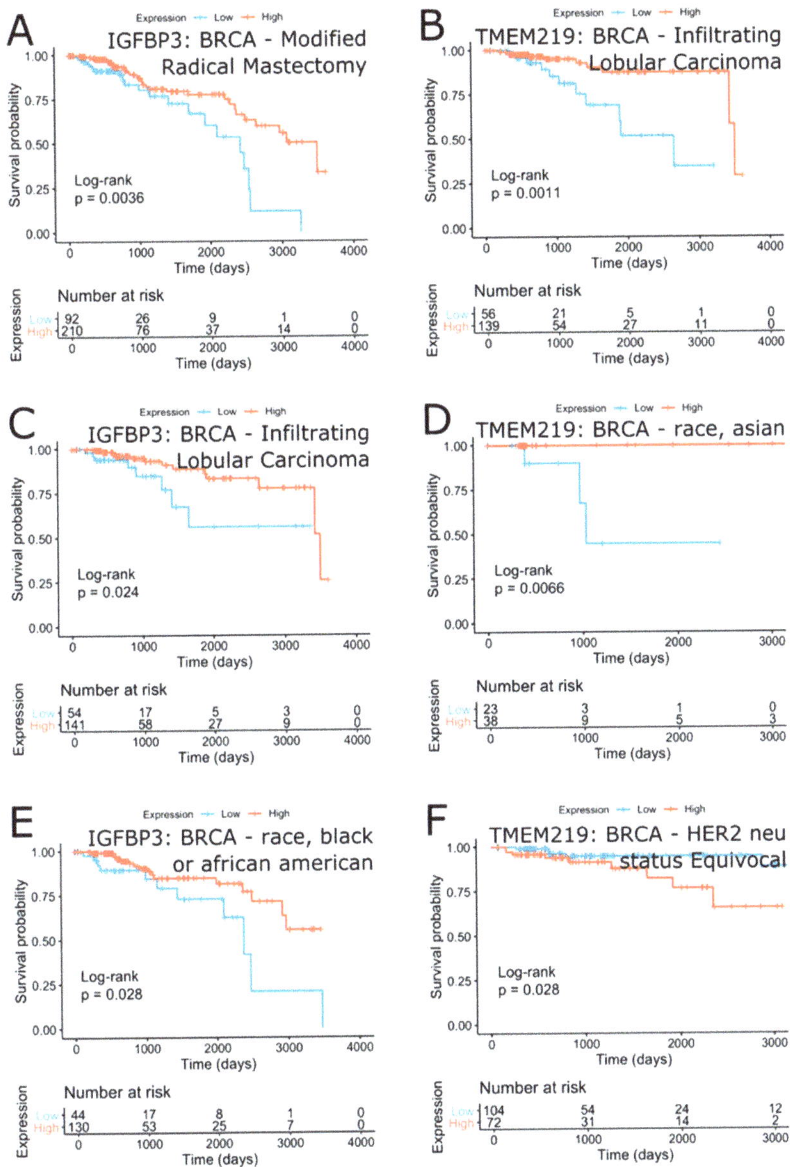

Figure 10. Survival effect of IGFBP-3 (**A**,**C**,**E**) and TMEM219 (**B**,**D**,**F**) in selected breast cancer subgroups. Unadjusted *p*-values are shown.

5. Conclusions

IGFBP-3 is a multifunctional protein and is involved in the pathophysiology of a variety of human diseases such as cancer, diabetes, fatty liver disease, ischemia, and Alzheimer's disease. Apart from the IGF/IGF-IR-dependent actions, IGFBP-3 exerts multiple biological activities through the IGF/IGF-IR-independent actions by interacting with distinct interacting proteins on the cell surface or within the cell. Much attention was given to identify a putative receptor for IGFBP-3 since early studies have demonstrated the anti-tumor function of IGFBP-3 in cancer. As described in this review, a few membrane proteins have been identified as "a putative IGFBP-3 receptor" and further characterized their functions with potential underlying mechanisms in cancer cells. Among them, TMEM219 appears to be the most critical IGFBP-3 receptor mediating anti-tumor and anti-metastatic activities of IGFBP-3. Given the fact that IGFBP-3/IGFBP-3R (TMEM219) axis is impaired and shown to have great impact on the survival outcome in specific cancers, IGFBP-3 and TMEM219 may serve as new diagnostic and prognostic biomarkers in specific cancers. Importantly, IGFBP-3R (TMEM219) agonists, in particular TMEM219 agonistic mAbs, are very attractive cancer therapeutics since these agonists would exhibit no other biological activities of IGFBP-3 induced by the interaction with other binding partners. Further characterization of specific gene regulation by TMEM219 activation and its crosstalk with other key signaling pathways will open a new avenue to treat many different types of cancer as a targeted monotherapy and a combination therapy with other chemotherapies.

Author Contributions: Writing, Q.C., M.D., and Y.O. Review and editing, M.D. and Y.O. All authors have read and agreed to the published version of the manuscript.

Funding: This work was supported, in part, by NIH-NCI Grant R21CA216685 (to Y.O).

Conflicts of Interest: The authors declare no conflicts of interest.

References

1. Wood, W.I.; Cachianes, G.; Henzel, W.J.; Winslow, G.A.; Spencer, S.A.; Hellmiss, R.; Martin, J.L.; Baxter, R.C. Cloning and expression of the growth hormone-dependent insulin-like growth factor-binding protein. *Mol. Endocrinol.* **1988**, *2*, 1176–1185. [CrossRef] [PubMed]
2. Hwa, V.; Oh, Y.; Rosenfeld, R.G. The insulin-like growth factor-binding protein (IGFBP) superfamily. *Endocr. Rev.* **1999**, *20*, 761–787. [CrossRef] [PubMed]
3. Jogie-Brahim, S.; Feldman, D.; Oh, Y. Unraveling insulin-like growth factor binding protein-3 actions in human disease. *Endocr. Rev.* **2009**, *30*, 417–437. [CrossRef] [PubMed]
4. Yamada, P.M.; Lee, K.W. Perspectives in mammalian IGFBP-3 biology: Local vs. systemic action. *Am. J. Physiol. Cell Physiol.* **2009**, *296*, C954–C976. [CrossRef]
5. Baxter, R.C. IGF binding proteins in cancer: Mechanistic and clinical insights. *Nat. Rev. Cancer* **2014**, *14*, 329–341. [CrossRef]
6. Johnson, M.A.; Firth, S.M. IGFBP-3: A cell fate pivot in cancer and disease. *Growth Horm. IGF Res.* **2014**, *24*, 164–173. [CrossRef]
7. Buckway, C.K.; Wilson, E.M.; Ahlsen, M.; Bang, P.; Oh, Y.; Rosenfeld, R.G. Mutation of three critical amino acids of the N-terminal domain of IGF-binding protein-3 essential for high affinity IGF binding. *J. Clin. Endocrinol. Metab.* **2001**, *86*, 4943–4950. [CrossRef]
8. Firth, S.M.; Baxter, R.C. Characterisation of recombinant glycosylation variants of insulin-like growth factor binding protein-3. *J. Endocrinol.* **1999**, *160*, 379–387. [CrossRef]
9. Hoeck, W.G.; Mukku, V.R. Identification of the major sites of phosphorylation in IGF binding protein-3. *J. Cell Biochem.* **1994**, *56*, 262–273. [CrossRef]
10. Bang, P.; Fielder, P.J. Human pregnancy serum contains at least two distinct proteolytic activities with the ability to degrade insulin-like growth factor binding protein-3. *Endocrinology* **1997**, *138*, 3912–3917. [CrossRef]
11. Firth, S.M.; Baxter, R.C. Cellular actions of the insulin-like growth factor binding proteins. *Endocr. Rev.* **2002**, *23*, 824–854. [CrossRef]
12. Schedlich, L.J.; Nilsen, T.; John, A.P.; Jans, D.A.; Baxter, R.C. Phosphorylation of insulin-like growth factor binding protein-3 by deoxyribonucleic acid-dependent protein kinase reduces ligand binding and enhances nuclear accumulation. *Endocrinology* **2003**, *144*, 1984–1993. [CrossRef] [PubMed]
13. Cobb, L.J.; Liu, B.; Lee, K.W.; Cohen, P. Phosphorylation by DNA-dependent protein kinase is critical for apoptosis induction by insulin-like growth factor binding protein-3. *Cancer Res.* **2006**, *66*, 10878–10884. [CrossRef] [PubMed]
14. Fowlkes, J.L.; Thrailkill, K.M.; Serra, D.M.; Suzuki, K.; Nagase, H. Matrix metalloproteinases as insulin-like growth factor binding protein-degrading proteinases. *Prog. Growth Factor Res.* **1995**, *6*, 255–263. [CrossRef]

15. Ingermann, A.R.; Yang, Y.F.; Han, J.; Mikami, A.; Garza, A.E.; Mohanraj, L.; Fan, L.; Idowu, M.; Ware, J.L.; Kim, H.S.; et al. Identification of a novel cell death receptor mediating IGFBP-3-induced anti-tumor effects in breast and prostate cancer. *J. Biol. Chem.* **2010**, *285*, 30233–30246. [CrossRef] [PubMed]

16. Spencer, E.M.; Chan, K. A 3-dimensional model for the insulin-like growth factor binding proteins (IGFBPs); supporting evidence using the structural determinants of the IGF binding site on IGFBP-3. *Prog. Growth Factor Res.* **1995**, *6*, 209–214. [CrossRef]

17. Ho, P.J.; Baxter, R.C. Characterization of truncated insulin-like growth factor-binding protein-2 in human milk. *Endocrinology* **1997**, *138*, 3811–3818. [CrossRef]

18. Devi, G.R.; Yang, D.H.; Rosenfeld, R.G.; Oh, Y. Differential effects of insulin-like growth factor (IGF)-binding protein-3 and its proteolytic fragments on ligand binding, cell surface association, and IGF-I receptor signaling. *Endocrinology* **2000**, *141*, 4171–4179. [CrossRef]

19. Galanis, M.; Firth, S.M.; Bond, J.; Nathanielsz, A.; Kortt, A.A.; Hudson, P.J.; Baxter, R.C. Ligand-binding characteristics of recombinant amino- and carboxyl-terminal fragments of human insulin-like growth factor-binding protein-3. *J. Endocrinol.* **2001**, *169*, 123–133. [CrossRef]

20. Fowlkes, J.L.; Serra, D.M. Characterization of glycosaminoglycan-binding domains present in insulin-like growth factor-binding protein-3. *J. Biol. Chem.* **1996**, *271*, 14676–14679. [CrossRef]

21. Smith, E.P.; Lu, L.; Chernausek, S.D.; Klein, D.J. Insulin-like growth factor-binding protein-3 (IGFBP-3) concentration in rat Sertoli cell-conditioned medium is regulated by a pathway involving association of IGFBP-3 with cell surface proteoglycans. *Endocrinology* **1994**, *135*, 359–364. [CrossRef] [PubMed]

22. Campbell, P.G.; Durham, S.K.; Hayes, J.D.; Suwanichkul, A.; Powell, D.R. Insulin-like growth factor-binding protein-3 binds fibrinogen and fibrin. *J. Biol. Chem.* **1999**, *274*, 30215–30221. [CrossRef] [PubMed]

23. Baxter, R.C.; Martin, J.L.; Beniac, V.A. High molecular weight insulin-like growth factor binding protein complex. Purification and properties of the acid-labile subunit from human serum. *J. Biol. Chem.* **1989**, *264*, 11843–11848. [PubMed]

24. Huq, A.; Singh, B.; Meeker, T.; Mascarenhas, D. The metal-binding domain of IGFBP-3 selectively delivers therapeutic molecules into cancer cells. *Anticancer Drugs* **2009**, *20*, 21–31. [CrossRef]

25. Schedlich, L.J.; Le Page, S.L.; Firth, S.M.; Briggs, L.J.; Jans, D.A.; Baxter, R.C. Nuclear import of insulin-like growth factor-binding protein-3 and -5 is mediated by the importin beta subunit. *J. Biol. Chem.* **2000**, *275*, 23462–23470. [CrossRef]

26. Lee, K.W.; Liu, B.; Ma, L.; Li, H.; Bang, P.; Koeffler, H.P.; Cohen, P. Cellular internalization of insulin-like growth factor binding protein-3: Distinct endocytic pathways facilitate re-uptake and nuclear localization. *J. Biol. Chem.* **2004**, *279*, 469–476. [CrossRef]

27. Mohan, S.; Baylink, D.J.; Pettis, J.L. Insulin-like growth factor (IGF)-binding proteins in serum—do they have additional roles besides modulating the endocrine IGF actions? *J. Clin. Endocrinol. Metab.* **1996**, *81*, 3817–3820. [CrossRef]

28. Rajaram, S.; Baylink, D.J.; Mohan, S. Insulin-like growth factor-binding proteins in serum and other biological fluids: Regulation and functions. *Endocr. Rev.* **1997**, *18*, 801–831.

29. Clemmons, D.R. Role of insulin-like growth factor binding proteins in controlling IGF actions. *Mol. Cell Endocrinol.* **1998**, *140*, 19–24. [CrossRef]

30. Kelley, K.M.; Oh, Y.; Gargosky, S.E.; Gucev, Z.; Matsumoto, T.; Hwa, V.; Ng, L.; Simpson, D.M.; Rosenfeld, R.G. Insulin-like growth factor-binding proteins (IGFBPs) and their regulatory dynamics. *Int. J. Biochem. Cell Biol.* **1996**, *28*, 619–637. [CrossRef]

31. Ferry, R.J., Jr.; Cerri, R.W.; Cohen, P. Insulin-like growth factor binding proteins: New proteins, new functions. *Horm. Res.* **1999**, *51*, 53–67. [CrossRef] [PubMed]

32. Collett-Solberg, P.F.; Cohen, P. Genetics, chemistry, and function of the IGF/IGFBP system. *Endocrine* **2000**, *12*, 121–136. [CrossRef]

33. Oh, Y.; Muller, H.L.; Lamson, G.; Rosenfeld, R.G. Insulin-like growth factor (IGF)-independent action of IGF-binding protein-3 in Hs578T human breast cancer cells. Cell surface binding and growth inhibition. *J. Biol. Chem.* **1993**, *268*, 14964–14971. [PubMed]

34. Rajah, R.; Valentinis, B.; Cohen, P. Insulin-like growth factor (IGF)-binding protein-3 induces apoptosis and mediates the effects of transforming growth factor-beta1 on programmed cell death through a p53-and IGF-independent mechanism. *J. Biol. Chem.* **1997**, *272*, 12181–12188. [CrossRef] [PubMed]

35. Leal, S.M.; Huang, S.S.; Huang, J.S. Interactions of high affinity insulin-like growth factor-binding proteins with the type V transforming growth factor-beta receptor in mink lung epithelial cells. *J. Biol. Chem.* **1999**, *274*, 6711–6717. [CrossRef]

36. Wu, H.B.; Kumar, A.; Tsai, W.C.; Mascarenhas, D.; Healey, J.; Rechler, M.M. Characterization of the inhibition of DNA synthesis in proliferating mink lung epithelial cells by insulin-like growth factor binding protein-3. *J. Cell Biochem.* **2000**, *77*, 288–297. [CrossRef]

37. Ikonen, M.; Liu, B.; Hashimoto, Y.; Ma, L.; Lee, K.W.; Niikura, T.; Nishimoto, I.; Cohen, P. Interaction between the Alzheimer's survival peptide humanin and insulin-like growth factor-binding protein 3 regulates cell survival and apoptosis. *Proc. Natl. Acad. Sci. USA* **2003**, *100*, 13042–13047. [CrossRef]

38. Chen, X.; Ferry, R.J., Jr. Novel actions of IGFBP-3 on intracellular signaling pathways of insulin-secreting cells. *Growth Horm. IGF Res.* **2006**, *16*, 41–48. [CrossRef]

39. Chang, K.H.; Chan-Ling, T.; McFarland, E.L.; Afzal, A.; Pan, H.; Baxter, L.C.; Shaw, L.C.; Caballero, S.; Sengupta, N.; Li Calzi, S.; et al. IGF binding protein-3 regulates hematopoietic stem cell and endothelial precursor cell function during vascular development. *Proc. Natl. Acad. Sci. USA* **2007**, *104*, 10595–10600. [CrossRef]

40. Grimberg, A.; Coleman, C.M.; Burns, T.F.; Himelstein, B.P.; Koch, C.J.; Cohen, P.; El-Deiry, W.S. p53-Dependent and p53-independent induction of insulin-like growth factor binding protein-3 by deoxyribonucleic acid damage and hypoxia. *J. Clin. Endocrinol. Metab.* **2005**, *90*, 3568–3574. [CrossRef]

41. Lofqvist, C.; Chen, J.; Connor, K.M.; Smith, A.C.; Aderman, C.M.; Liu, N.; Pintar, J.E.; Ludwig, T.; Hellstrom, A.; Smith, L.E. IGFBP3 suppresses retinopathy through suppression of oxygen-induced vessel loss and promotion of vascular regrowth. *Proc. Natl. Acad. Sci. USA* **2007**, *104*, 10589–10594. [CrossRef] [PubMed]

42. Williams, A.C.; Smartt, H.; H-Zadeh, A.; MMacfarlane, M.; Paraskeva, C.; Collard, T.J. Insulin-like growth factor binding protein 3 (IGFBP-3) potentiates TRAIL-induced apoptosis of human colorectal carcinoma cells through inhibition of NF-kappaB. *Cell Death Differ.* **2007**, *14*, 137–145. [CrossRef]

43. Mehta, H.H.; Gao, Q.; Galet, C.; Paharkova, V.; Wan, J.; Said, J.; Sohn, J.J.; Lawson, G.; Cohen, P.; Cobb, L.J.; et al. IGFBP-3 is a metastasis suppression gene in prostate cancer. *Cancer Res.* **2011**, *71*, 5154–5163. [CrossRef] [PubMed]

44. Mohanraj, L.; Kim, H.S.; Li, W.; Cai, Q.; Kim, K.E.; Shin, H.J.; Lee, Y.J.; Lee, W.J.; Kim, J.H.; Oh, Y. IGFBP-3 inhibits cytokine-induced insulin resistance and early manifestations of atherosclerosis. *PLoS ONE* **2013**, *8*, e55084. [CrossRef] [PubMed]

45. Min, H.K.; Maruyama, H.; Jang, B.K.; Shimada, M.; Mirshahi, F.; Ren, S.; Oh, Y.; Puri, P.; Sanyal, A.J. Suppression of IGF binding protein-3 by palmitate promotes hepatic inflammatory responses. *FASEB J.* **2016**, *30*, 4071–4082. [CrossRef] [PubMed]

46. Kim, M.; Cai, Q.; Oh, Y. Therapeutic potential of alpha-1 antitrypsin in human disease. *Ann. Pediatr. Endocrinol. Metab.* **2018**, *23*, 131–135. [CrossRef]

47. Arab, J.P.; Cabrera, D.; Sehrawat, T.S.; Jalan-Sakrikar, N.; Verma, V.K.; Simonetto, D.; Cao, S.; Yaqoob, U.; Leon, J.; Freire, M.; et al. Hepatic stellate cell activation promotes alcohol-induced steatohepatitis through Igfbp3 and SerpinA12. *J. Hepatol.* **2020**. [CrossRef]

48. Gucev, Z.S.; Oh, Y.; Kelley, K.M.; Rosenfeld, R.G. Insulin-like growth factor binding protein 3 mediates retinoic acid-and transforming growth factor beta2-induced growth inhibition in human breast cancer cells. *Cancer Res.* **1996**, *56*, 1545–1550.

49. Hembree, J.R.; Agarwal, C.; Beard, R.L.; Chandraratna, R.A.; Eckert, R. Retinoid X receptor-specific retinoids inhibit the ability of retinoic acid receptor-specific retinoids to increase the level of insulin-like growth factor binding protein-3 in human ectocervical epithelial cells. *Cancer Res.* **1996**, *56*, 1794–1799.

50. Rozen, F.; Zhang, J.; Pollak, M. Antiproliferative action of tumor necrosis factor-alpha on MCF-7 breastcancer cells is associated with increased insulin-like growth factor binding protein-3 accumulation. *Int. J. Oncol.* **1998**, *13*, 865–869. [CrossRef]

51. Colston, K.W.; Perks, C.M.; Xie, S.P.; Holly, J.M. Growth inhibition of both MCF-7 and Hs578T human breast cancer cell lines by vitamin D analogues is associated with increased expression of insulin-like growth factor binding protein-3. *J. Mol. Endocrinol.* **1998**, *20*, 157–162. [CrossRef] [PubMed]

52. Huynh, H.; Yang, X.; Pollak, M. Estradiol and antiestrogens regulate a growth inhibitory insulin-like growth factor binding protein 3 autocrine loop in human breast cancer cells. *J. Biol. Chem.* **1996**, *271*, 1016–1021. [CrossRef] [PubMed]

53. Martin, J.L.; Baxter, R.C. Transforming growth factor-beta stimulates production of insulin-like growth factor-binding protein-3 by human skin fibroblasts. *Endocrinology* **1991**, *128*, 1425–1433. [CrossRef] [PubMed]

54. Oh, Y.; Muller, H.L.; Ng, L.; Rosenfeld, R.G. Transforming growth factor-beta-induced cell growth inhibition in human breast cancer cells is mediated through insulin-like growth factor-binding protein-3 action. *J. Biol. Chem.* **1995**, *270*, 13589–13592. [CrossRef] [PubMed]

55. Zhou, W.; Wang, S.; Ying, Y.; Zhou, R.; Mao, P. miR-196b/miR-1290 participate in the antitumor effect of resveratrol via regulation of IGFBP3 expression in acute lymphoblastic leukemia. *Oncol. Rep.* **2017**, *37*, 1075–1083. [CrossRef] [PubMed]

56. Xia, Y.; Jin, L.; Zhang, B.; Xue, H.; Li, Q.; Xu, Y. The potentiation of curcumin on insulin-like growth factor-1 action in MCF-7 human breast carcinoma cells. *Life Sci.* **2007**, *80*, 2161–2169. [CrossRef] [PubMed]

57. Vijayababu, M.R.; Arunkumar, A.; Kanagaraj, P.; Arunakaran, J. Effects of quercetin on insulin-like growth factors (IGFs) and their binding protein-3 (IGFBP-3) secretion and induction of apoptosis in human prostate cancer cells. *J. Carcinog.* **2006**, *5*, 10. [CrossRef]

58. Buckbinder, L.; Talbott, R.; Velasco-Miguel, S.; Takenaka, I.; Faha, B.; Seizinger, B.R.; Kley, N. Induction of the growth inhibitor IGF-binding protein 3 by p53. *Nature* **1995**, *377*, 646–649. [CrossRef]

59. Bourdon, J.C.; Deguin-Chambon, V.; Lelong, J.C.; Dessen, P.; May, P.; Debuire, B.; May, E. Further characterisation of the p53 responsive element–identification of new candidate genes for trans-activation by p53. *Oncogene* **1997**, *14*, 85–94. [CrossRef]

60. Ludwig, R.L.; Bates, S.; Vousden, K.H. Differential activation of target cellular promoters by p53 mutants with impaired apoptotic function. *Mol. Cell Biol.* **1996**, *16*, 4952–4960. [CrossRef]

61. Barbieri, C.E.; Perez, C.A.; Johnson, K.N.; Ely, K.A.; Billheimer, D.; Pietenpol, J.A. IGFBP-3 is a direct target of transcriptional regulation by DeltaNp63alpha in squamous epithelium. *Cancer Res.* **2005**, *65*, 2314–2320. [CrossRef] [PubMed]

62. Chang, Y.S.; Wang, L.; Liu, D.; Mao, L.; Hong, W.K.; Khuri, F.R.; Lee, H.Y. Correlation between insulin-like growth factor-binding protein-3 promoter methylation and prognosis of patients with stage I non-small cell lung cancer. *Clin. Cancer Res.* **2002**, *8*, 3669–3675. [PubMed]

63. Wiley, A.; Katsaros, D.; Fracchioli, S.; Yu, H. Methylation of the insulin-like growth factor binding protein-3 gene and prognosis of epithelial ovarian cancer. *Int. J. Gynecol. Cancer* **2006**, *16*, 210–218. [CrossRef] [PubMed]

64. Ibanez de Caceres, I.; Cortes-Sempere, M.; Moratilla, C.; Machado-Pinilla, R.; Rodriguez-Fanjul, V.; Manguan-Garcia, C.; Cejas, P.; Lopez-Rios, F.; Paz-Ares, L.; de CastroCarpeno, J.; et al. IGFBP-3 hypermethylation-derived deficiency mediates cisplatin resistance in non-small-cell lung cancer. *Oncogene* **2010**, *29*, 1681–1690. [CrossRef] [PubMed]

65. Harada, A.; Jogie-Brahim, S.; Oh, Y. Tobacco specific carcinogen 4-(methylnitrosamino)-1-(3-pyridyl)-1-butanone suppresses a newly identified anti-tumor IGFBP-3/IGFBP-3R system in lung cancer cells. *Lung Cancer* **2013**, *80*, 270–277. [CrossRef] [PubMed]

66. Perez-Carbonell, L.; Balaguer, F.; Toiyama, Y.; Egoavil, C.; Rojas, E.; Guarinos, C.; Andreu, M.; Llor, X.; Castells, A.; Jover, R.; et al. IGFBP3 methylation is a novel diagnostic and predictive biomarker in colorectal cancer. *PLoS ONE* **2014**, *9*, e104285. [CrossRef] [PubMed]

67. Pernia, O.; Belda-Iniesta, C.; Pulido, V.; Cortes-Sempere, M.; Rodriguez, C.; Vera, O.; Soto, J.; Jimenez, J.; Taus, A.; Rojo, F.; et al. Methylation status of IGFBP-3 as a useful clinical tool for deciding on a concomitant radiotherapy. *Epigenetics* **2014**, *9*, 1446–1453. [CrossRef] [PubMed]

68. Perks, C.M.; Holly, J.M. Epigenetic regulation of insulin-like growth factor binding protein-3 (IGFBP-3) in cancer. *J. Cell Commun. Signal* **2015**, *9*, 159–166. [CrossRef]

69. Fu, T.; Pappou, E.P.; Guzzetta, A.A.; Calmon Mde, F.; Sun, L.; Herrera, A.; Li, F.; Wolfgang, C.L.; Baylin, S.B.; Iacobuzio-Donahue, C.A.; et al. IGFBP-3 Gene Methylation in Primary Tumor Predicts Recurrence of Stage II Colorectal Cancers. *Ann. Surg.* **2016**, *263*, 337–344. [CrossRef] [PubMed]

70. Oh, Y.; Muller, H.L.; Pham, H.; Rosenfeld, R.G. Demonstration of receptors for insulin-like growth factor binding protein-3 on Hs578T human breast cancer cells. *J. Biol. Chem.* **1993**, *268*, 26045–26048.

71. Huang, S.S.; Ling, T.Y.; Tseng, W.F.; Huang, Y.H.; Tang, F.M.; Leal, S.M.; Huang, J.S. Cellular growth inhibition by IGFBP-3 and TGF-beta1 requires LRP-1. *FASEB J.* **2003**, *17*, 2068–2081. [CrossRef] [PubMed]

72. Mishra, S.; Raz, A.; Murphy, L.J. Insulin-like growth factor binding protein-3 interacts with autocrine motility factor/phosphoglucose isomerase (AMF/PGI) and inhibits the AMF/PGI function. *Cancer Res.* **2004**, *64*, 2516–2522. [CrossRef] [PubMed]

73. Weinzimer, S.A.; Gibson, T.B.; Collett-Solberg, P.F.; Khare, A.; Liu, B.; Cohen, P. Transferrin is an insulin-like growth factor-binding protein-3 binding protein. *J. Clin. Endocrinol. Metab.* **2001**, *86*, 1806–1813. [CrossRef] [PubMed]

74. Leal, S.M.; Liu, Q.; Huang, S.S.; Huang, J.S. The type V transforming growth factor beta receptor is the putative insulin-like growth factor-binding protein 3 receptor. *J. Biol. Chem.* **1997**, *272*, 20572–20576. [CrossRef]

75. Huang, S.S.; Leal, S.M.; Chen, C.L.; Liu, I.H.; Huang, J.S. Identification of insulin receptor substrate proteins as key molecules for the TbetaR-V/LRP-1-mediated growth inhibitory signaling cascade in epithelial and myeloid cells. *FASEB J.* **2004**, *18*, 1719–1721. [CrossRef]

76. Lee, S.H.; Takahashi, M.; Honke, K.; Miyoshi, E.; Osumi, D.; Sakiyama, H.; Ekuni, A.; Wang, X.; Inoue, S.; Gu, J.; et al. Loss of core fucosylation of low-density lipoprotein receptor-related protein-1 impairs its function, leading to the upregulation of serum levels of insulin-like growth factor-binding protein 3 in Fut8-/-mice. *J. Biochem.* **2006**, *139*, 391–398. [CrossRef]

77. Mangasser-Stephan, K.; Gressner, A.M. Molecular and functional aspects of latent transforming growth factor-beta binding protein: Just a masking protein? *Cell Tissue Res.* **1999**, *297*, 363–370. [CrossRef]

78. Gui, Y.; Murphy, L.J. Interaction of insulin-like growth factor binding protein-3 with latent transforming growth factor-beta binding protein-1. *Mol. Cell Biochem.* **2003**, *250*, 189–195. [CrossRef]

79. Rifkin, D.B. Latent transforming growth factor-beta (TGF-beta) binding proteins: Orchestrators of TGF-beta availability. *J. Biol. Chem.* **2005**, *280*, 7409–7412. [CrossRef]

80. Nwosu, Z.C.; Ebert, M.P.; Dooley, S.; Meyer, C. Caveolin-1 in the regulation of cell metabolism: A cancer perspective. *Mol. Cancer* **2016**, *15*, 71. [CrossRef]

81. Oufattole, M.; Lin, S.W.; Liu, B.; Mascarenhas, D.; Cohen, P.; Rodgers, B.D. Ribonucleic acid polymerase II binding subunit 3 (Rpb3), a potential nuclear target of insulin-like growth factor binding protein-3. *Endocrinology* **2006**, *147*, 2138–2146. [CrossRef] [PubMed]

82. Wu, C.; Yao, G.; Zou, M.; Chen, G.; Wang, M.; Liu, J.; Wang, J.; Xu, D. N-Acetylgalactosaminyltransferase 14, a novel insulin-like growth factor binding protein-3 binding partner. *Biochem. Biophys. Res. Commun.* **2007**, *357*, 360–365. [CrossRef] [PubMed]

83. Li, C.; Harada, A.; Oh, Y. IGFBP-3 sensitizes antiestrogen-resistant breast cancer cells through interaction with GRP78. *Cancer Lett.* **2012**, *325*, 200–206. [CrossRef] [PubMed]

84. Grkovic, S.; O'Reilly, V.C.; Han, S.; Hong, M.; Baxter, R.C.; Firth, S.M. IGFBP-3 binds GRP78, stimulates autophagy and promotes the survival of breast cancer cells exposed to adverse microenvironments. *Oncogene* **2013**, *32*, 2412–2420. [CrossRef]

85. Liu, B.; Lee, H.Y.; Weinzimer, S.A.; Powell, D.R.; Clifford, J.L.; Kurie, J.M.; Cohen, P. Direct functional interactions between insulin-like growth factor-binding protein-3 and retinoid X receptor-alpha regulate transcriptional signaling and apoptosis. *J. Biol. Chem.* **2000**, *275*, 33607–33613. [CrossRef] [PubMed]

86. Schedlich, L.J.; O'Han, M.K.; Leong, G.M.; Baxter, R.C. Insulin-like growth factor binding protein-3 prevents retinoid receptor heterodimerization: Implications for retinoic acid-sensitivity in human breast cancer cells. *Biochem. Biophys. Res. Commun.* **2004**, *314*, 83–88. [CrossRef]

87. Lee, K.W.; Cobb, L.J.; Paharkova-Vatchkova, V.; Liu, B.; Milbrandt, J.; Cohen, P. Contribution of the orphan nuclear receptor Nur77 to the apoptotic action of IGFBP-3. *Carcinogenesis* **2007**, *28*, 1653–1658. [CrossRef]

88. Moreno-Santos, I.; Castellano-Castillo, D.; Lara, M.F.; Fernandez-Garcia, J.C.; Tinahones, F.J.; Macias-Gonzalez, M. IGFBP-3 Interacts with the Vitamin D Receptor in Insulin Signaling Associated with Obesity in Visceral Adipose Tissue. *Int. J. Mol. Sci.* **2017**, *18*, 2349. [CrossRef]

89. Chan, S.S.; Schedlich, L.J.; Twigg, S.M.; Baxter, R.C. Inhibition of adipocyte differentiation by insulin-like growth factor-binding protein-3. *Am. J. Physiol. Endocrinol. Metab.* **2009**, *296*, E654–E663. [CrossRef]

90. Njomen, E.; Evans, H.G.; Gedara, S.H.; Heyl, D.L. Humanin Peptide Binds to Insulin-Like Growth Factor-Binding Protein 3 (IGFBP3) and Regulates Its Interaction with Importin-beta. *Protein. Pept. Lett.* **2015**, *22*, 869–876. [CrossRef]

91. Wang, H.; Tachibana, K.; Zhang, Y.; Iwasaki, H.; Kameyama, A.; Cheng, L.; Guo, J.; Hiruma, T.; Togayachi, A.; Kudo, T.; et al. Cloning and characterization of a novel UDP-GalNAc:polypeptide

N-acetylgalactosaminyltransferase, pp-GalNAc-T14. *Biochem. Biophys. Res. Commun.* **2003**, *300*, 738–744. [CrossRef]

92. Kwon, O.S.; Oh, E.; Park, J.R.; Lee, J.S.; Bae, G.Y.; Koo, J.H.; Kim, H.; Choi, Y.L.; Choi, Y.S.; Kim, J.; et al. GalNAc-T14 promotes metastasis through Wnt dependent HOXB9 expression in lung adenocarcinoma. *Oncotarget* **2015**, *6*, 41916–41928. [CrossRef]

93. Yang, J.; Hu, Y.; Wu, J.; Kong, S. Effects of IGFBP-3 and GalNAc-T14 on proliferation and cell cycle of glioblastoma cells and its mechanism. *J. Pharm. Pharmacol.* **2020**, *72*, 218–226. [CrossRef] [PubMed]

94. Li, J.; Ni, M.; Lee, B.; Barron, E.; Hinton, D.R.; Lee, A.S. The unfolded protein response regulator GRP78/BiP is required for endoplasmic reticulum integrity and stress-induced autophagy in mammalian cells. *Cell Death Differ.* **2008**, *15*, 1460–1471. [CrossRef] [PubMed]

95. Bertolotti, A.; Zhang, Y.; Hendershot, L.M.; Harding, H.P.; Ron, D. Dynamic interaction of BiP and ER stress transducers in the unfolded-protein response. *Nat. Cell Biol.* **2000**, *2*, 326–332. [CrossRef]

96. Wang, M.; Wey, S.; Zhang, Y.; Ye, R.; Lee, A.S. Role of the unfolded protein response regulator GRP78/BiP in development, cancer, and neurological disorders. *Antioxid. Redox. Signal* **2009**, *11*, 2307–2316. [CrossRef] [PubMed]

97. Jing, G.; Wang, J.J.; Zhang, S.X. ER stress and apoptosis: A new mechanism for retinal cell death. *Exp. Diabetes Res.* **2012**, *2012*, 589589. [CrossRef]

98. Lee, K.W.; Ma, L.; Yan, X.; Liu, B.; Zhang, X.K.; Cohen, P. Rapid apoptosis induction by IGFBP-3 involves an insulin-like growth factor-independent nucleomitochondrial translocation of RXRalpha/Nur77. *J. Biol. Chem.* **2005**, *280*, 16942–16948. [CrossRef]

99. Kastner, P.; Mark, M.; Chambon, P. Nonsteroid nuclear receptors: What are genetic studies telling us about their role in real life? *Cell* **1995**, *83*, 859–869. [CrossRef]

100. Mangelsdorf, D.J.; Evans, R.M. The RXR heterodimers and orphan receptors. *Cell* **1995**, *83*, 841–850. [CrossRef]

101. Germain, P.; Staels, B.; Dacquet, C.; Spedding, M.; Laudet, V. Overview of nomenclature of nuclear receptors. *Pharmacol. Rev.* **2006**, *58*, 685–704. [CrossRef] [PubMed]

102. Hazel, T.G.; Nathans, D.; Lau, L.F. A gene inducible by serum growth factors encodes a member of the steroid and thyroid hormone receptor superfamily. *Proc. Natl. Acad. Sci. USA* **1988**, *85*, 8444–8448. [CrossRef] [PubMed]

103. Bhattacharyya, N.; Pechhold, K.; Shahjee, H.; Zappala, G.; Elbi, C.; Raaka, B.; Wiench, M.; Hong, J.; Rechler, M.M. Nonsecreted insulin-like growth factor binding protein-3 (IGFBP-3) can induce apoptosis in human prostate cancer cells by IGF-independent mechanisms without being concentrated in the nucleus. *J. Biol. Chem.* **2006**, *281*, 24588–24601. [CrossRef] [PubMed]

104. Zappala, G.; Elbi, C.; Edwards, J.; Gorenstein, J.; Rechler, M.M.; Bhattacharyya, N. Induction of apoptosis in human prostate cancer cells by insulin-like growth factor binding protein-3 does not require binding to retinoid X receptor-alpha. *Endocrinology* **2008**, *149*, 1802–1812. [CrossRef] [PubMed]

105. Yamanaka, Y.; Fowlkes, J.L.; Wilson, E.M.; Rosenfeld, R.G.; Oh, Y. Characterization of insulin-like growth factor binding protein-3 (IGFBP-3) binding to human breast cancer cells: Kinetics of IGFBP-3 binding and identification of receptor binding domain on the IGFBP-3 molecule. *Endocrinology* **1999**, *140*, 1319–1328. [CrossRef]

106. Ruan, W.; Becker, V.; Klingmuller, U.; Langosch, D. The interface between self-assembling erythropoietin receptor transmembrane segments corresponds to a membrane-spanning leucine zipper. *J. Biol. Chem.* **2004**, *279*, 3273–3279. [CrossRef]

107. Noordeen, N.A.; Carafoli, F.; Hohenester, E.; Horton, M.A.; Leitinger, B. A transmembrane leucine zipper is required for activation of the dimeric receptor tyrosine kinase DDR1. *J. Biol. Chem.* **2006**, *281*, 22744–22751. [CrossRef]

108. Ding, H.F.; Lin, Y.L.; McGill, G.; Juo, P.; Zhu, H.; Blenis, J.; Yuan, J.; Fisher, D.E. Essential role for caspase-8 in transcription-independent apoptosis triggered by p53. *J. Biol. Chem.* **2000**, *275*, 38905–38911. [CrossRef]

109. Zhao, H.; Ross, F.P.; Teitelbaum, S.L. Unoccupied alpha(v)beta3 integrin regulates osteoclast apoptosis by transmitting a positive death signal. *Mol. Endocrinol.* **2005**, *19*, 771–780. [CrossRef]

110. Baxter, R.C.; Martin, J.L. Structure of the Mr 140,000 growth hormone-dependent insulin-like growth factor binding protein complex: Determination by reconstitution and affinity-labeling. *Proc. Natl. Acad. Sci. USA* **1989**, *86*, 6898–6902. [CrossRef]

111. Han, J.; Jogie-Brahim, S.; Harada, A.; Oh, Y. Insulin-like growth factor-binding protein-3 suppresses tumor growth via activation of caspase-dependent apoptosis and cross-talk with NF-kappaB signaling. *Cancer Lett.* **2011**, *307*, 200–210. [CrossRef] [PubMed]

112. Oh, Y. Treatment of Diseases Related to IGFBP-3 and Its Receptor. International Application Number No. PCT/US2017/059244, International Publication Number WO/2018/085252, 11 May 2018.

113. Zhang, X.D.; Nguyen, T.; Thomas, W.D.; Sanders, J.E.; Hersey, P. Mechanisms of resistance of normal cells to TRAIL induced apoptosis vary between different cell types. *FEBS Lett.* **2000**, *482*, 193–199. [CrossRef]

114. Wagner, K.W.; Punnoose, E.A.; Januario, T.; Lawrence, D.A.; Pitti, R.M.; Lancaster, K.; Lee, D.; von Goetz, M.; Yee, S.F.; Totpal, K.; et al. Death-receptor O-glycosylation controls tumor-cell sensitivity to the proapoptotic ligand Apo2L/TRAIL. *Nat. Med.* **2007**, *13*, 1070–1077. [CrossRef] [PubMed]

115. Dubuisson, A.; Micheau, O. Antibodies and Derivatives Targeting DR4 and DR5 for Cancer Therapy. *Antibodies* **2017**, *6*, 16. [CrossRef] [PubMed]

116. Li, B.; Dewey, C. RSEM: Accurate transcript quantification from RNA-Seq data with or without a reference genome. *BMC Bioinform.* **2011**, *12*. [CrossRef] [PubMed]

117. R2: Genomics Analysis and Visualization Platform. Available online: http://r2.amc.nl (accessed on 12 May 2020).

118. Mihaly, Z.; Kormos, M.; Lanczky, A.; Dank, M.; Budczies, J.; Szasz, M.A.; Gyorffy, B. A meta-analysis of gene expression-based biomarkers predicting outcome after tamoxifen treatment in breast cancer. *Breast Cancer Res. Treat.* **2013**, *140*, 219–232. [CrossRef]

119. Benjamini, Y.; Hochberg, Y. Controlling the false discovery rate: A practical and powerful approach to multiple testing. *J. R. Stat. Soc. Ser. B* **1995**, 289–300. [CrossRef]

Review

Targeting the IGF-Axis for Cancer Therapy: Development and Validation of an IGF-Trap as a Potential Drug

Yinhsuan Michely Chen [1,2,†], Shu Qi [2,†], Stephanie Perrino [2], Masakazu Hashimoto [2,3] and Pnina Brodt [1,2,3,4,*]

[1] Department of Medicine, Division of Experimental Medicine, McGill University, Montreal, QC H3A 0G4, Canada; michely.chen@mail.mcgill.ca
[2] The Research Institute of the McGill University Health Center, Montreal, QC H4A 3J1, Canada; shu.qi@mail.mcgill.ca (S.Q.); stephanie.perrino@mail.mcgill.ca (S.P.); oita521@yahoo.co.jp (M.H.)
[3] Department of Surgery, McGill University, Montreal, QC H3A 0G4, Canada
[4] Department of Oncology, McGill University, Montreal, QC H3A 0G4, Canada
*   Correspondence: pnina.brodt@mcgill.ca; Tel.: +1-514-934-1934
†   These authors contributed equally to this work.

Received: 1 April 2020; Accepted: 23 April 2020; Published: 29 April 2020

Abstract: The insulin-like growth factor (IGF)-axis was implicated in cancer progression and identified as a clinically important therapeutic target. Several IGF-I receptor (IGF-IR) targeting drugs including humanized monoclonal antibodies have advanced to phase II/III clinical trials, but to date, have not progressed to clinical use, due, at least in part, to interference with insulin receptor signaling and compensatory signaling by the insulin receptor (IR) isoform A that can bind IGF-II and initiate mitogenic signaling. Here we briefly review the current state of IGF-targeting biologicals, discuss some factors that may be responsible for their poor performance in the clinic and outline the stepwise bioengineering and validation of an IGF-Trap—a novel anti-cancer therapeutic that could bypass these limitations. The IGF-Trap is a heterotetramer, consisting of the entire extracellular domain of the IGF-IR fused to the Fc portion of human IgG_1. It binds human IGF-I and IGF-II with a three-log higher affinity than insulin and could inhibit IGF-IR driven cellular functions such as survival, proliferation and invasion in multiple carcinoma cell models in vitro. In vivo, the IGF-Trap has favorable pharmacokinetic properties and could markedly reduce metastatic outgrowth of colon and lung carcinoma cells in the liver, outperforming IGF-IR and ligand-binding monoclonal antibodies. Moreover, IGF-Trap dose-response profiles correlate with their bio-availability profiles, as measured by the IGF kinase receptor-activation (KIRA) assay, providing a novel, surrogate biomarker for drug efficacy. Our studies identify the IGF-Trap as a potent, safe, anti-cancer therapeutic that could overcome some of the obstacles encountered by IGF-targeting biologicals that have already been evaluated in clinical settings.

Keywords: IGF-I receptor; signaling; targeted therapeutics; IGF-Trap

---

1. Background Information

1.1. The Insulin-Like Growth Factor (IGF)-Axis

The IGF-axis consists of two cell surface receptors (IGF-IR and IGF-IIR), the ligands IGF-I and IGF-II, high affinity binding proteins (IGFBP-1-6) and their proteases (reviewed in [1,2]. IGF-IR shares a 60% sequence homology with the insulin receptor (IR). It is synthesized as a polypeptide precursor that undergoes post translational modification (glycosylation, proteolytic cleavage and dimerization) to form a heterotetramer composed of two α and two β subunits linked by α–α and α–β disulphide bonds. The α subunits are extracellular and contain the ligand binding site, while the β subunits have an extracellular domain, a transmembrane domain and an intracellular portion that contains the tyrosine kinase domain [3].

Upon ligand binding, the tyrosine kinase domain in the β subunit is activated, inducing a conformational change that leads to autophosphorylation at Tyr950 that serves as a docking site for signalling substrates including the insulin receptor substrate (IRS) proteins IRS-1-4, and the activation of PI3-kinase (PI3K)/Akt/ mammalian target of rapamycin (mTOR) and Raf /MEK/ERK signaling. This leads to regulation of cell survival and protein synthesis on one hand, and gene expression, cellular proliferation and differentiation, on the other [4,5].

IGF-IR can also translocate to the nucleus in a ligand-dependent manner following SUMOlyation of three lysine residues on the β-subunit [6]. In the nucleus, IGF-IR can act as a transcriptional co-activator with LEF/TCF, increasing promoter activity of the downstream target genes cyclin D1 and Axin2, upregulating their expression and promoting cell cycle progression [7]. In several human malignancies including clear cell renal cancer, colorectal carcinoma and pediatric glioma, nuclear IGF-IR was associated with advanced disease and adverse prognosis [8–10]. Codony-Servat el al. showed that in colorectal carcinoma cells treated with IGF-IR-blocking antibodies, nuclear translocation increased, suggesting that nuclear sequestration of the receptor could contribute to therapy resistance [9].

1.2. Hybrid Receptors and Crosstalk with Other Receptors

Two IR isoforms, IR-A and IR-B, formed by the alternative splicing of exon 11, have been identified [11]. IR-A is expressed predominantly in embryonic and fetal tissues, in the central nervous system (CNS) and hematopoietic cells and is frequently upregulated in cancer cells, whereas IR-B is expressed mainly in the liver, fat and muscle where it binds insulin with high affinity and mediates its metabolic functions. IR-A can bind IGF-II and insulin with high affinities, and this can initiate mitogenic signaling and tumorigenesis. RNA sequencing data based on analysis of 6943 samples, representing 21 tumor types in the Cancer Genome Atlas, revealed IR-A expression in all tumor types analyzed, and IR-B expression was also detected in many tumor samples. However the IR-A/IR-B ratio is generally in favor of the IR-A isoform in many cancer types including breast, colon, and lung carcinomas (extensively reviewed in [11,12].

Since many cancer cells overexpress both the insulin and IGF-I receptors and due to the high sequence homology between these receptors, hybrid receptors consisting of one insulin αβ hemi-receptor and one IGF-IR αβ hemi-receptor can also form. The IR-A/IGF-IR hybrids bind insulin and both IGF-I and IGF-II with similar high affinities, while IR-B/IGF-IR hybrids bind IGF-I with high affinity, IGF-II with lower affinity and insulin with poor affinity [13]. The specific signaling and functions of the hybrid receptors remain largely unknown, as they can bind and be activated by all three ligands. In a study of human breast carcinoma specimens and cell lines, hybrid receptor levels exceeded those of IGF-IR in a large proportion of specimens and in cultured cells; hybrid receptor autophosphorylation in response to IGF-I exceeded IGF-IR autophosphorylation and could initiate growth signaling, suggesting that these receptors could contribute to ligand mediated signal transduction [14] (reviewed in [15]). In a recent study using inducible chimeric receptors in mammary carcinoma cells, both IGF-IR and the hybrid receptor were found to induce cell proliferation, but only IGF-IR had anti-apoptotic effects [16], suggesting that it activates distinct signaling pathways. The high expression of IR-A in many cancer types and its ability to initiate mitogenic signaling in response to IGF-II, as well as the presence of signaling-competent hybrid receptors may have been a major factor in the outcome of clinical trials for IGF-IR targeting antibodies and other inhibitors and has emerged as a major consideration in the design of IGF-axis targeting drugs.

Furthermore, IGF-IR/IR signaling is part of a complex network of receptor tyrosine kinase (RTK)-initiated pathways. The IGF-IR crosstalks with several RTKs including the epidermal growth factor receptor (EGFR), fibroblast growth factor receptor (FGFR), platelet-derived growth factor receptor (PDGFR) and human epidermal growth factor receptor 2 (HER-2), as well as with the steroid hormone receptors, estrogen receptor (ER) and androgen receptor. For example, crosstalk between ERα and IGF-IR was demonstrated in uterine cells, where IGF-IR signaling could be triggered by estradiol/ER-mediated induction of IGF-I synthesis in vitro. Conversely, ER transcriptional activity could be induced by IGF-I, in an estradiol-independent manner and IGF-I-induced ER transcriptional activity could be induced in the uteri of ovariectomized mice in vivo [17,18]. Crosstalk between IGF-IR and EGFR and compensatory actions between their signaling pathways have been identified as potential resistance mechanisms to drugs that target either of these axes. Thus, treatment of head and neck squamous cell carcinoma and non-small cell lung cancer cells with the fully humanized anti-IGF-IR monoclonal antibody (MAb), Cixutumumab, induced Akt and mammalian target of rapamycin (mTOR) activation, resulting in EGFR, Akt1, and survivin synthesis and EGFR pathway activation. This inter-dependence and compensatory RTK signaling has been identified as another obstacle to successful therapeutic targeting of the IGF-axis.

1.3. Targeting of the IGF-Axis for Cancer Therapy—The Rational

Increased expression of IGF-IR and/or its ligands has been documented in many human malignancies such as lung, breast, colon and prostate carcinoma, glioblastoma and melanoma, and high expression levels were shown to be associated with metastasis, shorter survival and poor prognosis [19], identifying this axis as a target for cancer therapy. High circulating IGF-I levels were identified as a predictive factor in several malignancies including lung, breast, colorectal and prostate carcinoma [20], and thought to contribute to cellular transformation and malignant progression [2]. The important role of circulating IGF-I in cancer development was demonstrated in vivo using liver-specific-IGF-I deficient (LID and iLID) mouse models where decreased mammary tumor incidence and progression and reduced colon adenocarcinoma growth and metastasis were documented [21,22]. This was also observed in other mouse models of reduced circulating IGF-I levels such as the lit/lit mice that have only 10% of normal circulating IGF-I levels due to reduced growth hormone (GH) production and in dw/dw dwarf mice that are deficient in GH and IGF-I production [22,23]. In addition to circulating IGF-I, tissue IGF-I levels that activate IGF-IR signaling in a paracrine or autocrine fashion were also shown to contribute to tumorigenesis in both animal models and human studies [24–26]. In a study of 125 primary non-small cell lung cancer compared to benign pulmonary lesions, high IGF-I and IGF-IR levels were associated with advanced-stage disease and expression of IGF-I correlated with tumor size and poor outcome [27]. However, in contrast to these findings, tumor IGF-I levels were found to be associated with better overall survival in studies of prostate and breast cancer tumors [28,29]. This may reflect the dual role of IGF-I as a proliferation and differentiation factor, depending on the cellular context [29–32]. As these studies were based on immunohistochemical evaluation or gene expression analyses performed on whole tumor tissue, the precise source of IGF-I in these studies cannot be definitively identified. The relative contributions of circulating and local IGF-I levels to malignant progression and the role of IGFBPs in modulating their effects remain an open question with implications for IGF-targeting and patient stratification [33].

The IGF ligands form complexes with six high-affinity IGFBPs that modulate their half-life and bioavailability [34]. Lower circulating IGFBP levels were found to be associated with increased risk for several cancers including premenopausal breast carcinoma, prostate carcinoma, colorectal carcinoma, lung cancer, endometrial cancer and bladder cancer [2]. The identification of all components of the axis as contributors to the development of malignant disease has spurred an intensive effort to design inhibitors and strategies for blockade of IGF-IR signaling. These inhibitors can be broadly divided into drugs that target the receptor (monoclonal antibodies and small molecule tyrosine kinase inhibitors (TKI)) and strategies that reduce ligand bioavailability to the cognate receptor. A brief summary of the experience with these drugs is provided below and in Table 1.

Table 1. Insulin-like growth factor (IGF) targeting strategies: the pre-clinical and clinical experience.

Target	Approach	Advantages	Disadvantages	Reference
IGF- insulin receptor (IR)	Nucleic acid approach	High specificity via mRNA degradation	Toxicity, challenges in drug delivery and uptake Compensatory signaling through IR-A Low translational potential	[35]
	Antibodies	Induce internalization and downregulation of IGF-IR	Adverse effects on glucose metabolism Hyperglycemia activation of IR-A by IGF-II nuclear translocation of IGF-IR Compensatory receptor tyrosine kinase (RTK) signaling	[36,37]
	Bispecific antibodies	Neutralizing two or more targets improved protein stability to oxidative and thermal stress Inhibit compensatory signaling by other RTKs	Steric hindrance large, reduced intra-tumoral penetration	[38,39]
	Tyrosine kinase inhibitors (TKI)	Cross reactivity with IR	Affects metabolic insulin signaling via IR-B hyperglycemia short half-life	[37,40]
IGF- ligands	Antibodies	Block IGF-IR and IR-A activation Low affinity for insulin minimizes adverse effects on glucose metabolism Reduced ligand bioavailability in the serum	Efficacy depends on IGF-IR expression levels Reduced plasma IGF levels may trigger compensatory feedback mechanisms	[41]
	Traps	Block IGF-IR and IR-A activation Low affinity for insulin minimizes adverse effects on glucose metabolism Reduce ligand bioavailability in the serum Fc fusion proteins increase serum half-life	Size may limit diffusion into the tumor site Oligomerization due to disulfide bonds may affect manufacturability Could potentially trigger a compensatory feedback mechanism upon long-term administration	[41,42]

1.3.1. IGF-Targeting for Cancer Management: The Current Landscape and Overall Clinical Experience

Targeting the IGF-I Receptor

Receptor-specific antibodies: IGF-IR antibodies can inhibit signaling by binding to the extracellular α subunits, blocking ligand binding and triggering receptor internalization. Several humanized or fully human neutralizing anti-IGF-IR antibodies have entered clinical trials. Included among them are cixutumumab (IMC-A12-ImClone, New York, NY, USA), Figitumumab (CP-751,871-Pfizer, New York, NY, USA), Dalotuzumab (MK-0646; h7C10-Pierre Fabre (Paris, France) and Merck (Kenilworth, NJ, USA)), ganitumab (AMG 479-Amgen Thousand Oaks, CA, USA), Teprotumumab (R1507-Genmab (Copenhagen, Denmark) and Roche (Basel, Switzerland)), Robatumumab (SCH 717454, 19D12-ImmunoGen (Waltham, MA, USA) and Sanofi (Paris, France)), Istiratumab (MM141-Merrimack (Cambridge, MA, USA)), BIIB022 (Biogen (Cambridge, MA, USA)), and AVE1642 (EM164-Biogen, Cambridge, MA, USA). Unfortunately, the use of most of these drugs in cancer therapy has been discontinued after several obstacles were identified [43]. IGF-IR blocking drugs could cause insulin resistance, hyperinsulinemia and mild hyperglycemia [43]. In addition, the therapeutic responses to the monoclonal antibodies were disappointing, and this was attributed to several potential factors including: (i) a compensatory feedback mechanism that leads to increased IGF production due to increased growth hormone release [44], (ii) IR-A signalling that can be initiated by IGF-II (the main plasma IGF-IR ligand in human) and leads to mitogenic signaling; and (iii) cancer cell resistance due to activation of compensatory RTK signaling [36,45,46]. Several of the anti-IGF-IR antibodies have also been tested in combination with chemotherapy or antibodies to other RTKs [47,48]. Despite pre-clinical data to suggest that these combinations could be effective in targeting resistant tumor subpopulations [49–56], the results of clinical trials have generally been disappointing, resulting in termination due to lack of demonstrable efficacy [43,57]. An exception may be teprotumumab that had a successful phase III clinical trial with thyroid eye diseases and has been U.S. Food and Drug Administration (FDA)-approved for the treatment of Graves' disease [58,59]. To address potential resistance due to activation of other RTKs, bispecific antibodies that target a second kinase have been generated. These include XGFR, a bispecific anti-IGF-IR/EGFR antibody that showed inhibition of tumor growth and enhanced immune activation in pancreatic cancer in vivo [39], and Istiratumab (MM-141) which co-targets IGF-IR and ErbB3. MM-141 was tested in combination with standard of care (SOC) chemotherapy in a phase II clinical trial for pancreatic cancer, but failed to show a survival advantage in comparison to SOC alone [60,61]. Of importance, however, are the reports that specific IGF-targeting drugs were generally well tolerated.

Several small molecule tyrosine kinase inhibitors were also developed to target IGF-IR signaling including Masoprocol (INSM-18, NDGA–InsMed (Bridgewater Township, NJ, USA)), Linsitinib (OSI-906–OSI (Farmingdale, NY, USA)), BMS-754807 (BMS (Montreal, QC, Canada)), AXL1717 (Picropodophyllin- Axelar AB (Solna, Sweden)) and XL-228 (Exelixis (Alameda, CA, USA)). A potential advantage of small TKIs is that they may also inhibit IR-A-initiated signaling due to the high homology between these receptors. However, this is a double-edged sword, as disruption of IR signaling can have deleterious effects on glucose metabolism and lead to hyperinsulinemia and hyperglycemia [36,37]. To date, no IGF-IR TKI has advanced to clinical use.

Targeting the IGF-Ligands

An alternative approach to blocking IGF-IR signaling is targeting the ligands to reduce their bioavailability to the receptor. An advantage of this approach is that while it can inhibit IGF-IR and IR-A-derived mitogenic signaling, it has no direct effect on insulin-mediated metabolic functions. Two dual IGF-I/IGF-II neutralizing antibodies, Dusigtumab (MEDI-573-MedImmune, Gaithersburg, MD, USA) and Xentuzumab (BI-836845-Boehringer-Ingelheim, Ingelheim am Rhein, Germany), have entered phase I clinical trials [62,63] and had minimal adverse effects. However, the efficacy of ligand-neutralizing antibodies may be limited by cell surface expression levels of IGF-IR on the cancer cells, as they determine maximal ligand binding capacity [41].

IGFBPs are naturally occurring molecules that modulate the bioavailability of IGF ligands. IGFBP-3, the predominant IGFBP in the circulation can also induce IGF-independent apoptosis by mediating the pro-apoptotic function of TGFβ, in an IGF-IR independent manner. In addition, IGFBP-3 plays a role in the DNA repair response to DNA-damaging therapy and was shown to co-translocate to the nucleus of breast cancer cells with EGFR and DNA-dependent protein kinase in response to DNA damage, to mediate this function [64,65]. Recombinant human rhIGFBP-3 was shown to potentiate the effect of Herceptin on Herceptin-resistant human breast cancer cells in vitro as well as in a xenograft model in vivo by reducing Akt and ERK signaling [66] and an exogenously administered protease-resistant IGFBP-2 was shown to inhibit the growth of breast cancer cells in vitro and in a xenograft model in vivo [67]. However, to date, IGFBPs have not advanced to clinical testing, possibly because of their short half-life in vivo.

2. Traps in the Clinic—Advantages and Challenges

An effective strategy for blocking the action of cell surface receptors is the use of soluble decoys that bind the ligand with high affinity, reducing its bioavailability to the cognate receptor in a highly specific manner [68–70]. The efficacy of such decoys can be significantly improved by the addition of an IgG Fc domain resulting in a more stable ligand known as "Trap". For example, a soluble tumor necrosis factor (TNF)-α receptor-Fc fusion protein (Etanercept, Enbrel®, approved in 1998) is currently in routine clinical use for the treatment of inflammatory conditions such as rheumatoid arthritis [71]; Interleukin (IL)-1-Trap (Rilonacept, Arcalyst®, approved in 2008) is used for the treatment of cryopyrin-associated periodic syndromes (CAPS) [72], and a VEGFR1/VEGFR2-Fc decoy (VEGF-TRAP-Aflibercept, Regeneron (Eastview, NY, USA)) was approved for the treatment of wet macular degeneration under the trade name Eylea and for metastatic colorectal cancer as Zaltrap [68]. Although the development of IGF-IR decoys for cancer treatment has been reported [73], to date, none have advanced into clinical use.

To construct high affinity and high efficacy ligand binding Traps, two or more distinct receptor domains have to be fused to the Fc molecule. This fusion strategy can result in highly potent therapeutic drugs. For example, Rilonacept was engineered with the extracellular domains of the IL-1 receptor (IL-1R1) and the IL-1R accessory protein (IL-1-RAcP) fused to the Fc domain of human IgG$_1$, resulting in a potent (IC$_{50}$ = 6.5 pM), high affinity (Kd = 1.5 pM) IL-1R antagonist [72,74]. Aflibercept is composed of the ligand-binding domains of VEGFR1 and VEGFR2 fused to the Fc domain of IgG$_1$ and has a higher affinity to multiple isoforms of VEGF than the VEGF targeting MAbs. It was consequently found to be more effective than the MAbs Ranibizumab and Bevacizumab in patients with marked loss of visual acuity [75], and a superior inhibitor of angiogenesis in a model of neuroblastoma, where it caused regression of coopted vascular structures at high doses [76]. Recently, an Fc-fusion EGFR decoy comprising the truncated extracellular domains of EGFR/ErbB-1 and ErbB-4 fused to Fc was shown to have high-affinity ligand binding to EGF-like growth factors and could inhibit the invasive growth and metastasis of mammary carcinoma cells [77].

An advantage of Fc fusion proteins is their increased stability and extended half-life in vivo that is mediated primarily through their binding to the neonatal Fc receptor (FcRn) and their reduced renal elimination [78]. This increase in half-life reduces the dosing frequency and immunogenicity of the fusion proteins with clear clinical benefits. This was shown for both Etanercept [75,79] in the treatment of rheumatic diseases and for Aflibercept in the treatment of age-related macular degeneration [75]. Other advantages conferred by the Fc portion of Trap proteins, particularly in the context of cancer treatment, are their ability to trigger antibody-dependent cell-mediated cytotoxicity (ADCC) and antibody-dependent cell-mediated phagocytosis through Fc binding to Fc-gamma receptors (FcγRs) [80], and the activation of complement-dependent cytotoxicity (CDC) by the binding of complement C1q, leading to tumor cell killing [81,82].

The development of Fc-based therapeutics can be challenging. In order to improve recombinant protein expression, protein folding and protein stability, various modifications to the protein are required that can result in undesirable consequences such as altered protein–protein interactions, high molecular weight complex formation and aggregation, resulting in decreased bioactivity and increased risk of immunogenicity [83]. For example, removal of terminal carbohydrate moieties is an efficient way to eliminate undesired effector functions, but de-glycosylation can lead to instability and protein aggregation [84]. Moreover, even small changes in the amino acid sequence can have a considerable effect on the stability and safety profile of a drug [83]. The rapid emergence of technologies for protein engineering and modification will necessitate careful assessment of the risk/benefit profile, before they are transitioned to clinical use.

3. The IGF-Trap—A Stepwise Bioengineering Venture

3.1. IGF-IR Decoys-Background Information

The identification of the IGF-axis as a target in cancer therapy has spurred many attempts to inhibit this axis through nucleic acid-based and protein-engineering strategies (Table 1). One early approach was the development, by several groups, of IGF-IR decoys that when secreted by the cancer cells, reduce ligand bioavailability to the cognate receptor and act as dominant negative receptor mutants. The Baserga group was first to report that transfection of a 486 amino acid (486/stop) truncated receptor into rat glioma C6 cells and subsequently, into human metastatic breast cancer MDA-MB-435 cells, resulted in the secretion of this receptor into the conditioned medium, inhibiting cancer cell invasion, increasing apoptosis and reducing colony formation in vitro. C6 cells expressing this decoy had reduced tumorigenesis in vivo, while MDA-MB-435 cells had reduced metastasis [73,85]. Sachdev et al. subsequently reported on the production of a C-terminal-truncated 262 bp IGF-IR decoy that retained the ligand binding domain but lacked the autophosphorylated tyrosine residues in the carboxyl terminus. They showed that LCC6 cells—a metastatic variant of breast carcinoma MDA-MB-435 cells—transfected with this truncated receptor lost their motility in response to IGF-I and the ability to metastasize in a xenograft model [86]. Min et al. analyzed the effects of two decoys of 482 and 950 amino acids in a xenograft model of human gastric cancer. Consistent with the above studies, they found that expression of these decoys suppressed tumorigenicity in vitro and in vivo, blocked ligand-induced Akt-1 activation and markedly increased the sensitivity of the cells to radiation and chemotherapy-induced apoptosis [87].

3.2. The Incremental Production/Validation Process for an IGF-Trap

Our laboratory used a stepwise approach to engineer an IGF-Trap with potent growth inhibitory activity against multiple aggressive carcinomas. Initially a truncated (t) IGF-IR was engineered consisting of the first 933 amino acids and spanning the entire extracellular domain of the native receptor (IGF-IR^{933}). This truncated receptor was expressed in highly metastatic murine lung carcinoma H-59 cells. We confirmed that these cells produced and secreted into the medium a (β^t–α–α–β^t) heterotetramer that neutralized exogenously added IGF-I and inhibited IGF-I-induced signaling and IGF-IR-mediated proliferation, invasion, and apoptosis resistance. Expression of this truncated receptor had a dramatic effect on the metastatic potential of H-59 cells, reducing hepatic metastases by 90% following their intrasplenic/portal inoculation and significantly extending the long-term, disease-free survival of the mice (Figure 1) [88]. These results identified the IGFIR^{933} as a potent anti-tumorigenic and anti-metastatic agent with potential applications for cancer therapy and prompted us to begin exploring the translational potential of this decoy as a biological therapeutic. Initially, two cell and gene therapy strategies were used. Namely, we genetically engineered autologous bone marrow stromal cells stably secreting the IGF-IR^{933} decoy and implanted them subcutaneously into mice to achieve sustained production of this decoy in vivo [89]. We confirmed that these cells were able to generate high plasma levels of sIGFIR for at least three weeks, with a longer duration in athymic nude mice, suggestive of immune-based elimination of the stromal cells in immunocompetent mice [89]. In mice implanted with IGF-IR[933]-producing stromal cells, a marked reduction in experimental hepatic metastases of colon and lung carcinoma cells was observed (Figure 2). Moreover, in hepatic micro-metastases, a significant reduction in intra-lesional angiogenesis and an increase in tumor cell apoptosis were seen, suggesting that the IGF-IR decoy impeded early events in the process of liver metastasis. The results showed that sustained delivery of a soluble IGF-IR decoy was highly effective in preventing the expansion of liver metastases. This was also confirmed when a second approach was used, namely when a gutless adenovirus expressing sIGFIR was injected into mice intravenously, leading to production of measurable sIGFIR plasma levels for up to 21 days and resulting in significant inhibition of experimental liver metastasis [90].

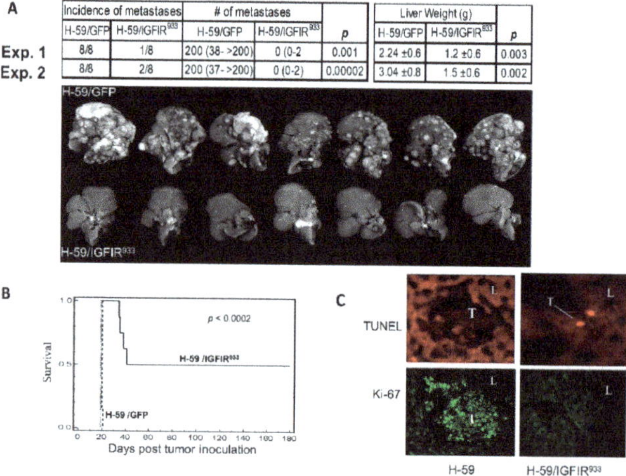

	Incidence of metastases		# of metastases			Liver Weight (g)		
	H-59/GFP	H-59/IGFIR[933]	H-59/GFP	H-59/IGFIR[933]	p	H-59/GFP	H-59/IGFIR[933]	p
Exp. 1	8/8	1/8	200 (38->200)	0 (0-2	0.001	2.24 ±0.6	1.2 ±0.6	0.003
Exp. 2	8/8	2/8	200 (37->200)	0 (0-2)	0.00002	3.04 ±0.8	1.5 ±0.6	0.002

Figure 1. Loss of metastatic potential in lung carcinoma cells expressing a soluble IGF-IR decoy (IGFIR[933]). Lewis lung carcinoma subline H-59 cells were transduced with retroparticles expressing the truncated 933 aa IGF-IR decoy (H-59/IGFIR[933]) or GFP only (H-59/GFP) and 10^5 tumor cells injected into syngeneic C57Bl/6 female mice via the intrasplenic/portal route to generate experimental liver metastases. Mice were sacrificed and visible metastases enumerated 14 days later. Shown in (**A**) (top) are the median numbers of metastases (and range) per liver based on eight animals per group in two separate experiments. Liver weights (means ± SD) are shown on the right, and representative livers from experiment (Exp.) 2 are shown on the bottom. Shown in (**B**) are survival data for mice inoculated in a similar manner ($p < 0.0002$) and in (**C**) terminal deoxynucleotidyl transferase (Tdt)-mediated nick end labeling (TUNEL) assay (top) and Ki-67 staining (bottom) performed on liver (L) cryostat sections prepared 5 days post tumor (T) injection (Mag. X135). Reproduced from [88].

Having observed marked reductions in experimental liver metastases in mice with sustained high plasma levels of an IGF-IR decoy, and in an effort to expedite potential translation of this technology to the clinic, we used recombinant technology to engineer and scale-up production of an IGF-Trap with potent anti-cancer activity. This was achieved in a two-stage process. Initially, we generated the soluble receptor decoy expressed in CHO cells downstream of a cumate-inducible promoter, using lentivirus particles. CHO cell clones identified as high producers were expanded and protein production initiated by the addition of 1 mg/mL cumate followed by a 7–8-day incubation, before the soluble protein was harvested and a stepwise purification of sIGF-IR performed. High binding affinity of the recombinant protein for hIGF-I and a 10^3-fold lower affinity for insulin were confirmed by surface plasmon resonance (SPR) and the biological activity of this protein was assessed and validated in multiple functional assays including IGF-initiated proliferation, invasion, anchorage independent growth and anoikis [91].

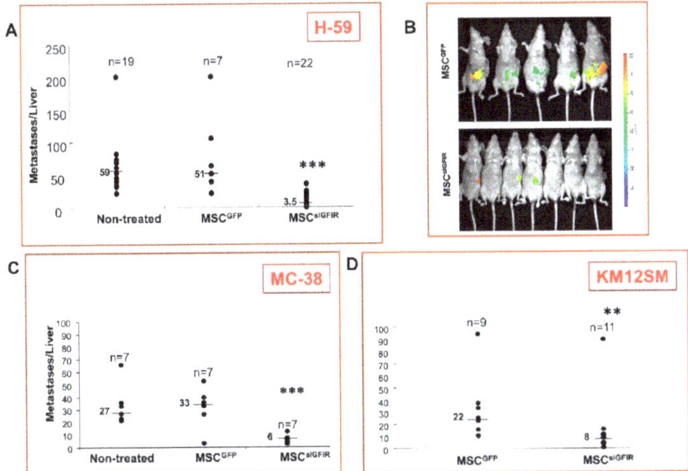

Figure 2. Bone marrow stromal cells producing a soluble IGF-IR inhibit experimental hepatic metastasis of lung and colon carcinoma. Syngeneic female C57Bl/6 (**A** and **C**) or nude (**B** and **D**) mice were implanted with 10^7 genetically engineered marrow-derived stromal cells (MSCs) expressing sIGFIR (MSC^sIGFIR) or control MSC (MSC^GFP) embedded in Matrigel. Fourteen days later (**A–D**), the mice were inoculated via the intrasplenic/portal route with 10^5 H-59 (**A** and **B**), 5×10^4 murine colon carcinoma MC-38 (**C**) or 10^6 human colon carcinoma KM12SM (**D**) cells. Mice were euthanized and liver metastases enumerated 14–16 (**A**), 18 (**C**) or 21 (**D**) days after or imaged using the IVIS 100 Xenogen 15 days (**B**) post tumor inoculation. Shown in (**A**) are the pooled data of three and in (**B–D**) individual experiments. Results of optical imaging are shown in (**B**). ** $p < 0.01$, *** $p < 0.001$, as determined by the non-parametric Mann–Whitney test. Reproduced from [89].

In order to improve the pharmacokinetic and potential therapeutic properties of this soluble receptor, thereby optimizing it for clinical translation, we then generated a sIGFIR–hFc–IgG$_1$ fusion protein—the IGF-Trap—that was produced in CHO cells using a similar production/scale-up strategy (Figure 3). We found that the addition of the Fc fragment did not alter the individual binding kinetics or overall affinity of the recombinant protein. The IGF-Trap bound hIGF-I with highest affinity and hIGF-II and murine IGF-I with moderately lower affinities, and had a three-log weaker affinity for insulin, confirming the high affinity and specificity of the IGF-Trap and a binding profile consistent with that observed with the cognate cell surface receptor [91]. Similar to sIGFIR, the IGF-Trap inhibited IGF-IR signaling and IGF-I and IGF-II- regulated cellular functions in several carcinoma cell types including breast, lung and colon carcinoma cells in vitro. It had a favorable pharmacokinetic profile in vivo with a half-life of 47.5 h as compared to 21.9 h for sIGFIR, confirming that the addition of the two Fc domains improved the stability of this protein in vivo [91]. Moreover, IGF-Trap treatment inhibited the growth of human and murine breast carcinoma cells and markedly reduced experimental liver metastasis of colon and lung carcinoma in vivo (representative data shown in Figure 3). Interestingly, we found that the IGF-Trap had superior therapeutic efficacy to an anti-IGF-IR antibody or IGF-binding protein-1 when used at similar or higher concentrations in a human breast cancer model and experimental murine colon cancer metastasis assays, respectively.

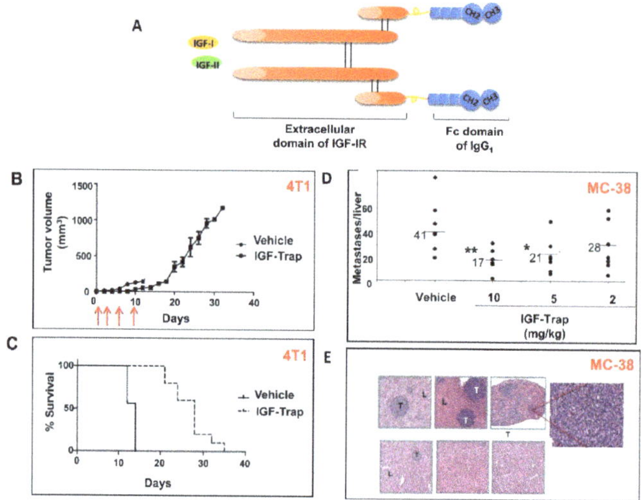

Figure 3. The IGF-Trap inhibits the orthotopic growth of mammary carcinoma and liver metastasis of colon carcinoma cells. Balb/c (**B** and **C**) or C57Bl/6 (**D** and **E**) mice were injected into the mammary fat pad (MFP) with 5×10^4 4T1 cells (**B** and **C**) or via the intrasplenic/portal route with 5×10^4 MC-38 cells (**D** and **E**). IGF-Trap injections were administered i.v. to 4T1 injected mice 4 h and 3, 6 and 10 days (arrows) post tumor inoculation (10 mg/kg for the first 2 injections and 5 mg/kg subsequently) and to MC-38 injected mice, 24 h and 4 and 7 days post tumor inoculation. Shown in (**A**) is a diagrammatic representation of the 2nd generation IGF-Trap. Shown in (**B**) are mean tumor volumes (±SD) and in (**C**) a Kaplan–Meier survival curve ($p < 0.01$ using Mantel-Cox or Gehan-Breslow-Wilcoxon Tests). Local MFP tumors grew rapidly in all untreated mice, causing morbidity by day 14, while in the treated mice, tumor growth was seen only after cessation of treatment. Shown in (**D**) are the numbers of visible liver metastases enumerated 18 days post tumor injection. Bars (and numbers) denote medians. Shown in (**E**) are representative hematoxylin and eosin-stained, formalin-fixed and paraffin-embedded sections obtained from different livers of MC-38-injected mice (magnification ×20; inset ×400). T: tumor; L: liver; * $p < 0.05$; ** $p < 0.01$. Reproduced from [91].

3.3. A 3rd Generation IGF-Trap—Properties, Bioactivity and Challenges

A problem frequently encountered with Fc-fusion proteins is the formation of high-molecular-weight (HMW) complexes due to oligomerization by irregular disulfide bonding between adjacent Fc fragments [92,93]. The IGF-Trap is a tetramer with two β subunits, each fused to one Fc domain of IgG₁, and this proximity of adjacent F_C domains lends itself to undesirable disulfide bonding and large complex formation. Indeed, we documented HMW protein species that migrated at the > 400 kDa range in the IGF-Trap preparations. We showed that these HMW species did not contribute significantly to the biologic activity of the Trap and could be minimized by step elution following Protein-A column purification [91]. In an effort to further improve the purity and manufacturability of the IGF-Trap, we therefore re-engineered the parent protein to eliminate such aberrant disulfide bonding by cysteine-to-serine substitutions in the hinge region of the human IgG₁ Fc fragment, as well as by incorporating a longer and more flexible linker between the IGF-IR ectodomain and the Fc domain. Four different modified Traps were produced, and two were selected for further evaluation, based on a polyacrylamide gel profile that confirmed the elimination of HMW species in these preparations. We found that the IGF-Trap in which Cys-Ser substitutions in the Fc hinge region were combined with the addition of a flexible linker (IGF-Trap 3.3) had a considerably improved pharmacokinetic profile with a marked increase in the area under the serum concentration-time curve. Moreover, this IGF-Trap had an enhanced therapeutic profile, as evaluated in an experimental colon carcinoma liver metastasis model and was superior to a ligand binding antibody used under the same conditions (Figure 4). This indicated that depletion of HMW species and the increased stability also improved the pharmacodynamic properties of the Trap.

The IGF kinase-receptor-activation (KIRA) assay measures ligand bioavailability by quantifying phosphorylated IGF-I receptor levels. While traditional end-point bioassays measure downstream effects of IGF-IR activation such as cell proliferation and survival, the KIRA assay is based on measuring receptor activation per se, thereby avoiding errors due to detection of other confounding signaling pathways. Moreover, naturally occurring IGFBPs and proteases in the circulation affect the bioavailability/bioactivity of IGF-I. While immune-based approaches such as enzyme linked immunoassays (ELISA) measure both total (BP-bound) and free ligand, the two-step KIRA assay provides a more accurate measure of bioactive ligands [94–97]. Using the KIRA assay, we found that IGF-I serum bioavailability correlated well with the IGF-Trap pharmacokinetic/pharmacodynamic profile, providing a novel, surrogate marker for its therapeutic efficacy [98].

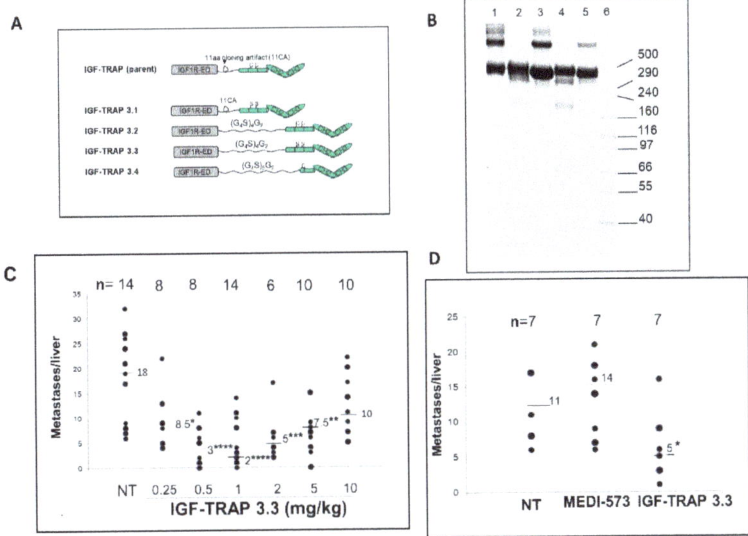

Figure 4. Cysteine-serine substitutions in the Fc domain of the IGF-Trap reduce high-molecular-weight (HMW) oligomers and improve pharmacodynamic properties. Shown in (**A**) is a schematic representation of the modifications engineered in the parent (2nd generation) IGF-Trap and in (**B**) results of SDS-PAGE performed on purified parental or modified IGFIR-hFc-IgG₁ proteins, using denaturing and non-reducing condition. Lanes: 1—parent IGF-Trap; 2—IGF-Trap 3.1; 3—IGF-Trap 3.2; 4—IGF-Trap 3.3; 5—IGF-Trap 3.4; 6—HMW protein standard (Invitrogen). Shown in (**C**) is the number of metastases enumerated in individual livers in three different experiments where mice were inoculated via the intrasplenic/portal route with 5×10^4 MC-38 cells, treated with IGF-Trap 3.3 at the indicated doses from day 1 and thereafter twice weekly, for a total of 5 injections and sacrificed 16–18 days later. The total number of mice per treatment group is indicated on the top. Shown in (**D**) are results of a separate experiment where one group of mice was treated with 1 mg/kg of the anti-ligand MAb MEDI-573. Horizontal bars denote medians. NT: non-treated. * $p < 0.05$, ** $p < 0.01$, *** $p < 0.005$, **** $p < 0.001$, as assessed by the non-parametric Mann–Whitney test. Reproduced from [98].

4. Targeting the IGF-IR in the Tumor Microenvironment

4.1. IGF-IR Is Expressed on Immune Cells and Plays a Role in Immunosuppression

The major immune cell subtypes (i.e., T and B lymphocytes), myeloid derived mononuclear cells and NK cells express the IGF-IR and are responsive to IGF ligands [99]. Although complex, there is compelling evidence that within a tumor microenvironment (TME), the IGF axis promotes an anti-inflammatory, immunosuppressive response that enables cancer expansion. Thus, IGF-I was shown to negatively regulate DC activation, impair their antigen-presenting function [100] and stimulate the proliferation of immunosuppressive regulatory T cells (Treg) [101,102]. IGF-IR activation was also linked to macrophage polarization to the pro-tumorigenic M2 phenotype [103,104]. Treatment of DC with the IGF inhibitor NVP-AEW541 restored DC-mediated antigen presentation and anti-tumor immunity [105]. A deficit in IGF-I signaling in macrophages was associated with a decreased capacity to induce the M2 state and an increased responsiveness to the pro-inflammatory cytokine IFNγ [104]. Moreover, the inhibitor NT157 that targets both the IGF-IR and STAT3 inhibited expression of pro-tumorigenic cytokines, chemokines and growth factors including IL-6, CCL2, CCL5 and TGFβ [106]. IGF-I was also shown to play a role in the survival of neutrophils by blocking Fas-mediated apoptosis [99]. Of interest, in patients treated with a MAb to IGF-IR (AMG 479), high levels of antibody binding to neutrophils were documented [107]. Finally, IGF-I may also play a role in the tumor-promoting effect of myeloid derived suppressor cells (MDSC) [108]. Collectively, these studies identify the IGF axis as a contributor to a pro-tumorigenic TME, suggesting that in addition to their direct positive effect on tumor cell survival and proliferation, the IGFs also potentiate escape mechanisms from immune-mediated tumor cell destruction.

4.2. Multiple Effects of the IGF-Trap on the Tumor Microenvironment

In addition to directly targeting IGF-signaling in the cancer cells, the IGF-Trap also had indirect effects on metastatic expansion by targeting the pro-metastatic microenvironment of the liver. As shown above, treatment with the IGF-Trap inhibited neovascularization in the early stages of metastases [91], suggesting that it affected endothelial cell migration and/or proliferation. Moreover, we have shown that IGF-I regulates hepatic stellate cell (HSC) activation in both cancer metastasis and cancer-free (CCl_4-induced liver injury) models, and the IGF-Trap caused a significant reduction in HSC activation in response to metastatic colon cancer cells [109]. When analyzing neutrophil phenotypes in a colon cancer liver metastasis model, we also observed a reduction in CXCL4^{high}/ICAM-1^{low} N2 polarized neutrophils in IGF-Trap treated mice that may potentially be mediated through regulation of TGFβ expression levels [110]. Finally, we have shown that IGF signaling regulates type IV collagen production in metastatic cancer cells, thereby promoting their growth in the liver [111,112]. Given the critical role that the extracellular matrix (ECM) plays in the TME [113], the IGF-Trap may therefore also impede metastatic expansion in this organ by altering the tumor-associated ECM. Thus, the IGF-Trap can have a multi-pronged effect on metastatic expansion, particularly in the liver, by impeding cancer cell proliferation, while also rendering the TME less hospitable to their expansion.

4.3. Future Prospective: The Case for Combinatorial Therapy with the IGF-Trap

The TME in primary or secondary sites can either promote or suppress the progression of malignant disease. The nature of the immune response engendered within the TME is a major factor determining the balance between these opposing outcomes [114–117]. Recent advances in immunotherapy, based on targeting immune checkpoints such as PD-1 and CTLA-4 have yielded promising therapeutic results in several aggressive and treatment-refractory cancers such as malignant melanoma, small cell lung cancer and renal cell carcinoma [118–120]. To date, however, immunotherapy has failed to show promise in the treatment of malignancies such as colorectal carcinoma and pancreatic ductal adenocarcinoma that metastasize to the liver [121,122]. This may be due, at least in part, to the presence of immunosuppressive cells such MDSC and M2 macrophages that impede T cell mediated cytotoxicity.

Thus, therapeutic approaches that can target an immunosuppressive TME and enhance the efficacy of immunotherapy are currently being sought [122,123]. As reviewed above, the major innate and adaptive immune cell subtypes express IGF-IR and are responsive to IGF ligands [99]. Although the role of IGF-IR in the development and function of immune cells is complex, there is compelling evidence that within the TME, the IGF axis promotes an anti-inflammatory, immunosuppressive response that enables cancer expansion. Thus, IGF targeting was shown to alter the tumor immune ME in colon cancer, reducing anti-inflammatory cytokines [106] and our own data identified IGF-IR on neutrophils and HSCs as a contributor to liver metastasis [109,110]. Collectively, these data provide a compelling rationale for combinatorial immunotherapy using immune checkpoint inhibitors together with IGF-targeting drugs. These combinations may be particularly effective for malignancies of the gastrointestinal track that metastasize to the liver, an organ with an innate immune hyporeactivity and the site of IGF-I production. Our bio-distribution studies have identified the liver as a major site for IGF-Trap accumulation, possibly due to the high local level of IGF-I [98]. This suggests that the IGF-Trap may be particularly well suited for combinatorial immunotherapy in liver-metastatic diseases.

5. Conclusions

Clinical trials with IGF-targeting biologicals exposed several obstacles to their successful use in cancer therapy. Due to the homology and crosstalk between IGF-IR and IR, several inhibitors of IGF-IR signaling (including tyrosine kinase inhibitors) were found to also disrupt IR signaling, resulting in undesirable side effects such as hyperinsulinemia and hyperglycemia. The responses to more specific drugs, such as anti-IGF-IR antibodies, were also disappointing, and this has been attributed to several potential factors, including increased GH release, IGF-II/IR-A signaling, rescue signaling by alternate RTKs and increased IGF-IR nuclear translocation. Recently it was proposed that IGF-IR targeting by antibodies or kinase inhibitors may result in alternative, kinase-independent ERK signaling mediated via recruitment of interacting proteins such as β-arrestins, limiting the effectiveness of these inhibitors (reviewed in [124,125]).

The IGF-Trap offers key advantages over receptor targeting antibodies and small-molecule inhibitors. With high specificity for IGF-I and IGF-II, and poor affinity for insulin, the deleterious effects on the physiological functions of insulin can be minimized. Since the IGF-Trap binds circulating ligands, penetration and diffusion into solid tumors are not major obstacles to efficacy, although uptake at the tumor site, if achieved, could have the added benefit of neutralizing locally produced ligands. Moreover, the high binding affinity of the IGF-Trap for IGF-II should reduce IGF-II bioavailability for IR-A activation, bypassing one of the major resistance mechanisms to IGF-IR targeting drugs. In addition, the potential of anti-IGF-IR antibodies to act as natural agonists and activate alternate IGF-IR signaling can be circumvented with the use of an IGF-Trap [125], and targeting of the ligands rather than a cell surface receptor should minimize non-desirable side effects due to antibody-dependent cellular cytotoxicity (ADCC) that can be mediated by the Fc portion of cell bound antibodies [126]. Finally, our evidence suggests that the IGF-Trap, by reducing ligand bioavailability can target several components of the tumor microenvironment, further enhancing its inhibitory activity on tumor cell growth. Having established the utility of the KIRA for monitoring IGF-Trap efficacy in vivo, our data suggest that it could provide a surrogate marker for response evaluation and a potential tool for patient stratification. Collectively, there is therefore a compelling rationale for transitioning this technology to the clinic for treatment of malignant disease, either alone or in combination with other treatment modalities.

Author Contributions: All the authors contributed literature research and texts to this review. Y.M.C. wrote the introductory section on IGF-targeting strategies, S.Q. wrote the review on Trap technology. S.P. contributed to the section on IGF-Trap development and M.H. contributed to the section on the role of the IGF-axis in the tumor microenvironment. P.B. collated, revised and edited the manuscript. All authors have read and agreed to the published version of the manuscript.

Funding: This research was funded by a Terry Fox Frontiers Initiative Grant of the National Cancer Institute of Canada, grants MOP-77677 and PP2-141724 from the Canadian Institute for Health Research, grant IT04483 from MITACS, and research support from Amorchem Inc.

Acknowledgments: The bioengineering, production and optimization of the IGF-Trap were made possible by generous grants from the Quebec Ministry of Economic Development and MSBiv, The Terry Fox Research Institute, grants MOP-77677 and PP2-141724 from the Canadian Institute for Health Research, grant IT04483 from MITACS and funding by Amorchem Inc. The development of the IGF-Trap was a collaborative endeavor. The many contributions of the protein engineering team of the Human Health Therapeutics Research Centre, the National Research Council of Canada, Montreal, Quebec, Canada under the leadership of Bernard Massie are gratefully acknowledged.

Conflicts of Interest: The authors declare no conflicts of interest.

References

1. Butler, A.A.; Yakar, S.; Gewolb, I.H.; Karas, M.; Okubo, Y.; LeRoith, D. Insulin-like growth factor-I receptor signal transduction: At the interface between physiology and cell biology. *Comp. Biochem. Physiol. B Biochem. Mol. Biol.* **1998**, *121*, 19–26. [CrossRef]

2. Samani, A.A.; Yakar, S.; LeRoith, D.; Brodt, P. The role of the IGF system in cancer growth and metastasis: Overview and recent insights. *Endocr. Rev.* **2007**, *28*, 20–47. [CrossRef] [PubMed]

3. De Meyts, P.; Whittaker, J. Structural biology of insulin and IGF1 receptors: Implications for drug design. *Nat. Rev. Drug Discov.* **2002**, *1*, 769–783. [CrossRef] [PubMed]

4. Hakuno, F.; Takahashi, S.-I. 40 years of IGF1: IGF1 receptor signaling pathways. *J. Mol. Endocrinol.* **2018**, *61*, T69–T86. [CrossRef] [PubMed]

5. Li, J.; Choi, E.; Yu, H.; Bai, X.-c. Structural basis of the activation of type 1 insulin-like growth factor receptor. *Nat. Commun.* **2019**, *10*, 1–11. [CrossRef]

6. Sehat, B.; Tofigh, A.; Lin, Y.; Trocmé, E.; Liljedahl, U.; Lagergren, J.; Larsson, O. SUMOylation mediates the nuclear translocation and signaling of the IGF-1 receptor. *Sci. Signal.* **2010**, *3*, ra10. [CrossRef]

7. Warsito, D.; Sjöström, S.; Andersson, S.; Larsson, O.; Sehat, B. Nuclear IGF1R is a transcriptional co-activator of LEF1/TCF. *Embo Rep.* **2012**, *13*, 244–250. [CrossRef]

8. Aleksic, T.; Chitnis, M.M.; Perestenko, O.V.; Gao, S.; Thomas, P.H.; Turner, G.D.; Protheroe, A.S.; Howarth, M.; Macaulay, V.M. Type 1 insulin-like growth factor receptor translocates to the nucleus of human tumor cells. *Cancer Res.* **2010**, *70*, 6412–6419. [CrossRef]

9. Codony-Servat, J.; Cuatrecasas, M.; Asensio, E.; Montironi, C.; Martínez-Cardús, A.; Marín-Aguilera, M.; Horndler, C.; Martínez-Balibrea, E.; Rubini, M.; Jares, P. Nuclear IGF-1R predicts chemotherapy and targeted therapy resistance in metastatic colorectal cancer. *Br. J. Cancer* **2017**, *117*, 1777–1786. [CrossRef]

10. Clément, F.; Martin, A.; Venara, M.; de Luján Calcagno, M.; Mathó, C.; Maglio, S.; Lombardi, M.G.; Bergadá, I.; Pennisi, P.A. Type 1 IGF Receptor Localization in Paediatric Gliomas: Significant Association with WHO Grading and Clinical Outcome. *Horm. Cancer* **2018**, *9*, 205–214. [CrossRef]

11. Belfiore, A.; Malaguarnera, R.; Vella, V.; Lawrence, M.C.; Sciacca, L.; Frasca, F.; Morrione, A.; Vigneri, R. Insulin Receptor Isoforms in Physiology and Disease: An Updated View. *Endocr Rev.* **2017**, *38*, 379–431. [CrossRef]

12. Vella, V.; Milluzzo, A.; Scalisi, N.M.; Vigneri, P.; Sciacca, L. Insulin Receptor Isoforms in Cancer. *Int. J. Mol. Sci.* **2018**, *19*, 3615. [CrossRef]

13. Pandini, G.; Frasca, F.; Mineo, R.; Sciacca, L.; Vigneri, R.; Belfiore, A. Insulin/insulin-like growth factor I hybrid receptors have different biological characteristics depending on the insulin receptor isoform involved. *J. Biol. Chem.* **2002**, *277*, 39684–39695. [CrossRef]

14. Pandini, G.; Vigneri, R.; Costantino, A.; Frasca, F.; Ippolito, A.; Fujita-Yamaguchi, Y.; Siddle, K.; Goldfine, I.D.; Belfiore, A. Insulin and insulin-like growth factor-I (IGF-I) receptor overexpression in breast cancers leads to insulin/IGF-I hybrid receptor overexpression: Evidence for a second mechanism of IGF-I signaling. *Clin. Cancer Res.* **1999**, *5*, 1935–1944. [PubMed]

15. Belfiore, A. The role of insulin receptor isoforms and hybrid insulin/IGF-I receptors in human cancer. *Curr. Pharm. Des.* **2007**, *13*, 671–686. [CrossRef] [PubMed]

16. Chen, J.; Nagle, A.M.; Wang, Y.F.; Boone, D.N.; Lee, A.V. Controlled dimerization of insulin-like growth factor-1 and insulin receptors reveals shared and distinct activities of holo and hybrid receptors. *J. Biol. Chem.* **2018**, *293*, 3700–3709. [CrossRef] [PubMed]

17. Klotz, D.M.; Hewitt, S.C.; Ciana, P.; Raviscioni, M.; Lindzey, J.K.; Foley, J.; Maggi, A.; DiAugustine, R.P.; Korach, K.S. Requirement of estrogen receptor-α in insulin-like growth factor-1 (IGF-1)-induced uterine responses and in vivo evidence for IGF-1/estrogen receptor cross-talk. *J. Biol. Chem.* **2002**, *277*, 8531–8537. [CrossRef]

18. Janssen, J.A.; Varewijck, A.J. IGF-IR targeted therapy: Past, present and future. *Front. Endocrinol.* **2014**, *5*, 224. [CrossRef]

19. Sun, Y.; Sun, X.; Shen, B. Molecular imaging of IGF-1R in cancer. *Mol. Imaging* **2017**, *16*, 1536012117736648. [CrossRef]

20. Shanmugalingam, T.; Bosco, C.; Ridley, A.J.; Van Hemelrijck, M. Is there a role for IGF-1 in the development of second primary cancers? *Cancer Med.* **2016**, *5*, 3353–3367. [CrossRef]

21. Wu, Y.; Brodt, P.; Sun, H.; Mejia, W.; Novosyadlyy, R.; Nunez, N.; Chen, X.; Mendoza, A.; Hong, S.H.; Khanna, C.; et al. Insulin-like growth factor-I regulates the liver microenvironment in obese mice and promotes liver metastasis. *Cancer Res.* **2010**, *70*, 57–67. [CrossRef] [PubMed]

22. Yakar, S.; Leroith, D.; Brodt, P. The role of the growth hormone/insulin-like growth factor axis in tumor growth and progression: Lessons from animal models. *Cytokine Growth Factor Rev.* **2005**, *16*, 407–420. [CrossRef] [PubMed]

23. Chhabra, Y.; Waters, M.J.; Brooks, A.J. Role of the growth hormone–IGF-1 axis in cancer. *Expert Rev. Endocrinol. Metab.* **2011**, *6*, 71–84. [CrossRef] [PubMed]

24. De Ostrovich, K.K.; Lambertz, I.; Colby, J.K.; Tian, J.; Rundhaug, J.E.; Johnston, D.; Conti, C.J.; DiGiovanni, J.; Fuchs-Young, R. Paracrine overexpression of insulin-like growth factor-1 enhances mammary tumorigenesis in vivo. *Am. J. Pathol.* **2008**, *173*, 824–834. [CrossRef]

25. DiGiovanni, J.; Kiguchi, K.; Frijhoff, A.; Wilker, E.; Bol, D.K.; Beltrán, L.; Moats, S.; Ramirez, A.; Jorcano, J.; Conti, C. Deregulated expression of insulin-like growth factor 1 in prostate epithelium leads to neoplasia in transgenic mice. *Proc. Natl. Acad. Sci. USA* **2000**, *97*, 3455–3460. [CrossRef]

26. Kim, W.-Y.; Jin, Q.; Oh, S.-H.; Kim, E.S.; Yang, Y.J.; Lee, D.H.; Feng, L.; Behrens, C.; Prudkin, L.; Miller, Y.E. Elevated epithelial insulin-like growth factor expression is a risk factor for lung cancer development. *Cancer Res.* **2009**, *69*, 7439–7448. [CrossRef]

27. Fu, S.; Tang, H.; Liao, Y.; Xu, Q.; Liu, C.; Deng, Y.; Wang, J.; Wang, J.; Fu, X. Expression and clinical significance of insulin-like growth factor 1 in lung cancer tissues and perioperative circulation from patients with non-small-cell lung cancer. *Curr. Oncol.* **2016**, *23*, 12. [CrossRef]

28. Mancarella, C.; Casanova-Salas, I.; Calatrava, A.; García-Flores, M.; Garofalo, C.; Grilli, A.; Rubio-Briones, J.; Scotlandi, K.; López-Guerrero, J.A. Insulin-like growth factor 1 receptor affects the survival of primary prostate cancer patients depending on TMPRSS2-ERG status. *BMC Cancer* **2017**, *17*, 367. [CrossRef]

29. Mu, L.; Tuck, D.; Katsaros, D.; Lu, L.; Schulz, V.; Perincheri, S.; Menato, G.; Scarampi, L.; Harris, L.; Yu, H. Favorable outcome associated with an IGF-1 ligand signature in breast cancer. *Breast Cancer Res. Treat.* **2012**, *133*, 321–331. [CrossRef]

30. Baserga, R.; Peruzzi, F.; Reiss, K. The IGF-1 receptor in cancer biology. *Int J. Cancer* **2003**, *107*, 873–877. [CrossRef]

31. Valentinis, B.; Baserga, R. IGF-I receptor signalling in transformation and differentiation. *Mol. Pathol.* **2001**, *54*, 133–137. [CrossRef] [PubMed]

32. Valentinis, B.; Romano, G.; Peruzzi, F.; Morrione, A.; Prisco, M.; Soddu, S.; Cristofanelli, B.; Sacchi, A.; Baserga, R. Growth and differentiation signals by the insulin-like growth factor 1 receptor in hemopoietic cells are mediated through different pathways. *J. Biol. Chem.* **1999**, *274*, 12423–12430. [CrossRef] [PubMed]

33. Llanos, A.A.; Brasky, T.M.; Dumitrescu, R.G.; Marian, C.; Makambi, K.H.; Kallakury, B.V.; Spear, S.L.; Perry, D.J.; Convit, R.J.; Platek, M.E. Plasma IGF-1 and IGFBP-3 may be imprecise surrogates for breast concentrations: An analysis of healthy women. *Breast Cancer Res. Treat.* **2013**, *138*, 571–579. [CrossRef] [PubMed]

34. Haywood, N.J.; Slater, T.A.; Matthews, C.J.; Wheatcroft, S.B. The insulin like growth factor and binding protein family: Novel therapeutic targets in obesity & diabetes. *Mol. Metab.* **2019**, *19*, 86–96.

35. Bohula, E.A.; Playford, M.P.; Macaulay, V.M. Targeting the type 1 insulin-like growth factor receptor as anti-cancer treatment. *Anti-Cancer Drugs* **2003**, *14*, 669–682. [CrossRef]
36. Buck, E.; Gokhale, P.C.; Koujak, S.; Brown, E.; Eyzaguirre, A.; Tao, N.; Rosenfeld-Franklin, M.; Lerner, L.; Chiu, M.I.; Wild, R.; et al. Compensatory insulin receptor (IR) activation on inhibition of insulin-like growth factor-1 receptor (IGF-1R): Rationale for cotargeting IGF-1R and IR in cancer. *Mol. Cancer Ther.* **2010**, *9*, 2652–2664. [CrossRef]
37. Osher, E.; Macaulay, V.M. Therapeutic targeting of the IGF axis. *Cells* **2019**, *8*, 895. [CrossRef]
38. Runcie, K.; Budman, D.R.; John, V.; Seetharamu, N. Bi-specific and tri-specific antibodies-the next big thing in solid tumor therapeutics. *Mol. Med.* **2018**, *24*, 50. [CrossRef]
39. Schanzer, J.M.; Wartha, K.; Moessner, E.; Hosse, R.J.; Moser, S.; Croasdale, R.; Trochanowska, H.; Shao, C.; Wang, P.; Shi, L.; et al. XGFR*, a novel affinity-matured bispecific antibody targeting IGF-1R and EGFR with combined signaling inhibition and enhanced immune activation for the treatment of pancreatic cancer. *mAbs* **2016**, *8*, 811–827. [CrossRef]
40. Quinn, D.I.; Baudin, E.; Demeure, M.J.; Fassnacht, M.; Hammer, G.D.; Poondru, S.; Fleege, T.; Rorig, R.; Berruti, A. International randomized, double-blind, placebo-controlled, phase 3 study of linsitinib (OSI-906, L) in patients (pts) with locally advanced or metastatic adrenocortical carcinoma (ACC). *J. Clin. Oncol.* **2014**, *32*, 4507. [CrossRef]
41. Tian, D.; Kreeger, P.K. Analysis of the quantitative balance between insulin-like growth factor (IGF)-1 ligand, receptor, and binding protein levels to predict cell sensitivity and therapeutic efficacy. *BMC Syst. Biol.* **2014**, *8*, 98. [CrossRef] [PubMed]
42. Beck, A.; Reichert, J.M. Therapeutic Fc-fusion proteins and peptides as successful alternatives to antibodies. *mAbs* **2011**, *3*, 415–416. [CrossRef] [PubMed]
43. Pollak, M. The insulin and insulin-like growth factor receptor family in neoplasia: An update. *Nat. Rev. Cancer* **2012**, *12*, 159–169. [CrossRef] [PubMed]
44. Gualberto, A.; Pollak, M. Emerging role of insulin-like growth factor receptor inhibitors in oncology: Early clinical trial results and future directions. *Oncogene* **2009**, *28*, 3009–3021. [CrossRef] [PubMed]
45. Avnet, S.; Sciacca, L.; Salerno, M.; Gancitano, G.; Cassarino, M.F.; Longhi, A.; Zakikhani, M.; Carboni, J.M.; Gottardis, M.; Giunti, A.; et al. Insulin receptor isoform A and insulin-like growth factor II as additional treatment targets in human osteosarcoma. *Cancer Res.* **2009**, *69*, 2443–2452. [CrossRef]
46. Frasca, F.; Pandini, G.; Scalia, P.; Sciacca, L.; Mineo, R.; Costantino, A.; Goldfine, I.D.; Belfiore, A.; Vigneri, R. Insulin receptor isoform A, a newly recognized, high-affinity insulin-like growth factor II receptor in fetal and cancer cells. *Mol. Cell. Biol.* **1999**, *19*, 3278–3288. [CrossRef]
47. Haluska, P.; Worden, F.; Olmos, D.; Yin, D.; Schteingart, D.; Batzel, G.N.; Paccagnella, M.L.; de Bono, J.S.; Gualberto, A.; Hammer, G.D. Safety, tolerability, and pharmacokinetics of the anti-IGF-1R monoclonal antibody figitumumab in patients with refractory adrenocortical carcinoma. *Cancer Chemother. Pharmacol.* **2010**, *65*, 765–773. [CrossRef]
48. Ramalingam, S.S.; Spigel, D.R.; Chen, D.; Steins, M.B.; Engelman, J.A.; Schneider, C.P.; Novello, S.; Eberhardt, W.E.; Crino, L.; Habben, K.; et al. Randomized phase II study of erlotinib in combination with placebo or R1507, a monoclonal antibody to insulin-like growth factor-1 receptor, for advanced-stage non-small-cell lung cancer. *J. Clin. Oncol. Off. J. Am. Soc. Clin. Oncol.* **2011**, *29*, 4574–4580. [CrossRef]
49. Cohen, B.D.; Baker, D.A.; Soderstrom, C.; Tkalcevic, G.; Rossi, A.M.; Miller, P.E.; Tengowski, M.W.; Wang, F.; Gualberto, A.; Beebe, J.S.; et al. Combination therapy enhances the inhibition of tumor growth with the fully human anti-type 1 insulin-like growth factor receptor monoclonal antibody CP-751,871. *Clin. Cancer Res.* **2005**, *11*, 2063–2073. [CrossRef]
50. Frogne, T.; Jepsen, J.S.; Larsen, S.S.; Fog, C.K.; Brockdorff, B.L.; Lykkesfeldt, A.E. Antiestrogen-resistant human breast cancer cells require activated protein kinase B/Akt for growth. *Endocr. -Relat. Cancer* **2005**, *12*, 599–614. [CrossRef]
51. Ye, J.J.; Liang, S.J.; Guo, N.; Li, S.L.; Wu, A.M.; Giannini, S.; Sachdev, D.; Yee, D.; Brunner, N.; Ikle, D.; et al. Combined effects of tamoxifen and a chimeric humanized single chain antibody against the type I IGF receptor on breast tumor growth in vivo. *Horm. Metab Res.* **2003**, *35*, 836–842. [PubMed]
52. Gee, J.M.; Robertson, J.F.; Gutteridge, E.; Ellis, I.O.; Pinder, S.E.; Rubini, M.; Nicholson, R.I. Epidermal growth factor receptor/HER2/insulin-like growth factor receptor signalling and oestrogen receptor activity in clinical breast cancer. *Endocr. Relat. Cancer* **2005**, *12* (Suppl. 1), S99–S111. [CrossRef] [PubMed]

53. Jones, H.E.; Gee, J.M.; Taylor, K.M.; Barrow, D.; Williams, H.D.; Rubini, M.; Nicholson, R.I. Development of strategies for the use of anti-growth factor treatments. *Endocr. -Relat. Cancer* **2005**, *12* (Suppl. 1), S173–S182. [CrossRef] [PubMed]

54. Nahta, R.; Yuan, L.X.; Zhang, B.; Kobayashi, R.; Esteva, F.J. Insulin-like growth factor-I receptor/human epidermal growth factor receptor 2 heterodimerization contributes to trastuzumab resistance of breast cancer cells. *Cancer Res.* **2005**, *65*, 11118–11128. [CrossRef] [PubMed]

55. Camirand, A.; Zakikhani, M.; Young, F.; Pollak, M. Inhibition of insulin-like growth factor-1 receptor signaling enhances growth-inhibitory and proapoptotic effects of gefitinib (Iressa) in human breast cancer cells. *Breast Cancer Res. BCR* **2005**, *7*, R570–R579. [CrossRef]

56. Chakravarti, A.; Loeffler, J.S.; Dyson, N.J. Insulin-like growth factor receptor I mediates resistance to anti-epidermal growth factor receptor therapy in primary human glioblastoma cells through continued activation of phosphoinositide 3-kinase signaling. *Cancer Res.* **2002**, *62*, 200–207.

57. Gualberto, A.; Pollak, M. Clinical development of inhibitors of the insulin-like growth factor receptor in oncology. *Curr. Drug Targets* **2009**, *10*, 923–936. [CrossRef]

58. Douglas, R.S.; Kahaly, G.J.; Patel, A.; Sile, S.; Thompson, E.H.Z.; Perdok, R.; Fleming, J.C.; Fowler, B.T.; Marcocci, C.; Marino, M.; et al. Teprotumumab for the Treatment of Active Thyroid Eye Disease. *N. Engl. J. Med.* **2020**, *382*, 341–352. [CrossRef]

59. Markham, A. Teprotumumab: First Approval. *Drugs* **2020**, *10*, 1–4. [CrossRef]

60. Ko, A.H.; Cubillo, A.; Kundranda, M.; Zafar, S.F.; Meiri, E.; Bendell, J.; Alguel, H.; Rivera Herrero, F.; Ahn, E.; Watkins, D.; et al. CARRIE: A randomized, double-blind, placebo-controlled phase II study of istiratumab (MM-141) plus nab-paclitaxel and gemcitabine versus nab-paclitaxel and gemcitabine in front-line metastatic pancreatic cancer. *Ann. Oncol.* **2018**, *29* (Suppl. 8), viii720. [CrossRef]

61. Kundranda, M.; Gracian, A.C.; Zafar, S.F.; Meiri, E.; Bendell, J.; Algul, H.; Rivera, F.; Ahn, E.R.; Watkins, D.; Pelzer, U.; et al. Randomized, double-blind, placebo-controlled phase II study of istiratumab (MM-141) plus nab-paclitaxel and gemcitabine versus nab-paclitaxel and gemcitabine in front-line metastatic pancreatic cancer (CARRIE). *Ann. Oncol.* **2020**, *31*, 79–87. [CrossRef] [PubMed]

62. de Bono, J.; Lin, C.C.; Chen, L.T.; Corral, J.; Michalarea, V.; Rihawi, K.; Ong, M.; Lee, J.H.; Hsu, C.H.; Yang, J.C.; et al. Two first-in-human studies of xentuzumab, a humanised insulin-like growth factor (IGF)-neutralising antibody, in patients with advanced solid tumours. *Br. J. Cancer* **2020**, *10*, 1–9. [CrossRef] [PubMed]

63. Iguchi, H.; Nishina, T.; Nogami, N.; Kozuki, T.; Yamagiwa, Y.; Yagawa, K. Phase I dose-escalation study evaluating safety, tolerability and pharmacokinetics of MEDI-573, a dual IGF-I/II neutralizing antibody, in Japanese patients with advanced solid tumours. *Invest. New Drugs* **2015**, *33*, 194–200. [CrossRef] [PubMed]

64. Rajah, R.; Valentinis, B.; Cohen, P. Insulin-like growth factor (IGF)-binding protein-3 induces apoptosis and mediates the effects of transforming growth factor-β1 on programmed cell death through a p53-and IGF-independent mechanism. *J. Biol. Chem.* **1997**, *272*, 12181–12188. [CrossRef]

65. Lin, M.; Marzec, K.; Martin, J.; Baxter, R. The role of insulin-like growth factor binding protein-3 in the breast cancer cell response to DNA-damaging agents. *Oncogene* **2014**, *33*, 85–96. [CrossRef]

66. Jerome, L.; Alami, N.; Belanger, S.; Page, V.; Yu, Q.; Paterson, J.; Shiry, L.; Pegram, M.; Leyland-Jones, B. Recombinant human insulin-like growth factor binding protein 3 inhibits growth of human epidermal growth factor receptor-2–overexpressing breast tumors and potentiates Herceptin activity in vivo. *Cancer Res.* **2006**, *66*, 7245–7252. [CrossRef]

67. Soh, C.L.; McNeil, K.; Owczarek, C.M.; Hardy, M.P.; Fabri, L.J.; Pearse, M.; Delaine, C.A.; Forbes, B.E. Exogenous administration of protease-resistant, non-matrix-binding IGFBP-2 inhibits tumour growth in a murine model of breast cancer. *Br. J. Cancer* **2014**, *110*, 2855–2864. [CrossRef]

68. Holash, J.; Davis, S.; Papadopoulos, N.; Croll, S.D.; Ho, L.; Russell, M.; Boland, P.; Leidich, R.; Hylton, D.; Burova, E.; et al. VEGF-Trap: A VEGF blocker with potent antitumor effects. *Proc. Natl. Acad. Sci. USA* **2002**, *99*, 11393–11398. [CrossRef]

69. Trieu, Y.; Wen, X.Y.; Skinnider, B.F.; Bray, M.R.; Li, Z.; Claudio, J.O.; Masih-Khan, E.; Zhu, Y.X.; Trudel, S.; McCart, J.A.; et al. Soluble interleukin-13Ralpha2 decoy receptor inhibits Hodgkin's lymphoma growth in vitro and in vivo. *Cancer Res.* **2004**, *64*, 3271–3275. [CrossRef]

70. Tseng, J.F.; Farnebo, F.A.; Kisker, O.; Becker, C.M.; Kuo, C.J.; Folkman, J.; Mulligan, R.C. Adenovirus-mediated delivery of a soluble form of the VEGF receptor Flk1 delays the growth of murine and human pancreatic adenocarcinoma in mice. *Surgery* **2002**, *132*, 857–865. [CrossRef]

71. Messori, A.; Santarlasci, B.; Vaiani, M. New drugs for rheumatoid arthritis. *N. Engl. J. Med.* **2004**, *351*, 937–938. [CrossRef] [PubMed]

72. Hoffman, H.M.; Throne, M.L.; Amar, N.J.; Sebai, M.; Kivitz, A.J.; Kavanaugh, A.; Weinstein, S.P.; Belomestnov, P.; Yancopoulos, G.D.; Stahl, N.; et al. Efficacy and safety of rilonacept (interleukin-1 Trap) in patients with cryopyrin-associated periodic syndromes: Results from two sequential placebo-controlled studies. *Arthritis Rheum.* **2008**, *58*, 2443–2452. [CrossRef] [PubMed]

73. D'Ambrosio, C.; Ferber, A.; Resnicoff, M.; Baserga, R. A soluble insulin-like growth factor I receptor that induces apoptosis of tumor cells in vivo and inhibits tumorigenesis. *Cancer Res.* **1996**, *56*, 4013–4020.

74. Dubois, E.A.; Rissmann, R.; Cohen, A.F. Rilonacept and canakinumab. *Br. J. Clin. Pharm.* **2011**, *71*, 639–641. [CrossRef] [PubMed]

75. Pechtner, V.; Karanikas, C.A.; García-Pérez, L.E.; Glaesner, W. A new approach to drug therapy: Fc-fusion technology. *Prim Health Care* **2017**, 7.

76. Kim, E.S.; Serur, A.; Huang, J.; Manley, C.A.; McCrudden, K.W.; Frischer, J.S.; Soffer, S.Z.; Ring, L.; New, T.; Zabski, S.; et al. Potent VEGF blockade causes regression of coopted vessels in a model of neuroblastoma. *Proc. Natl. Acad. Sci. USA* **2002**, *99*, 11399–11404. [CrossRef] [PubMed]

77. Lindzen, M.; Carvalho, S.; Starr, A.; Ben-Chetrit, N.; Pradeep, C.R.; Kostler, W.J.; Rabinkov, A.; Lavi, S.; Bacus, S.S.; Yarden, Y. A recombinant decoy comprising EGFR and ErbB-4 inhibits tumor growth and metastasis. *Oncogene* **2012**, *31*, 3505–3515. [CrossRef] [PubMed]

78. Strohl, W.R. Fusion Proteins for Half-Life Extension of Biologics as a Strategy to Make Biobetters. *BioDrugs* **2015**, *29*, 215–239. [CrossRef]

79. Marotte, H.; Cimaz, R. Etanercept - TNF receptor and IgG1 Fc fusion protein: Is it different from other TNF blockers? *Expert Opin. Biol. Ther.* **2014**, *14*, 569–572. [CrossRef]

80. Ravetch, J.V.; Bolland, S. IgG Fc receptors. *Annu. Rev. Immunol.* **2001**, *19*, 275–290. [CrossRef]

81. Czajkowsky, D.M.; Hu, J.; Shao, Z.; Pleass, R.J. Fc-fusion proteins: New developments and future perspectives. *Embo Mol. Med.* **2012**, *4*, 1015–1028. [CrossRef]

82. Idusogie, E.E.; Wong, P.Y.; Presta, L.G.; Gazzano-Santoro, H.; Totpal, K.; Ultsch, M.; Mulkerrin, M.G. Engineered antibodies with increased activity to recruit complement. *J. Immunol.* **2001**, *166*, 2571–2575. [CrossRef] [PubMed]

83. Kimchi-Sarfaty, C.; Schiller, T.; Hamasaki-Katagiri, N.; Khan, M.A.; Yanover, C.; Sauna, Z.E. Building better drugs: Developing and regulating engineered therapeutic proteins. *Trends Pharm. Sci* **2013**, *34*, 534–548. [CrossRef] [PubMed]

84. Yang, C.; Gao, X.; Gong, R. Engineering of Fc Fragments with Optimized Physicochemical Properties Implying Improvement of Clinical Potentials for Fc-Based Therapeutics. *Front. Immunol.* **2017**, *8*, 1860. [CrossRef] [PubMed]

85. Dunn, S.E.; Ehrlich, M.; Sharp, N.J.; Reiss, K.; Solomon, G.; Hawkins, R.; Baserga, R.; Barrett, J.C. A dominant negative mutant of the insulin-like growth factor-I receptor inhibits the adhesion, invasion, and metastasis of breast cancer. *Cancer Res.* **1998**, *58*, 3353–3361.

86. Sachdev, D.; Hartell, J.S.; Lee, A.V.; Zhang, X.; Yee, D. A dominant negative type I insulin-like growth factor receptor inhibits metastasis of human cancer cells. *J. Biol. Chem.* **2004**, *279*, 5017–5024. [CrossRef]

87. Min, Y.; Adachi, Y.; Yamamoto, H.; Imsumran, A.; Arimura, Y.; Endo, T.; Hinoda, Y.; Lee, C.T.; Nadaf, S.; Carbone, D.P.; et al. Insulin-like growth factor I receptor blockade enhances chemotherapy and radiation responses and inhibits tumour growth in human gastric cancer xenografts. *Gut* **2005**, *54*, 591–600. [CrossRef]

88. Samani, A.A.; Chevet, E.; Fallavollita, L.; Galipeau, J.; Brodt, P. Loss of tumorigenicity and metastatic potential in carcinoma cells expressing the extracellular domain of the type 1 insulin-like growth factor receptor. *Cancer Res.* **2004**, *64*, 3380–3385. [CrossRef]

89. Wang, N.; Fallavollita, L.; Nguyen, L.; Burnier, J.; Rafei, M.; Galipeau, J.; Yakar, S.; Brodt, P. Autologous bone marrow stromal cells genetically engineered to secrete an igf-I receptor decoy prevent the growth of liver metastases. *Mol. Ther.* **2009**, *17*, 1241–1249. [CrossRef]

90. Wang, N.; Lu, Y.; Pinard, M.; Pilotte, A.; Gilbert, R.; Massie, B.; Brodt, P. Sustained production of a soluble IGF-I receptor by gutless adenovirus-transduced host cells protects from tumor growth in the liver. *Cancer Gene* **2013**, *20*, 229–236. [CrossRef]

91. Wang, N.; Rayes, R.F.; Elahi, S.M.; Lu, Y.; Hancock, M.A.; Massie, B.; Rowe, G.E.; Aomari, H.; Hossain, S.; Durocher, Y.; et al. The IGF-Trap: Novel Inhibitor of Carcinoma Growth and Metastasis. *Mol. Cancer Ther.* **2015**, *14*, 982–993. [CrossRef] [PubMed]

92. Liu, L. Pharmacokinetics of monoclonal antibodies and Fc-fusion proteins. *Protein Cell* **2018**, *9*, 15–32. [CrossRef] [PubMed]

93. Strand, J.; Huang, C.T.; Xu, J. Characterization of Fc-fusion protein aggregates derived from extracellular domain disulfide bond rearrangements. *J. Pharm Sci.* **2013**, *102*, 441–453. [CrossRef] [PubMed]

94. Sadick, M.D.; Intintoli, A.; Quarmby, V.; McCoy, A.; Canova-Davis, E.; Ling, V. Kinase receptor activation (KIRA): A rapid and accurate alternative to end-point bioassays. *J. Pharm. Biomed. Anal.* **1999**, *19*, 883–891. [CrossRef]

95. Chen, J.W.; Ledet, T.; Orskov, H.; Jessen, N.; Lund, S.; Whittaker, J.; De Meyts, P.; Larsen, M.B.; Christiansen, J.S.; Frystyk, J. A highly sensitive and specific assay for determination of IGF-I bioactivity in human serum. *Am. J. Physiol. Endocrinol. Metab* **2003**, *284*, E1149–E1155. [CrossRef] [PubMed]

96. Espelund, U.S.; Bjerre, M.; Hjortebjerg, R.; Rasmussen, T.R.; Lundby, A.; Hoeflich, A.; Folkersen, B.H.; Oxvig, C.; Frystyk, J. Insulin-Like Growth Factor Bioactivity, Stanniocalcin-2, Pregnancy-Associated Plasma Protein-A, and IGF-Binding Protein-4 in Pleural Fluid and Serum From Patients With Pulmonary Disease. *J. Clin. Endocrinol Metab* **2017**, *102*, 3526–3534. [CrossRef] [PubMed]

97. Frystyk, J. Quantification of the GH/IGF-axis components: Lessons from human studies. *Domest Anim. Endocrinol.* **2012**, *43*, 186–197. [CrossRef]

98. Vaniotis, G.; Moffett, S.; Sulea, T.; Wang, N.; Elahi, S.M.; Lessard, E.; Baardsnes, J.; Perrino, S.; Durocher, Y.; Frystyk, J.; et al. Enhanced anti-metastatic bioactivity of an IGF-TRAP re-engineered to improve physicochemical properties. *Sci. Rep.* **2018**, *8*, 17361. [CrossRef]

99. Smith, T.J. Insulin-like growth factor-I regulation of immune function: A potential therapeutic target in autoimmune diseases? *Pharm. Rev.* **2010**, *62*, 199–236. [CrossRef]

100. Xuan, N.T.; Hoang, N.H.; Nhung, V.P.; Duong, N.T.; Ha, N.H.; Hai, N.V. Regulation of dendritic cell function by insulin/IGF-1/PI3K/Akt signaling through klotho expression. *J. Recept. Signal. Transduct.* **2016**, 1–7. [CrossRef]

101. Bilbao, D.; Luciani, L.; Johannesson, B.; Piszczek, A.; Rosenthal, N. Insulin-like growth factor-1 stimulates regulatory T cells and suppresses autoimmune disease. *Embo Mol. Med.* **2014**, *6*, 1423–1435. [CrossRef] [PubMed]

102. Chellappa, S.; Hugenschmidt, H.; Hagness, M.; Line, P.D.; Labori, K.J.; Wiedswang, G.; Tasken, K.; Aandahl, E.M. Regulatory T cells that co-express RORgammat and FOXP3 are pro-inflammatory and immunosuppressive and expand in human pancreatic cancer. *Oncoimmunology* **2016**, *5*, e1102828. [CrossRef] [PubMed]

103. Higashi, Y.; Sukhanov, S.; Shai, S.Y.; Danchuk, S.; Tang, R.; Snarski, P.; Li, Z.; Lobelle-Rich, P.; Wang, M.; Wang, D.; et al. Insulin-Like Growth Factor-1 Receptor Deficiency in Macrophages Accelerates Atherosclerosis and Induces an Unstable Plaque Phenotype in Apolipoprotein E-Deficient Mice. *Circulation* **2016**, *133*, 2263–2278. [CrossRef] [PubMed]

104. Barrett, J.P.; Minogue, A.M.; Falvey, A.; Lynch, M.A. Involvement of IGF-1 and Akt in M1/M2 activation state in bone marrow-derived macrophages. *Exp. Cell Res.* **2015**, *335*, 258–268. [CrossRef]

105. Huang, C.T.; Chang, M.C.; Chen, Y.L.; Chen, T.C.; Chen, C.A.; Cheng, W.F. Insulin-like growth factors inhibit dendritic cell-mediated anti-tumor immunity through regulating ERK1/2 phosphorylation and p38 dephosphorylation. *Cancer Lett* **2015**, *359*, 117–126. [CrossRef]

106. Sanchez-Lopez, E.; Flashner-Abramson, E.; Shalapour, S.; Zhong, Z.; Taniguchi, K.; Levitzki, A.; Karin, M. Targeting colorectal cancer via its microenvironment by inhibiting IGF-1 receptor-insulin receptor substrate and STAT3 signaling. *Oncogene* **2016**, *35*, 2634–2644. [CrossRef]

107. Tolcher, A.W.; Sarantopoulos, J.; Patnaik, A.; Papadopoulos, K.; Lin, C.C.; Rodon, J.; Murphy, B.; Roth, B.; McCaffery, I.; Gorski, K.S.; et al. Phase I, pharmacokinetic, and pharmacodynamic study of AMG 479, a fully human monoclonal antibody to insulin-like growth factor receptor 1. *J. Clin. Oncol. Off. J. Am. Soc. Clin. Oncol.* **2009**, *27*, 5800–5807. [CrossRef]

108. Shaw, A.K.; Pickup, M.W.; Chytil, A.; Aakre, M.; Owens, P.; Moses, H.L.; Novitskiy, S.V. TGFbeta signaling in myeloid cells regulates mammary carcinoma cell invasion through fibroblast interactions. *PLoS ONE* **2015**, *10*, e0117908. [CrossRef]

109. Fernandez, M.C.; Rayes, R.; Ham, B.; Wang, N.; Bourdeau, F.; Milette, S.; Lllemann, M.; Bird, N.; Majeed, A.; Xu, J.; et al. The type I insulin-like growth factor regulates the liver stromal response to metastatic colon carcinoma cells. *Oncotarget* **2017**, *8*, 52281–52293. [CrossRef] [PubMed]

110. Rayes, R.F.; Milette, S.; Fernandez, M.C.; Ham, B.; Wang, N.; Bourdeau, F.; Perrino, S.; Yakar, S.; Brodt, P. Loss of neutrophil polarization in colon carcinoma liver metastases of mice with an inducible, liver-specific IGF-I deficiency. *Oncotarget* **2018**, *9*, 15691–15704. [CrossRef] [PubMed]

111. Burnier, J.V.; Wang, N.; Michel, R.P.; Hassanain, M.; Li, S.; Lu, Y.; Metrakos, P.; Antecka, E.; Burnier, M.N.; Ponton, A.; et al. Type IV collagen-initiated signals provide survival and growth cues required for liver metastasis. *Oncogene* **2011**, *30*, 3766–3783. [CrossRef] [PubMed]

112. Vaniotis, G.; Rayes, R.F.; Qi, S.; Milette, S.; Wang, N.; Perrino, S.; Bourdeau, F.; Nystrom, H.; He, Y.; Lamarche-Vane, N.; et al. Collagen IV-conveyed signals can regulate chemokine production and promote liver metastasis. *Oncogene* **2018**, *37*, 3790–3805. [CrossRef] [PubMed]

113. Lu, P.; Weaver, V.M.; Werb, Z. The extracellular matrix: A dynamic niche in cancer progression. *J. Cell Biol.* **2012**, *196*, 395–406. [CrossRef] [PubMed]

114. Brodt, P. Role of the Microenvironment in Liver Metastasis: From Pre- to Prometastatic Niches. *Clin. Cancer Res.* **2016**, *22*, 5971–5982. [CrossRef] [PubMed]

115. Joyce, J.A.; Pollard, J.W. Microenvironmental regulation of metastasis. *Nat. Reviews. Cancer* **2009**, *9*, 239–252. [CrossRef]

116. Lindau, D.; Gielen, P.; Kroesen, M.; Wesseling, P.; Adema, G.J. The immunosuppressive tumour network: Myeloid-derived suppressor cells, regulatory T cells and natural killer T cells. *Immunology* **2013**, *138*, 105–115. [CrossRef]

117. Milette, S.; Sicklick, J.K.; Lowy, A.M.; Brodt, P. Molecular Pathways: Targeting the Microenvironment of Liver Metastases. *Clin. Cancer Res.* **2017**, *23*, 6390–6399. [CrossRef]

118. Franklin, C.; Livingstone, E.; Roesch, A.; Schilling, B.; Schadendorf, D. Immunotherapy in melanoma: Recent advances and future directions. *Eur J. Surg Oncol* **2017**, *43*, 604–611. [CrossRef]

119. Muto, A.; Gridelli, C. New immunotherapy in the treatment of advanced renal cancer. *Expert Opin. Emerg. Drugs* **2019**. [CrossRef]

120. Osmani, L.; Askin, F.; Gabrielson, E.; Li, Q.K. Current WHO guidelines and the critical role of immunohistochemical markers in the subclassification of non-small cell lung carcinoma (NSCLC): Moving from targeted therapy to immunotherapy. *Semin. Cancer Biol.* **2018**, *52*, 103–109. [CrossRef]

121. Nevala-Plagemann, C.; Hidalgo, M.; Garrido-Laguna, I. From state-of-the-art treatments to novel therapies for advanced-stage pancreatic cancer. *Nat. Rev. Clin. Oncol.* **2019**. [CrossRef] [PubMed]

122. Upadhrasta, S.; Zheng, L. Strategies in Developing Immunotherapy for Pancreatic Cancer: Recognizing and Correcting Multiple Immune "Defects" in the Tumor Microenvironment. *J. Clin. Med.* **2019**, *8*, 1472. [CrossRef] [PubMed]

123. Aroldi, F.; Zaniboni, A. Immunotherapy for pancreatic cancer: Present and future. *Immunotherapy* **2017**, *9*, 607–616. [CrossRef] [PubMed]

124. Crudden, C.; Ilic, M.; Suleymanova, N.; Worrall, C.; Girnita, A.; Girnita, L. The dichotomy of the Insulin-like growth factor 1 receptor: RTK and GPCR: Friend or foe for cancer treatment? *Growth Horm. IGF Res.* **2015**, *25*, 2–12. [CrossRef]

125. Girnita, L.; Worrall, C.; Takahashi, S.; Seregard, S.; Girnita, A. Something old, something new and something borrowed: Emerging paradigm of insulin-like growth factor type 1 receptor (IGF-1R) signaling regulation. *Cell Mol. Life Sci.* **2014**, *71*, 2403–2427. [CrossRef]

126. Monteverde, M.; Milano, G.; Strola, G.; Maffi, M.; Lattanzio, L.; Vivenza, D.; Tonissi, F.; Merlano, M.; Nigro, C.L. The relevance of ADCC for EGFR targeting: A review of the literature and a clinically-applicable method of assessment in patients. *Crit. Rev. Oncol./Hematol.* **2015**, *95*, 179–190. [CrossRef]

Review

New Insights from IGF-IR Stimulating Activity Analyses: Pathological Considerations [†]

Joseph A.M.J.L. Janssen

Department of Internal Medicine, Erasmus MC, Room Rg 527 Dr. Molewaterplein 40, 3015 GD Rotterdam, The Netherlands; j.a.m.j.l.janssen@erasmusmc.nl; Tel.: +31-06-5003-2421 or +31-10-7040704
† Abbreviated Title: The Complexity of IGF-IR Stimulating Activity.

Received: 5 March 2020; Accepted: 1 April 2020; Published: 2 April 2020

Abstract: Insulin-like growth factor-I (IGF-I) and insulin-like growth factor-II (IGF-II) play a crucial factor in the growth, differentiation and survival of cells in health and disease. IGF-I and IGF-II primarily activate the IGF-I receptor (IGF-IR), which is present on the cell surface. Activation of the IGF-IR stimulates multiple pathways which finally results in multiple biological effects in a variety of tissues and cells. In addition, activation of the IGF-IR has been found to be essential for the growth of cancers. The conventional view in the past was that the IGF-IR was exclusively a tyrosine kinase receptor and that phosphorylation of tyrosine residues, after binding of IGF-I to the IGF-IR, started a cascade of post-receptor events. Recent research has shown that this view was too simplistic. It has been found that the IGF-IR also has kinase-independent functions and may even emit signals in the unoccupied state through some yet-to-be-defined non-canonical pathways. The IGF-IR may further form hybrids with the insulin receptors but also with receptor tyrosine kinases (RTKs) outside the insulin-IGF system. In addition, the IGF-IR has extensive cross-talk with many other receptor tyrosine kinases and their downstream effectors. Moreover, there is now emerging evidence that the IGF-IR utilizes parts of the G-protein coupled receptor (GPCR) pathways: the IGF-IR can be considered as a functional RTK/GPCR hybrid, which integrates the kinase signaling with some IGF-IR mediated canonical GPCR characteristics. Like the classical GPCRs the IGF-IR can also show homologous and heterologous desensitization. Recently, it has been found that after activation by a ligand, the IGF-IR may be translocated into the nucleus and function as a transcriptional cofactor. Thus, in recent years, it has become clear that the IGF-IR signaling pathways are much more complex than first thought. Therefore a big challenge for the (near) future will be how all the new knowledge about IGF-IR signaling can be translated into the clinical practice and improve diagnosis and treatment of diseases.

Keywords: IGF-I; IGF-II; insulin; IGF-IR; IRs; tyrosine kinase receptor; GPCRs; hybrids; phosphorylation; G-proteins; β-arrestins; functional RTK/GPCR hybrid; nuclear translocation

1. Introduction

The insulin-IGF system is formed by insulin, two insulin-like growth factors (IGF-I and IGF-II), four cell-membrane receptors (insulin receptor-A (IR-A), insulin receptor-B (IR-B), insulin-like growth factor-I receptor (IGF-IR) and insulin-like growth factor receptor-II (IGF-II-R)) and six IGF-binding proteins (IGFBP-1-6), several IGFBP- related proteins and IGFBP proteases [1–4]. All IGFBPs can bind both IGF-I and IGF-II (however with different binding affinity for some) [5]. Only the unbound forms of IGFs are thought to interact with the IGF-IR and the IGF-II receptor [6].

The IGF-I gene comprises a highly conserved sequence and contains six exons, which give rise to heterogeneous mRNA transcripts by a combination of multiple transcription initiation sites and alternative splicing [7]. These multiple transcripts code in humans for different precursor IGF-I polypeptides, namely the IGF-IEa, IGF-IEb and IGF-IEc isoforms, which also undergo posttranslational modifications, such as proteolytic processing and glycosylation [7]. Differential biological activities have been reported for the different IGF-I isoforms and thus both common and unique or complementary pathways exist for the IGF-I isoforms to promote biological effects [7].

As IGFs and insulin as well as the IGF-IR and the IRs share high sequence homology, they are able to bind and activate each other's cognate receptors but with considerably lower avidity. The IGF-IR can bind IGF-I and IGF-II with equally high affinity (10^{-10} M) whereas its affinity for insulin (10^{-8} M) is much lower [8]. In the past it was thought that the IGFs and the IGF-IR predominantly mediated growth-promoting effects whereas insulin and the IRs predominantly mediated metabolic effects [9,10]. However, in certain circumstances IGF-I and insulin can mediate very similar responses [11]. Nevertheless, IGF-I and IGF-II play a crucial factor in the regulation of growth, proliferation, differentiation, migration and survival of cells. In addition, activation of IGF-IR and its intracellular pathways has been found to be essential for growth of cancers [12].

IGF-IIR regulates the amount of circulating and tissue IGF-II by transporting IGF-II into the cell and degrading it [13]. IGF-II can also bind to the IGF-IR with high affinity [13].

Due to alternative splicing of exon 11 of the IR gene, two IR transcripts are generated in the human body: IR-A (lacking exon 11) and IR-B (with exon 11) [14–16]. The IR-A is predominantly expressed in fetal tissues, the central nervous system, hematopoietic cells and in cancer tissues [14]. The IR-B is predominantly expressed in the liver, muscles and fat cells, the major target tissues for the metabolic effects of insulin [14]. The binding of insulin to IR-B will mainly induce metabolic effects (glucose uptake, glycogen synthesis, glycolysis and fatty acid synthesis) in liver, muscles and adipocytes [14]. Binding of insulin to the IR-A will predominantly induce growth-promoting effects in fetal tissues and tumors. In contrast to IR-B, the IR-A may also bind IGF-II with high affinity and thereby stimulate growth-promoting effects [14].

Although the liver is the main producer of the circulating IGFs, the IGFs are synthesized in almost all tissues of the body [3].

The Insulin-like Growth Factor Binding Proteins (IGFBPs) are a family of six proteins with high affinity for the IGFs. They are widely expressed in most tissues and are flexible endocrine and autocrine/paracrine regulators of IGF activity, which is essential for this important physiological system [3]. IGFBPs may affect cells in both an IGF-dependent and -independent manner [5]. Although IGFBPs often inhibit IGF actions in many circumstances, in some conditions they may also potentiate IGF actions [5].

2. IGF-I and the IGF-I Receptor

The IGF-IR is displayed on the cell surface and expressed by nearly all human tissues and cell types [5,9,17]. Surface density of the IGF-IR represents an important determinant of the magnitude of responses to IGF-I and the signaling pattern it provokes [18]. The IGF-IR is a heterotetrameric transmembrane protein composed of two alpha and two beta subunits which are linked by disulfide bonds [19,20]. The beta subunit of the IGF-IR consists of a short extracellular domain which is involved in linkage to the alpha subunits, a transmembrane domain and a cytoplasmic domain containing tyrosine kinase activity [21]. The beta subunit contains a consensus ATP-binding sequence and multiple tyrosine residues that are phosphorylated following ligand binding to the alpha subunit [21]. Binding of IGF-I or another ligand to the alpha subunit of the IGF-IR, induces a closer proximity of regions within the transmembrane domain resulting in autophosphorylation of three intracellular tyrosine residues (Tyr^{1131}, Tyr^{1135}, and Tyr^{1136}) within the beta subunit [21–23].

The conventional view was that the IGF-IR was exclusively a tyrosine kinase receptor and that the binding of IGF-I to the IGF-IR was essential to start the intracellular downstream signal cascade (Figure 1). In this model, the activated receptor recruited and phosphorylated intracellularly substrates as the insulin receptor substrate proteins (IRSs) and SH2 containing collagen-related proteins (SHC) (Figure 1). Tyrosine phosphorylation of the IRSs in turn activated then the phosphatidylinositol 3-kinase (PI3K-Akt) pathway and its various biological responses, while tyrosine phosphorylation of SHC induced downstream signaling activation through the Ras/Raf/MEK/Erk pathway [24,25] (Figure 1).

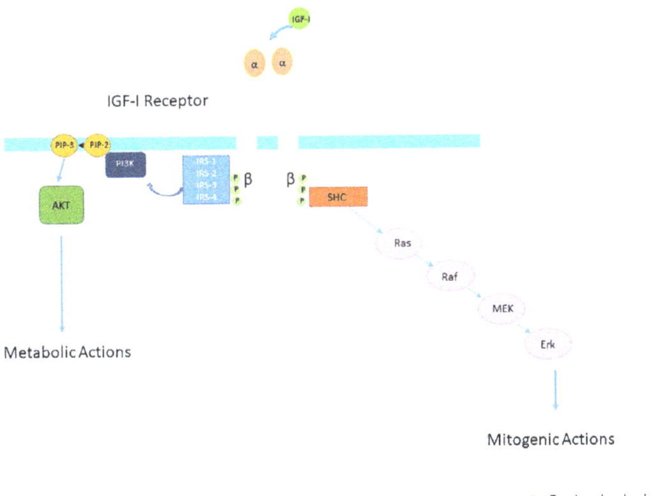

Figure 1. The Insulin-like Growth Factor-I (IGF-IR) is a transmembrane protein composed of two alpha (α) and two beta (β) subunits. The conventional view was that the IGF-IR was exclusively a tyrosine kinase receptor and that the binding of IGF-I to the IGF-IR started the intracellular downstream signal cascade. In this model IGF-I or IGF-II binding to the IGF-IR promotes tyrosine kinase activity and autophosporylation of the beta subunit of the IGF-IR. Intracellularly the activated IGF-IR receptor recruits phosphorylated substrates Insulin receptor substrates (IRSs) and SH2 containing collagen-related proteins (SHC). Tyrosine phosphorylation of IRSs and SHC proteins induces downstream signaling activation through the PI3K-AKT and Ras/Raf/MEK/Erk pathways. It was further thought that activation of the PI3K-AKT pathway had predominantly metabolic effects whereas activation of the Ras/Raf/MEK/Erk pathway had predominantly mitogenic effects.

3. The IGF-IR and Endocytosis

Many signaling receptors internalize via clathrin-coated pits [26]. Endocytosis of signaling receptors is widely recognized to confer control on cellular signaling responsiveness [27]. Ligand-induced activation typically increases receptor endocytic rate, and internalized receptors engage molecular sorting machineries that specify subsequent transport via divergent lysosomal and recycling routes [27]. These events, in turn, determine the degree to which cellular ligand responsiveness is attenuated ("down-regulated") or sustained ("re-sensitized") under conditions of prolonged or repeated ligand exposure [27].

The molecular basis for the close interactions between IGF-IR endocytosis and its signaling components is still poorly understood. Recently, it has been suggested that the ability of insulin receptor substrate-1 (IRS-1) to interact with the clathrin adapter protein AP2, which is essential for endocytosis, plays an important role in IGF-IR internalization [28]. Overexpression of IRS-I resulted in the accumulation of activated IGF-IR at the cellular membrane [28]. Conversely, knockdown of IRS-I induced faster internalization of IGF-IRs [28]. These data suggest that IRS-1 inhibits the recruitment of IGF-IR into clathrin-coated structures; the ability of IRS-I to bind to AP-2 avoids rapid endocytosis of the IGF-IR and prolongs its activity at the cell surface in HEK293T cells [28] (Figure 2A). In contrast, accelerating IGF-IR endocytosis via IRS-1 depletion induces the shift from sustained to transient Akt signaling [28] (Figure 2B). Thus, independent of its classic role as an adaptor in IGF-I receptor signaling,

IRS-1 has a role as an endocytic regulator of IGF-I receptor that ensures sustained IGF bioactivity, while IRS-1 degradation could be a trigger to internalize the IGF-IR [29].

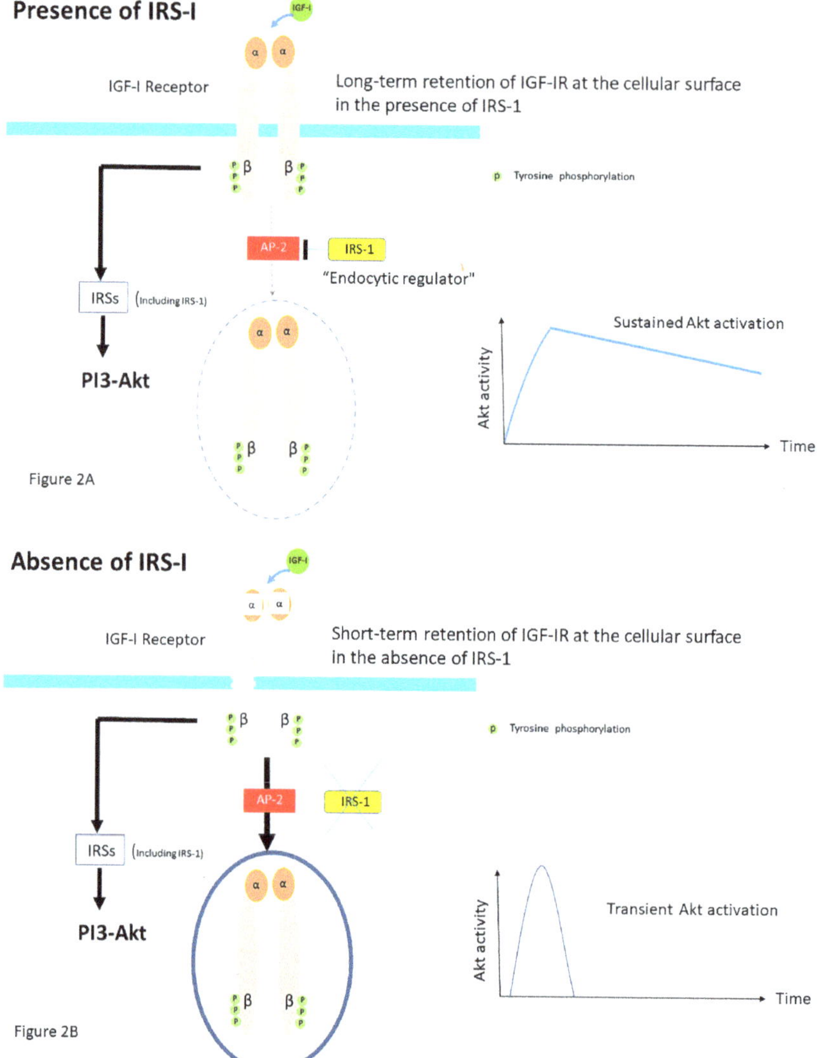

Figure 2. Proposed role of Insulin receptor substrate-1 (IRS-1). IRS-1 modulates how long ligand-activated IGF-IR remains at the cell surface before undergoing endocytosis in mammalian cells. IRS-1 interacts with the clathrin adaptor complex AP2. (**A**) In the presence of the IRS-1/AP2-complex in the cell IGF-IR endocytosis after the ligand stimulation is delayed. Mechanistically, IRS-1 inhibits the recruitment of IGF-IR into clathrin-coated structures; for this reason, IGF-IR avoids rapid endocytosis and prolongs its activity on the cell surface and this results in sustained activation of the AKT pathway. (**B**) In absence of IRS-1/AP2- complex in the cell, there is only short-term retention of the IGF-IR at the cell surface and IGF-IR endocytosis is accelerated. This results in a transient activation of the AKT pathway (Modified from Yoneyama et al. IRS-1 acts as an endocytic regulator of IGF-I receptor to facilitate sustained IGF signaling. eLIFE, 2018; 7. pii: e32893).

For the IGF-IR, ubiquitination also increases upon ligand binding [30]. The IGF-IR has been demonstrated to be a substrate for three ubiquitin ligases: Mdm2, (in human malignant melanoma cells), c-Cbl (HEK293 cells and human osteosarcoma cell lines U2OS and SAOS2) and Nedd4 (in mouse embryo fibroblasts). [30,31]. Mdm2 was originally described to control IGF-IR ubiquitination and thereby causing its degradation by the proteasome system [32]. Subsequently β-arrestins, known to be involved in the regulation of G-protein-coupled receptors (GPCRs), have also been identified as adaptor proteins to bring the oncoprotein Mdm2 to the IGF-IR in mouse embryo fibroblasts [33,34]. In addition, while removing the IGF-IR from the cell surface and inhibiting the "classical" kinase signaling pathway, β-arrestins may redirect the signaling wave through ERK [33] Ubiquitination may thus induce receptor internalization and degradation, but also enhance IGF-IR signaling [35] (which will be further addressed in the paragraph "the complexity of the post-receptor IGF-IR/IR pathways" below).

4. Structural Differences and Overlap between the IGF-IR and the IRs

It is hypothesized that the IGF-IR and IRs are created by gene duplication of common precursor receptor molecule [36]. Due to structural and functional homology, IGF-I and insulin can bind to (and activate) both IGF-IR and the IRs, as discussed above [37]. IGF-IR and IRs show 48% amino acid sequence homology [20]. Structural differences between the beta-subunit and kinase domains of the IGF-IR and the IRs leading to differences in substrate interactions may be (partly) responsible for IGF-I and insulin specificity as has been found in various cell types (rat-1 fibroblasts, murine skin keratinocytes and in NIH-3T3-fibroblasts) [38]. However, the signal transduction by the receptors may not be limited to its activation at the cell surface [39].

In addition to signaling through the classical tyrosine kinase pathways, it has been found that the IGF-IRs and IRs (in cells derived from C57Bl/6 mice) can emit signals in the unoccupied state through some yet-to-be-defined non-canonical pathways [40]. Boucher et al. demonstrated that cells lacking the IGF-IR and IR exhibit a major decrease in expression of multiple imprinted genes and microRNAs [40].

Although the IGF-IR and IRs have both distinct and overlapping functions, it has been suggested that in vivo specificity of the IGFs and insulin are at least in part reflected by the timing of the expression of the IGF-IR and IRs in target tissues in combination with ligand concentration and availability [41]. It has been further suggested that IGF-I and the IRs act as identical portals for the regulation of gene expression and that the differences between IGF-I and insulin effects are due to a modulation of the amplitude of the signal created by the specific ligand receptor interaction [41].

5. The IGF-IR and the IRs May Form Hybrids in the Human Body

In cells and tissues where both significant levels of the IGF-IRs and IRs are present, hybrids may be formed consisting of an alpha-beta subunit of the IGF-IR linked by disulfide bonds to an alpha-beta subunit of the IR [42] (Figure 3A). They are formed in the endoplasmic reticulum before they reach the cell surface [43]. Two splice variants of the hybrid receptors exist for the IR because the IR is expressed (as above discussed) either with (IR-B) or without 12 amino acids encoded by exon 11 (IR-A) [44]. Thus both IR-A/IGF-IR (Hybrid A) and IR-B/IGF-IR (Hybrid B) receptors can be formed. Although the biological functions of these hybrid receptors is still unclear, it has been suggested that hybrid receptors may play a role in the overlapping functions of IGF-I and insulin [21]. Several studies (in baby hamster kidney cells, NIH3T3 cells overexpressing IGFR and CHO cells overexpressing IR-B) have suggested that cells by increasing the relative expression level of the IGF-IR above that of the IR lose their insulin sensitivity because hybrid receptors bind insulin with low affinity [45,46]. In addition, binding of insulin to the alpha-beta subunit of the IR which is part of a hybrid, may result in autophosphorylation of its ow beta subunit and, following transphosphorylation of the beta subunit of the IGF-IR, may result in a signal for growth, [9]. In contrast, when IGF-I binds to the alpha-beta subunit of the IGF-IR, this may activate the beta subunit of the IRs by the same mechanisms and thereby activate growth (IR-A) or metabolism (IR-B) [9]. Although this latter mechanism could explain why hybrids may stimulate metabolic functions when stimulated by IGF-I, most functional studies have found that hybrid receptors behave more like IGF-IRs than IRs [47]. It has been hypothesized that this prioritization of hybrid receptors to IGF-I results from the ability of IGF-I to activate monomeric IGF-IR whereas, in contrast, dimerization of the IR has been considered necessary to induce a response to insulin [45].

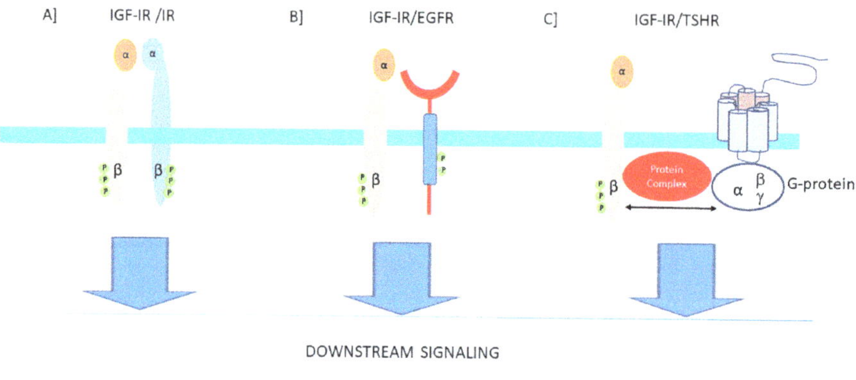

Figure 3. The IGF-I receptor may form hybrids with the insulin receptor, many other tyrosine kinase receptors outside the insulin-IGF system and G-protein coupled receptors. The figure shows three examples: (**A**) Hybrids may be formed consisting of an alpha-beta subunit of the IGF-IR linked by disulfide bonds to an alpha-beta subunit of the IR. Downstream signaling of both receptors converge via the canonical PI3K-Akt and ERK signaling pathways. Most functional studies have found that hybrid receptors behave more like IGF-IRs than IRs (See also text). (**B**) Hybrids may be formed consisting of an alpha-beta subunit of the IGF-IR linked and a monomer of the epidermal growth factor receptor (EGFR) which is also a tyrosine kinase receptor. Downstream signaling of both receptors converge via the canonical PI3K-Akt and ERK signaling pathways. Therefore, inhibition of one receptor of these hybrids may shift the signaling pathway in favor of the other available counterpart receptor. (**C**) The Thyroid Stimulating Hormone Receptor (TSH receptor), a typical G-protein coupled receptor, may form functional hybrids with the IGF-IR in the cellular membrane by forming a common protein complex. Bidirectional crosstalk between the IGF-IR and TSHR has been demonstrated. Stimulation of the IGF-IR by IGF-I/IGF-IR agonists may trigger the classical signaling pathway of the IGF IR, leading to downstream kinase-cascade signaling activation. In addition, stimulation of the IGF-IR by IGF-I/IGF-IR agonists may also utilize components of G-protein coupled receptor (GPCR) signaling and activate pathways conventionally used by TSHR.

In cells and tissues where both significant levels of the IGF-IRs and IRs are present, hybrids may be formed consisting of an alpha-beta subunit of the IGF-IR linked by disulfide bonds to an alpha-beta subunit of the IR [42] (Figure 3A). They are formed in the endoplasmic reticulum before they reach the cell surface [43]. Two splice variants of the hybrid receptors exist for the IR because the IR is expressed (as above discussed) either with (IR-B) or without 12 amino acids encoded by exon 11 (IR-A) [44]. Thus both IR-A/IGF-IR (Hybrid A) and IR-B/IGF-IR (Hybrid B) receptors can be formed. Although the biological functions of these hybrid receptors is still unclear, it has been suggested that hybrid receptors may play a role in the overlapping functions of IGF-I and insulin [21]. Several studies (in baby hamster kidney cells, NIH3T3 cells overexpressing IGFR and CHO cells overexpressing IR-B) have suggested that cells by increasing the relative expression level of the IGF-IR above that of the IR lose their insulin sensitivity because hybrid receptors bind insulin with low affinity [45,46]. In addition, binding of insulin to the alpha-beta subunit of the IR which is part of a hybrid, may result in autophosphorylation of its ow beta subunit and, following transphosphorylation of the beta subunit of the IGF-IR, may result in a signal for growth, [9]. In contrast, when IGF-I binds to the alpha-beta subunit of the IGF-IR, this may activate the beta subunit of the IRs by the same mechanisms and thereby activate growth (IR-A) or metabolism (IR-B) [9]. Although this latter mechanism could explain why hybrids may stimulate metabolic functions when stimulated by IGF-I, most functional studies have found that hybrid receptors behave more like IGF-IRs than IRs [47]. It has been hypothesized that this prioritization of hybrid receptors to IGF-I results from the ability of IGF-I to activate monomeric IGF-IR whereas, in contrast, dimerization of the IR has been considered necessary to induce a response to insulin [45].

It has been further hypothesized that IGF-IR/IR hybrids may affect tumor biology [48]. Specific downregulation of the IGF-IR by agents solely targeting the IGF-IR diminishes hybrid formation and this thereby enhances holo-IR formation [48]. An enhanced holo-IR formation results in an increase of insulin sensitivity [48]. As the IR, especially the IR-A, may also activate (post-receptor) signaling pathways involved in growth similar to the IGF-IR, the development of agents simultaneous targeting both IR-A and IGF-IR may be necessary to disrupt the malignant phenotype of cancers cells that are influenced by actions of the insulin-IGF system [48].

The IGF-IR may also heterodimerize with receptor tyrosine kinases (RTKs) outside the insulin-IGF system [49]. Heterodimerization of the IGF-IR with the EGFR is well-established [50] (Figure 3B). Downstream signaling of both receptors converge via the canonical PI3K-Akt and ERK signaling pathways. Therefore inhibition of one receptor in these hybrids may shift the signaling pathway in favor of the other available counterpart receptor [49,51,52]. Thus the compensatory signaling may be bidirectional [53]. Moreover, evidence exists that IGF-IR can activate independently downstream EGFR pathways and this may subsequently result in EFGR tyrosine kinase inhibitor (TKI) resistance [52]. The IGF-IR signaling pathway shows also cross-talk with the growth hormone receptor (GHR), thyroid stimulating hormone receptor (TSHR) (Figure 3C), estrogen receptor (ER), androgen receptor (AR) and human epidermal growth factor receptor 2 (HER-2) signaling pathways [54–59].

6. The Functional Relationship between Insulin/IGF Signaling and Discoidin Domain Receptors

In addition to its canonical role as a collagen receptor, it has recently been suggested that the discoidin domain receptor-1 (DDR1), a tyrosine kinase receptor, plays an important role in the regulation of the insulin-IGF system [60]. In contrast to most other RTKs, DDR1 is not activated by soluble growth factors but instead by various type of collagens [60]. While most other RTKs are fully activated in minutes, maximal activation of DDR1 occurs several hours after initial stimulation with collagen [60]. The DDR1 and the insulin-IGF system are linked by a feed-forward mechanism by which insulin and the IGFs induce DDR1 upregulation which in turn enhances expression and activity of the IRs and the IGF-IR [60]. The mechanisms by which DDR1 may affect downstream signaling of the IRs and IGF-IR are as yet not fully understood. It has been found that increasing DDR1 expression favors the expression of the more mitogenic IR-A isoform over the metabolic IR-B isoform and thus one of the functional consequences of this DDR1 upregulation may be increased IGF-II signaling through the IR-A [60]. This in turn may favor dedifferentiation and stem-like features [60]. It has been further hypothesized that inhibition of DDR1 may be a way to downregulate the tumor-inducing actions of the insulin/IGF system in human cancer cells while simultaneously inducing differentiation of cells without affecting the IR-B mediated metabolic actions [60]. In favor of this latter hypothesis, in vitro DDR1 silencing or downregulation blocked the IGF-2/IR-A autocrine loop in poorly differentiated thyroid cancer cells and induced cellular differentiation [61]. Although at present no clinical studies have shown that this strategy provides any clinical benefits for patients with tumors overexpressing the DRR1 and the insulin-IGF-I system, these results may help to develop novel therapeutic approaches for cancer.

7. The Complexity of the Post-Receptor IGF-IR/IR Pathways

In recent years, it has been become clear that the downstream complexity of the IGF-IR/IR pathways was grossly underestimated in the past. In this section we will discuss some important novel insights regarding this complexity. In the classical model, as discussed above, the IGF-IR was traditionally described as a tyrosine kinase receptor with an ON/OFF (active/inactive) system. In this system IGF-I binding to the IGF-IR stabilized the ON state (active) and this exclusively mediated kinase-dependent signaling activation of both the PI3-AKT and ERK pathways [62,63] (Figure 4A). Ubiquitin-mediated receptor downregulation and degradation was originally described as a response to ligand/receptor interaction and thus inseparable from kinase signaling activation [62].

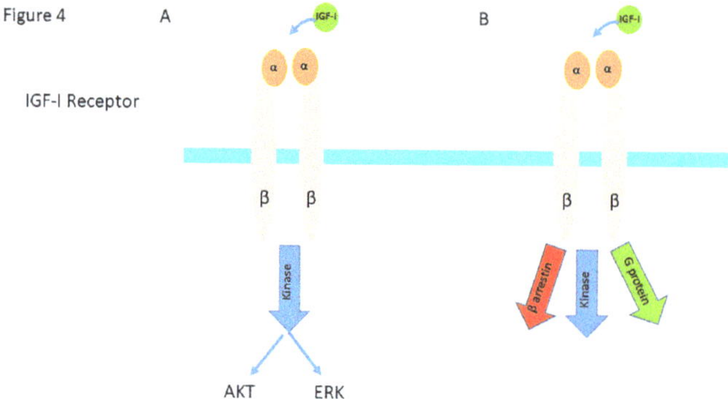

Figure 4. (**A**) In the classical model IGF-IR activation was triggered exclusively by ligand binding and signaling was exclusively mediated by a kinase cascade through phosphorylation. The ligand-activated IGF-IR was thought to lead to a balanced stimulation of the AKT/ERK pathways. (Abbreviation AKT= protein kinase B; ERK= extracellular signal –regulated kinase) (**B**) In the current model, binding of a ligand to the IGF-IR results not only in stimulating of the kinase cascade through phosphorylation of IRS-1, PI3K and AKT but also in activation and signaling by G-proteins and β-arrestins, as well as desensitization and internalization by β-arrestins. In this model ligand binding results in a balanced activation and signaling of the kinase cascade, G-proteins and β-arrestins, as well as desensitization and internalization by β-arrestins. (Modified from Girnita et al. Something old, something new and something borrowed: emerging paradigm of insulin-like growth factor type 1 receptor (IGF-1R) signaling regulation. Cell Mol Life Sci. 2014; 71:2403-27).

Almost 25 years ago, a study from Nobel Prize winner Dr. Robert Lefkowitz's laboratory reported that IGF-IR-dependent activation of the Mitogen-Activated Protein Kinase (MAPK) signaling pathway was inhibited by the Gαi-inhibitor pertussis toxin [64]. The last years there is emerging evidence that many RTKs can also utilize heterotetrameric G proteins to subserve some of their biological actions [65–67]. Recently, more extensive evidence has been published that the IGF-IR and IRs are also engaged in in G-protein coupled receptor (GPCR) signaling [62,63,65–67]. As GPCR phosphorylation by GPCR-kinases (GRKs) governs interactions of the receptors with β-arrestins, Zheng et al. investigated the regulatory roles of the four widely expressed GRKs on IGF-IR signaling/degradation [68]. They found that lowering GRK5/6 abolished IGF-I-mediated ERK and AKT activation, whereas GRK2 inhibition increased ERK activation and partially inhibited AKT signaling [68]. In addition, β-arrestin-mediated ERK signaling was enhanced by overexpression of GRK6 and diminished by GRK2. Similarly, they demonstrated opposing effects of GRK2 and -6 on IGF-IR degradation: GRK2 decreased whereas GRK6 enhanced ligand-induced degradation [68]. GRK2 and GRK6 co-immunoprecipitated with IGF-IR and increased IGF-IR serine phosphorylation, promoting β-arrestin1 association. Thus this study demonstrated distinct roles for GRK2 and GRK6 in IGF-IR signaling through β-arrestin binding with divergent functional outcomes [68].

Based on the insight that IGF-IR may also "borrow" components of GPCR signaling, including β-arrestins and G-protein-coupled-receptor kinases (GRKs), a new paradigm has emerged for the IGF-IR [62]. In this new paradigm, the IGF-IR is considered to be a functional RTK/GPCR hybrid, which integrates the kinase signaling with some IGF-IR mediated canonical GPCR characteristics [62]. Binding of IGF-I to the IGF-IR thus not only leads to balanced phosphorylation-dependent Akt/ERK signaling intracellularly, but results simultaneously also in activation of signaling by G-proteins and β-arrestins [62] (Figures 4B and 5A).

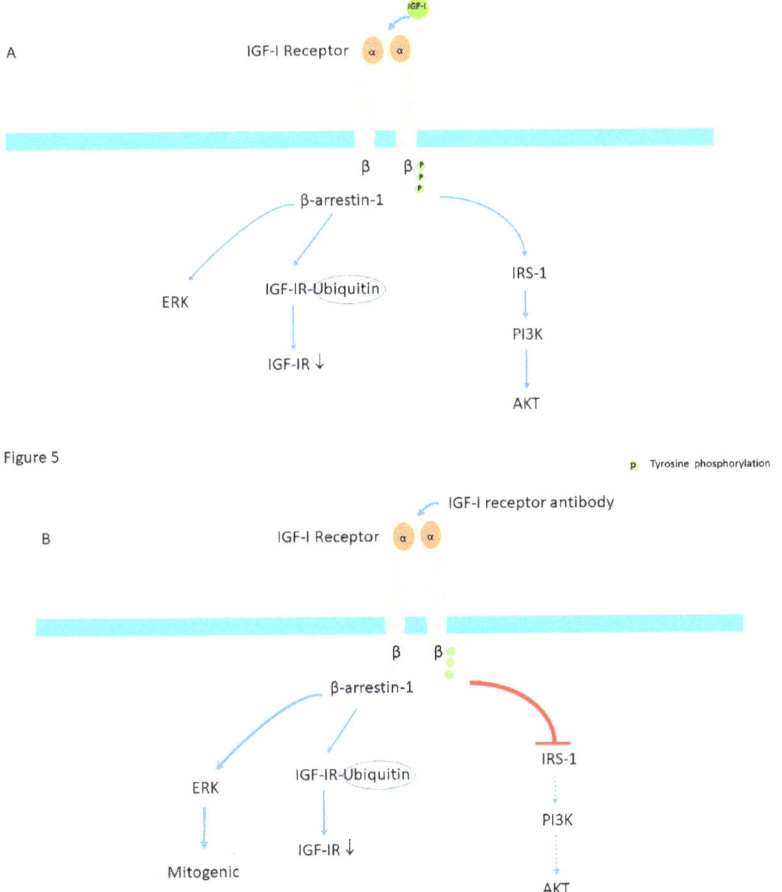

Figure 5. (**A**) Mechanisms of balanced agonism; activation of the IGF-IR stimulates not only the AKT pathway by phosphorylation of IRS-I and PI3K, but in addition, stimulates the β-arrestin-1 (β-arr1) pathway which leads to proteasomal degradation of the IGF-IR through an ubiquitin (Ub)-mediated mechanism and ERK activation. (**B**) Beta-arrestin-1 biased agonism. Binding of monoclonal (blocking) antibodies directed against the IGF-IR block the kinase cascade pathway (by blocking phosphorylation of IRS-1, PI3K and AKT) but simultaneously activate the β-arrestin-1 (β-arr1) pathway which induces enhanced IGF-IR receptor internalization (ubiquitination) and activation of ERK signaling pathway. (Modified from Salisbury & Tomblin. Insulin/Insulin-like growth factors in cancer: new roles for the aryl hydrocarbon receptor, tumor resistance mechanisms, and new blocking strategies. Front Endocrinol (Lausanne), 2015; 6:12).

The IRs have been shown also to interact with G-proteins and β-arrestin-1 [69]. However, the IGF-IR and IRs engage in different G-proteins for downstream signaling [65]. This possibly provides a mechanism that is responsible for the signaling specificity of these two receptors [65].

Also another new paradigm, the paradigm of biased signaling, has been proposed for IGF-IR and IR signaling [67,70]. The paradigm of biased signaling also originates from the GPCR signaling field [71]. The regulatory process which was discovered as the means by which classical GPCRs "desensitized" or tuned off, has been also found to be active for the IGF-IR [71]. In this paradigm the ligand is biased towards a specific signaling i.e., the signal mediated by binding of a ligand to a receptor is no longer balanced but the ligand elicits one response of the ligand over another compared with the classical ligand [53,63]. The IGF-IR has been extensively studied as an anti-cancer target However, monotherapy trials with IGF-IR targeted antibodies, have, overall, been very disappointing [12]. The anti-IGF-IR antibody Figitumumab (CP-751,871; CP) was designed as an antagonist to prevent ligand-receptor interaction [72]. Although it was found that CP blocked the kinase cascade pathway (by blocking phosphorylation of IRS-1, PI3K and AKT), as with all anti-IGF-IR antibodies, it simultaneously induced IGF-IR/β-arrestin-1 association with dual functional outcome: receptor ubiquitination and degradation and decrease in cell viability and β-arrestin-1-dependent ERK signaling activation [72].

Thus despite blocking all the "classical" tyrosine kinase-mediated effects of the IGF-IR, a blocking antibody directed against the IGF-IR may function as an IGF-IR/β-arrestin-1/ERK agonist and favor β-arrestin-1/ERK signaling [73] (Figure 5B). Another example of biased signaling of the IGF-IR signaling is mediated by LL-37, a newly recognized bacterial peptide, which after binding to the IGF-IR may function as an IGF-IR agonist by increasing β-arrestin-1 signaling and electively activating the ERK pathway but without affecting simultaneously the PI3K-AKT pathway [74].

One should make a distinction between homologous and heterologous desensitization. Homologous desensitization occurs within a receptor system when it alters its own responsiveness, for example the loss of responsiveness (desensitization) that can occur upon binding of insulin to the IR [75]. It is considered to limit or restrain a cell's responses to certain stimuli; it leaves a cell (transiently) less- or unresponsive to a ligand that activates the desensitized receptor but not to ligands that activate other receptors. In contrast, in heterologous desensitization the responsiveness of one receptor system is regulated (positively or negatively) by activation of another receptor system (i.e., "cross-talk") [75]. For example, insulin after binding to the IR may induce heterologous desensitization of the signaling of the IGF-IR by downregulating β-arrestin-1 and inhibiting of IGF-I-stimulated MAP kinase phosphorylation [76]. However, it has been found that this latter effect could be substantially rescued by ectopic expression of wild-type β-arrestin-1, consistent with the view that the decrease in cellular β-arrestin-1 content is a major mechanism for the observed desensitization effects of insulin on IGF-IR mediated signaling [76].

8. The Nuclear Translocation of the IGF-IR and IRs and Its Significance

The IGF-IR and IRs not only function at the cell surface. When after binding of IGF-I to the IGF-IR and the IGF-I/IGF-IR complex has been internalized into the cell, there are three potential outcomes for the internalized IGF-IR: it can go back to the cellular surface, it can be degraded or it can go to the nucleus [77].

It has been documented that both the IGF-IR and IRs can be translocated to the nucleus [78,79]. Nuclear transport of IGF-IR is enhanced by IGF-I and IGF-II but only modestly by insulin [79]. This transport correlated directly with the magnitude of ligand-induced receptor phosphorylation of the IGF-IR with these ligands [79]. In addition, it has been found that ligand-mediated phosphorylation of the IGF-IR is essential for nuclear trafficking [78].

IGF-IR nuclear import and chromatin binding can be blocked by an IGF-IR kinase inhibitor, indicating that indeed IGF-IR kinase activity is required for the IGF-IR to enter the nucleus [79].

The IGF-IR can undergo both caveolin- and clathrin mediated endocytosis [80,81]. Consistent with clathrin-mediated endocytosis, nuclear IGF-IR translocation can be blocked by the inhibitors of clathrin-dependent endocytosis (dansylcadaverine and the dynamin-1 inhibitor dynasore), but not by caveolin-1 depletion [79]. Nevertheless, the exact mechanisms responsible for nuclear import of the IGF-IR and IRs are still unclear [78]. Sehat et al. found that the α subunit (native size, 120 kD) together with the β subunit (native size, 95 kD) was present in the nuclear fraction, suggesting that nuclear IGF-IR was an intact receptor [82]. Data of Aleksic et al. also suggested that full-length IGF-IR translocates to the nucleus [79].

Both the IGF-IR and IRs are present in the perinuclear and nucleolar area of the nucleus in a small ubiquitin-like modifier SUMOylated form [78]. Receptor SUMOylation occurs in a ligand dependent fashion and it has been demonstrated that SUMOylation plays a crucial role in the nuclear translocation of the IGF-IR [82,83] (Figure 6). The SUMO-modified IGF-IR is deSUMOylated after passage across the nuclear membrane [82]. Nuclear IGF-IR binds to enhancer regions and activates transcription [84]. It is able to autoregulate expression of its own gene leading to an increase in IGF-IR promoter activity and IGF-IR expression [78] (Figure 6).

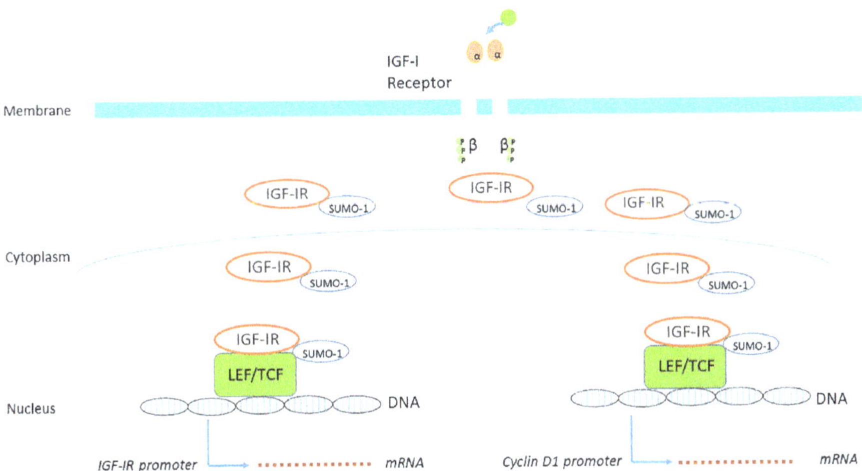

Figure 6. Studies have confirmed that ligand-mediated phosphorylation of the IGF-IR is essential for nuclear trafficking. Following binding of IGF-I to the IGF-IR at the cell surface, the IGF-IR is transported into the cell and further translocated from the cytoplasm into the nucleus. SUMOylation (SUMO-1) in the cytoplasm of the IGF-IR also plays a crucial role in the nuclear translocation of the IGF-IR from the cytoplasm. When the SUMOylated IGF-IR translocates to the nucleus, it is thought to be involved in the transcriptional enhancement of specific target genes. Nuclear IGF-IR is able to autoregulate expression of its own gene leading to an increase in IGF-IR promoter activity and IGF-IR expression (left). Nuclear IGF-IR may also bind to Cyclin D1 (and additional) promoters with ensuing target gene activation (right). (Modified from Sarfstein & Werner. Minireview: nuclear insulin and insulin-like growth factor-1 receptors: a novel paradigm in signal transduction. Endocrinology, 2013; 154:1672-9).

It has been further suggested that nuclear IGF-IR has biological significance in cancer; prognosis was less good and survival was shorter in patients whose tumor showed intense and/or widespread nuclear IGF-IR [79]. It has been reported that nuclear IGF-IR is a feature of pre-invasive lesions and invasive cancers including prostate, renal and breast cancers, and an association between nuclear IGF-IR and adverse prognosis was identified in renal cancer [79]. Subsequent data did associate nuclear IGF-IR with proliferation, tumorigenicity, resistance to EGFR inhibition and clinical response to therapeutic anti-IGF-IR antibodies, which suggests that IGF-IR nuclear import has biological significance and may contribute directly to IGF-IR function [79,85–90].

When the IGF-IR translocates to the nucleus, it is thought to be involved in the transcriptional enhancement of specific target genes [79,82,91]. It has been demonstrated that IGF-IR in the nucleus binds to the transcription factor lymphoid enhancer factor-1 (LEF1), leading to elevated protein levels of cyclin D1 and axin2 [84] (Figure 6). This might be an additional molecular mechanism by which IGF-IR promotes uncontrolled cell proliferation and contribute to the neoplastic transformation of cells [84].

When investigating the impact of IGF-IR levels on IGF-IR biosynthesis in estrogen receptor positive (ER+) and estrogen receptor depleted (ER-) breast cancer cells, it was found that in ER+ cell and ER- cells regulation of the IGF-IR gene and IGF-IR protein differed at the level of transcription; the IGF-IR protein was able to stimulate IGF-IR gene expression in ER- cells but not in ER+ cells [92]. Similarly to the IGF-IR, it was found that the IR was also translocated to the nucleus and to bind to the IGF-IR promoter. However, this was only observed in ER- cells but not in ER+ cells [92].

In addition, it has been found that transcription factors IGF-IR and IR display diametrically opposite activities in the context of IGF-IR gene regulation; in contrast to the IGF-IR, IR inhibited IGF-IR promoter activity [92]. Thus nuclear IGF-IR acted as a transcriptional activator of its own promoter, while nuclear IR functioned as a negative regulator of IGF-IR promoter activity [92]. Nevertheless, the authors of this latter paper concluded that the clinical implications of their findings—in particular the impact of IGF-IR/IR nuclear localization on targeted therapy—are at present unclear and require further investigation [92].

9. Conclusions

Although until recently the conventional view was that phosphorylation of tyrosine residues played a major role in the activation of the IGF-IR and initiated all downstream signaling, there is increasing evidence showing that this view was too simplistic and grossly underestimated the downstream complexity of the IGF-IR pathways. The IGF-IR has not only extensive cross-talk with many other receptors, but that the IGF-IR can be also considered as a functional RTK/GPCR hybrid, which integrates the kinase signaling with some IGF-IR mediated canonical GPCR characteristics. Like classical GPCRs the IGF-IR show homologous and heterologous desensitization. In addition, after activation by a ligand, the IGF-IR signaling can be translocated to the nucleus and function as a transcriptional cofactor. For example nuclear IGF-IR is able to autoregulate expression of its own gene.

Thus, it has become clear in recent years that the IGF-IR signaling pathway is far more complex than previously thought. It contains many points of regulation and shows signal divergence and cross-talk with many other signaling pathways at the receptor and post-receptor level. However, a big challenge for the (near) future will be how all this new knowledge about the IGF-IR signaling pathways can be translated into clinical practice and improve diagnosis and treatment of diseases.

Funding: This research received no external funding.

Acknowledgments: I thank Haim Werner for inviting me to agree to contribute a research article Special Issue on "Insulin-like Growth Factors in Development, Cancers and Aging" in the journal Cells. The author is grateful to Michael P. Brugts for comments and proofreading the manuscript.

Conflicts of Interest: The author declares no conflict of interest.

References

1. Collett-Solberg, P.F.; Cohen, P. The role of the insulin-like growth factor binding proteins and the igfbp proteases in modulating igf action. *Endocrinol. Metab. Clin. N. Am.* **1996**, *25*, 591–614. [CrossRef]
2. Janssen, J.; Varewijck, A.J.; Brugts, M.P. The insulin-like growth factor-i receptor stimulating activity (irsa) in health and disease. *Growth Horm. Igf Res.* **2019**, *48–49*, 16–28. [CrossRef] [PubMed]
3. Bach, L.A. Igf-binding proteins. *J. Mol. Endocrinol.* **2018**, *61*, T11–T28. [CrossRef] [PubMed]
4. Clemmons, D.R. Insulin-like growth factor binding proteins and their role in controlling igf actions. *Cytokine Growth Factor Rev.* **1997**, *8*, 45–62. [CrossRef]
5. Jones, J.I.; Clemmons, D.R. Insulin-like growth factors and their binding proteins: Biological actions. *Endocr. Rev.* **1995**, *16*, 3–34.
6. Janssen, J.A.; Lamberts, S.W. Is the measurement of free igf-i more indicative than that of total igf-i in the evaluation of the biological activity of the gh/igf-i axis? *J. Endocrinol. Invest.* **1999**, *22*, 313–315. [CrossRef]
7. Philippou, A.; Maridaki, M.; Pneumaticos, S.; Koutsilieris, M. The complexity of the igf1 gene splicing, posttranslational modification and bioactivity. *Mol. Med.* **2014**, *20*, 202–214. [CrossRef]
8. Steele-Perkins, G.; Turner, J.; Edman, J.C.; Hari, J.; Pierce, S.B.; Stover, C.; Rutter, W.J.; Roth, R.A. Expression and characterization of a functional human insulin-like growth factor i receptor. *J. Biol. Chem.* **1988**, *263*, 11486–11492.

9. LeRoith, D.; Werner, H.; Roberts, C.T. Molecular and Cellular Biology of the Insulin-Like Growth Factors. In *Molecular Endocrinology, Basic Concepts and Clinical Correlations*; Weintraub, B.D., Ed.; Raven Press: New York, NY, USA, 1995; pp. 181–193.

10. Siddle, K.; Urso, B.; Niesler, C.A.; Cope, D.L.; Molina, L.; Surinya, K.H.; Soos, M.A. Specificity in ligand binding and intracellular signalling by insulin and insulin-like growth factor receptors. *Biochem. Soc. Trans.* **2001**, *29*, 513–525. [CrossRef]

11. Froesch, E.; Zenobi, P.; Zapf, J. Metabolic Effects of Insulin-Like Growth Factor-i and Possible Therapeutic Aspects in Diabetes. John Wiley: Chichester, UK, 1993; pp. 109–129.

12. Janssen, J.A.; Varewijck, A.J. Igf-ir targeted therapy: Past, present and future. *Front. Endocrinol. (Lausanne)* **2014**, *5*, 224. [CrossRef]

13. Hassan, A.B. Keys to the hidden treasures of the mannose 6-phosphate/insulin-like growth factor 2 receptor. *Am. J. Pathol.* **2003**, *162*, 3–6. [CrossRef]

14. Belfiore, A.; Frasca, F.; Pandini, G.; Sciacca, L.; Vigneri, R. Insulin receptor isoforms and insulin receptor/insulin-like growth factor receptor hybrids in physiology and disease. *Endocr. Rev.* **2009**, *30*, 586–623. [CrossRef] [PubMed]

15. Ullrich, A.; Bell, J.R.; Chen, E.Y.; Herrera, R.; Petruzzelli, L.M.; Dull, T.J.; Gray, A.; Coussens, L.; Liao, Y.C.; Tsubokawa, M. Human insulin receptor and its relationship to the tyrosine kinase family of oncogenes. *Nature* **1985**, *313*, 756–761. [CrossRef] [PubMed]

16. Ebina, Y.; Ellis, L.; Jarnagin, K.; Edery, M.; Graf, L.; Clauser, E.; Ou, J.H.; Masiarz, F.; Kan, Y.W.; Goldfine, I.D. The human insulin receptor cdna: The structural basis for hormone-activated transmembrane signalling. *Cell* **1985**, *40*, 747–758. [CrossRef]

17. Varewijck, A.J.; Janssen, J.A. Insulin and its analogues and their affinities for the igf1 receptor. *Endocr. Relat. Cancer* **2012**, *19*, F63–F75. [CrossRef]

18. Roberts, C.T., Jr. Control of insulin-like growth factor (igf) action by regulation of igf-i receptor expression. *Endocr. J.* **1996**, *43*, S49–S55. [CrossRef]

19. Rechler, M.M.; Nissley, S.P. The nature and regulation of the receptors for insulin-like growth factors. *Annu. Rev. Physiol.* **1985**, *47*, 425–442. [CrossRef]

20. Ullrich, A.; Gray, A.; Tam, A.W.; Yang-Feng, T.; Tsubokawa, M.; Collins, C.; Henzel, W.; Le Bon, T.; Kathuria, S.; Chen, E. Insulin-like growth factor i receptor primary structure: Comparison with insulin receptor suggests structural determinants that define functional specificity. *EMBO J.* **1986**, *5*, 2503–2512. [CrossRef]

21. LeRoith, D.; Werner, H.; Beitner-Johnson, D.; Roberts, C.T., Jr. Molecular and cellular aspects of the insulin-like growth factor i receptor. *Endocr. Rev.* **1995**, *16*, 143–163. [CrossRef]

22. Lopaczynski, W.; Terry, C.; Nissley, P. Autophosphorylation of the insulin-like growth factor i receptor cytoplasmic domain. *Biochem. Biophys. Res. Commun.* **2000**, *279*, 955–960. [CrossRef]

23. Favelyukis, S.; Till, J.H.; Hubbard, S.R.; Miller, W.T. Structure and autoregulation of the insulin-like growth factor 1 receptor kinase. *Nat. Struct. Biol.* **2001**, *8*, 1058–1063. [CrossRef] [PubMed]

24. Myers, M.G., Jr.; Zhang, Y.; Aldaz, G.A.; Grammer, T.; Glasheen, E.M.; Yenush, L.; Wang, L.M.; Sun, X.J.; Blenis, J.; Pierce, J.H. Ymxm motifs and signaling by an insulin receptor substrate 1 molecule without tyrosine phosphorylation sites. *Mol. Cell Biol.* **1996**, *16*, 4147–4155. [CrossRef] [PubMed]

25. White, M.F. Irs proteins and the common path to diabetes. *Am. J. Physiol. Endocrinol. Metab.* **2002**, *283*, E413–E422. [CrossRef] [PubMed]

26. Traub, L.M.; Bonifacino, J.S. Cargo recognition in clathrin-mediated endocytosis. *Cold Spring Harb. Perspect. Biol.* **2013**, *5*, a016790. [CrossRef]

27. Irannejad, R.; Tsvetanova, N.G.; Lobingier, B.T.; von Zastrow, M. Effects of endocytosis on receptor-mediated signaling. *Curr. Opin. Cell Biol.* **2015**, *35*, 137–143. [CrossRef]

28. Yoneyama, Y.; Lanzerstorfer, P.; Niwa, H.; Umehara, T.; Shibano, T.; Yokoyama, S.; Chida, K.; Weghuber, J.; Hakuno, F.; Takahashi, S.I. Irs-1 acts as an endocytic regulator of igf-i receptor to facilitate sustained igf signaling. *eLIFE* **2018**, *7*, e32893. [CrossRef]

29. Takahashi, S.I. Igf research 2016–2018. *Growth Horm. Igf Res.* **2019**, *48–49*, 65–69. [CrossRef]

30. Worrall, C.; Nedelcu, D.; Serly, J.; Suleymanova, N.; Oprea, I.; Girnita, A.; Girnita, L. Novel mechanisms of regulation of igf-1r action: Functional and therapeutic implications. *Pediatr. Endocrinol. Rev.* **2013**, *10*, 473–484.

31. Vecchione, A.; Marchese, A.; Henry, P.; Rotin, D.; Morrione, A. The grb10/nedd4 complex regulates ligand-induced ubiquitination and stability of the insulin-like growth factor i receptor. *Mol. Cell Biol.* **2003**, *23*, 3363–3372. [CrossRef]

32. Girnita, L.; Girnita, A.; Larsson, O. Mdm2-dependent ubiquitination and degradation of the insulin-like growth factor 1 receptor. *Proc. Natl. Acad. Sci. USA* **2003**, *100*, 8247–8252. [CrossRef]

33. Girnita, L.; Shenoy, S.K.; Sehat, B.; Vasilcanu, R.; Vasilcanu, D.; Girnita, A.; Lefkowitz, R.J.; Larsson, O. Beta-arrestin and mdm2 mediate igf-1 receptor-stimulated erk activation and cell cycle progression. *J. Biol. Chem.* **2007**, *282*, 11329–11338. [CrossRef]

34. Girnita, L.; Shenoy, S.K.; Sehat, B.; Vasilcanu, R.; Girnita, A.; Lefkowitz, R.J.; Larsson, O. {beta}-arrestin is crucial for ubiquitination and down-regulation of the insulin-like growth factor-1 receptor by acting as adaptor for the mdm2 e3 ligase. *J. Biol. Chem.* **2005**, *280*, 24412–24419. [CrossRef] [PubMed]

35. Girnita, L.; Takahashi, S.I.; Crudden, C.; Fukushima, T.; Worrall, C.; Furuta, H.; Yoshihara, H.; Hakuno, F.; Girnita, A. Chapter seven - when phosphorylation encounters ubiquitination: A balanced perspective on igf-1r signaling. *Prog. Mol. Biol. Transl. Sci.* **2016**, *141*, 277–311. [PubMed]

36. De Meyts, P.; Sajid, W.; Palsgaard, J.; Theede, A.-M.; Gaugain, L.; Aladdin, H.; Whittaker, J. Insulin and igf-i Receptor Structure and Binding Proteins. In *Mechanisms of Insulin Action*; E.Pessin, A.R.S.J., Ed.; Landes Bioscience: Austin, TX, USA, 2007; pp. 1–32.

37. Zapf, J.; Schmid, C.; Froesch, E.R. Biological and immunological properties of insulin-like growth factors (igf) i and ii. *Clin Endocrinol. Metab.* **1984**, *13*, 3–30. [CrossRef]

38. Dupont, J.; LeRoith, D. Insulin and insulin-like growth factor i receptors: Similarities and differences in signal transduction. *Horm. Res.* **2001**, *55* (suppl. 2), 22–26. [CrossRef]

39. Bevan, P. Insulin signalling. *J. Cell Sci.* **2001**, *114*, 1429–1430.

40. Boucher, J.; Charalambous, M.; Zarse, K.; Mori, M.A.; Kleinridders, A.; Ristow, M.; Ferguson-Smith, A.C.; Kahn, C.R. Insulin and insulin-like growth factor 1 receptors are required for normal expression of imprinted genes. *Proc. Natl. Acad. Sci. USA* **2014**, *111*, 14512–14517. [CrossRef]

41. Boucher, J.; Tseng, Y.H.; Kahn, C.R. Insulin and insulin-like growth factor-1 receptors act as ligand-specific amplitude modulators of a common pathway regulating gene transcription. *J. Biol. Chem.* **2010**, *285*, 17235–17245. [CrossRef]

42. Slaaby, R. Specific insulin/igf1 hybrid receptor activation assay reveals igf1 as a more potent ligand than insulin. *Sci. Rep.* **2015**, *5*, 7911. [CrossRef]

43. Adams, T.E.; Epa, V.C.; Garrett, T.P.; Ward, C.W. Structure and function of the type 1 insulin-like growth factor receptor. *Cell Mol. Life Sci.* **2000**, *57*, 1050–1093. [CrossRef]

44. Mosthaf, L.; Grako, K.; Dull, T.J.; Coussens, L.; Ullrich, A.; McClain, D.A. Functionally distinct insulin receptors generated by tissue-specific alternative splicing. *Embo. J.* **1990**, *9*, 2409–2413. [CrossRef]

45. Slaaby, R.; Schaffer, L.; Lautrup-Larsen, I.; Andersen, A.S.; Shaw, A.C.; Mathiasen, I.S.; Brandt, J. Hybrid receptors formed by insulin receptor (ir) and insulin-like growth factor i receptor (igf-ir) have low insulin and high igf-1 affinity irrespective of the ir splice variant. *J. Biol. Chem.* **2006**, *281*, 25869–25874. [CrossRef] [PubMed]

46. Benyoucef, S.; Surinya, K.H.; Hadaschik, D.; Siddle, K. Characterization of insulin/igf hybrid receptors: Contributions of the insulin receptor l2 and fn1 domains and the alternatively spliced exon 11 sequence to ligand binding and receptor activation. *Biochem. J.* **2007**, *403*, 603–613. [CrossRef] [PubMed]

47. Pandini, G.; Frasca, F.; Mineo, R.; Sciacca, L.; Vigneri, R.; Belfiore, A. Insulin/insulin-like growth factor i hybrid receptors have different biological characteristics depending on the insulin receptor isoform involved. *J. Biol. Chem.* **2002**, *277*, 39684–39695. [CrossRef] [PubMed]

48. Zhang, H.; Pelzer, A.M.; Kiang, D.T.; Yee, D. Down-regulation of type i insulin-like growth factor receptor increases sensitivity of breast cancer cells to insulin. *Cancer Res.* **2007**, *67*, 391–397. [CrossRef] [PubMed]

49. King, E.R.; Wong, K.K. Insulin-like growth factor: Current concepts and new developments in cancer therapy. *Recent Pat. Anticancer Drug Discov.* **2012**, *7*, 14–30. [CrossRef]

50. Morgillo, F.; Woo, J.K.; Kim, E.S.; Hong, W.K.; Lee, H.Y. Heterodimerization of insulin-like growth factor receptor/epidermal growth factor receptor and induction of survivin expression counteract the antitumor action of erlotinib. *Cancer Res.* **2006**, *66*, 10100–10111. [CrossRef]

51. van der Veeken, J.; Oliveira, S.; Schiffelers, R.M.; Storm, G.; van Bergen En Henegouwen, P.M.; Roovers, R.C. Crosstalk between epidermal growth factor receptor- and insulin-like growth factor-1 receptor signaling: Implications for cancer therapy. *Curr. Cancer Drug Targets* **2009**, *9*, 748–760. [CrossRef]

52. Guix, M.; Faber, A.C.; Wang, S.E.; Olivares, M.G.; Song, Y.; Qu, S.; Rinehart, C.; Seidel, B.; Yee, D.; Arteaga, C.L. Acquired resistance to egfr tyrosine kinase inhibitors in cancer cells is mediated by loss of igf-binding proteins. *J. Clin. Investig.* **2008**, *118*, 2609–2619. [CrossRef]

53. Liefers-Visser, J.A.L.; Meijering, R.A.M.; Reyners, A.K.L.; van der Zee, A.G.J.; de Jong, S. Igf system targeted therapy: Therapeutic opportunities for ovarian cancer. *Cancer Treat. Rev.* **2017**, *60*, 90–99. [CrossRef]

54. Gee, J.M.; Robertson, J.F.; Gutteridge, E.; Ellis, I.O.; Pinder, S.E.; Rubini, M.; Nicholson, R.I. Epidermal growth factor receptor/her2/insulin-like growth factor receptor signalling and oestrogen receptor activity in clinical breast cancer. *Endocr. Relat. Cancer* **2005**, *12* (suppl. 1), S99–S111. [CrossRef]

55. Pandini, G.; Mineo, R.; Frasca, F.; Roberts, C.T., Jr.; Marcelli, M.; Vigneri, R.; Belfiore, A. Androgens up-regulate the insulin-like growth factor-i receptor in prostate cancer cells. *Cancer Res.* **2005**, *65*, 1849–1857. [CrossRef] [PubMed]

56. Coppola, D.; Ferber, A.; Miura, M.; Sell, C.; D'Ambrosio, C.; Rubin, R.; Baserga, R. A functional insulin-like growth factor i receptor is required for the mitogenic and transforming activities of the epidermal growth factor receptor. *Mol. Cell Biol.* **1994**, *14*, 4588–4595. [CrossRef] [PubMed]

57. Gan, Y.; Zhang, Y.; Buckels, A.; Paterson, A.J.; Jiang, J.; Clemens, T.L.; Zhang, Z.Y.; Du, K.; Chang, Y.; Frank, S.J. Igf-1r modulation of acute gh-induced stat5 signaling: Role of protein tyrosine phosphatase activity. *Mol. Endocrinol.* **2013**, *27*, 1969–1979. [CrossRef] [PubMed]

58. Tsui, S.; Naik, V.; Hoa, N.; Hwang, C.J.; Afifiyan, N.F.; Sinha Hikim, A.; Gianoukakis, A.G.; Douglas, R.S.; Smith, T.J. Evidence for an association between thyroid-stimulating hormone and insulin-like growth factor 1 receptors: A tale of two antigens implicated in graves' disease. *J. Immunol.* **2008**, *181*, 4397–4405. [CrossRef] [PubMed]

59. Smith, T.J.; Janssen, J. Insulin-like growth factor-i receptor and thyroid-associated ophthalmopathy. *Endocr. Rev.* **2019**, *40*, 236–267. [CrossRef] [PubMed]

60. Vella, V.; Malaguarnera, R.; Nicolosi, M.L.; Morrione, A.; Belfiore, A. Insulin/igf signaling and discoidin domain receptors: An emerging functional connection. *Biochim. Biophys. Acta Mol. Cell Res.* **2019**, *1866*, 118522. [CrossRef] [PubMed]

61. Vella, V.; Nicolosi, M.L.; Cantafio, P.; Massimino, M.; Lappano, R.; Vigneri, P.; Ciuni, R.; Gangemi, P.; Morrione, A.; Malaguarnera, R. Ddr1 regulates thyroid cancer cell differentiation via igf-2/ir-a autocrine signaling loop. *Endocr. Relat. Cancer* **2019**, *26*, 197–214. [CrossRef]

62. Girnita, L.; Worrall, C.; Takahashi, S.; Seregard, S.; Girnita, A. Something old, something new and something borrowed: Emerging paradigm of insulin-like growth factor type 1 receptor (igf-1r) signaling regulation. *Cell Mol. Life Sci.* **2014**, *71*, 2403–2427. [CrossRef]

63. Crudden, C.; Ilic, M.; Suleymanova, N.; Worrall, C.; Girnita, A.; Girnita, L. The dichotomy of the insulin-like growth factor 1 receptor: Rtk and gpcr: Friend or foe for cancer treatment? *Growth Horm. Igf Res.* **2015**, *25*, 2–12. [CrossRef]

64. Luttrell, L.M.; van Biesen, T.; Hawes, B.E.; Koch, W.J.; Touhara, K.; Lefkowitz, R.J. G beta gamma subunits mediate mitogen-activated protein kinase activation by the tyrosine kinase insulin-like growth factor 1 receptor. *J. Biol. Chem.* **1995**, *270*, 16495–16498. [CrossRef] [PubMed]

65. Dalle, S.; Ricketts, W.; Imamura, T.; Vollenweider, P.; Olefsky, J.M. Insulin and insulin-like growth factor i receptors utilize different g protein signaling components. *J. Biol. Chem.* **2001**, *276*, 15688–15695. [CrossRef] [PubMed]

66. Imamura, T.; Vollenweider, P.; Egawa, K.; Clodi, M.; Ishibashi, K.; Nakashima, N.; Ugi, S.; Adams, J.W.; Brown, J.H.; Olefsky, J.M. G alpha-q/11 protein plays a key role in insulin-induced glucose transport in 3t3-l1 adipocytes. *Mol. Cell Biol.* **1999**, *19*, 6765–6774. [CrossRef] [PubMed]

67. Lin, F.T.; Daaka, Y.; Lefkowitz, R.J. Beta-arrestins regulate mitogenic signaling and clathrin-mediated endocytosis of the insulin-like growth factor i receptor. *J. Biol. Chem.* **1998**, *273*, 31640–31643. [CrossRef] [PubMed]

68. Zheng, H.; Worrall, C.; Shen, H.; Issad, T.; Seregard, S.; Girnita, A.; Girnita, L. Selective recruitment of g protein-coupled receptor kinases (grks) controls signaling of the insulin-like growth factor 1 receptor. *Proc. Natl. Acad. Sci. USA* **2012**, *109*, 7055–7060. [CrossRef] [PubMed]

69. Usui, I.; Imamura, T.; Huang, J.; Satoh, H.; Shenoy, S.K.; Lefkowitz, R.J.; Hupfeld, C.J.; Olefsky, J.M. Beta-arrestin-1 competitively inhibits insulin-induced ubiquitination and degradation of insulin receptor substrate 1. *Mol. Cell Biol.* **2004**, *24*, 8929–8937. [CrossRef]
70. Povsic, T.J.; Kohout, T.A.; Lefkowitz, R.J. Beta-arrestin1 mediates insulin-like growth factor 1 (igf-1) activation of phosphatidylinositol 3-kinase (pi3k) and anti-apoptosis. *J. Biol. Chem.* **2003**, *278*, 51334–51339. [CrossRef]
71. Lefkowitz, R.J.; Shenoy, S.K. Transduction of receptor signals by beta-arrestins. *Science* **2005**, *308*, 512–517. [CrossRef]
72. Zheng, H.; Shen, H.; Oprea, I.; Worrall, C.; Stefanescu, R.; Girnita, A.; Girnita, L. Beta-arrestin-biased agonism as the central mechanism of action for insulin-like growth factor 1 receptor-targeting antibodies in ewing's sarcoma. *Proc. Natl. Acad. Sci. USA* **2012**, *109*, 20620–20625. [CrossRef]
73. Salisbury, T.B.; Tomblin, J.K. Insulin/insulin-like growth factors in cancer: New roles for the aryl hydrocarbon receptor, tumor resistance mechanisms, and new blocking strategies. *Front. Endocrinol. (Lausanne)* **2015**, *6*, 12. [CrossRef]
74. Girnita, A.; Zheng, H.; Gronberg, A.; Girnita, L.; Stahle, M. Identification of the cathelicidin peptide ll-37 as agonist for the type i insulin-like growth factor receptor. *Oncogene* **2012**, *31*, 352–365. [CrossRef] [PubMed]
75. Berg, K.A.; Maayani, S.; Clarke, W.P. Interactions between effectors linked to serotonin receptors. *Ann. N.Y. Acad. Sci.* **1998**, *861*, 111–120. [CrossRef] [PubMed]
76. Dalle, S.; Imamura, T.; Rose, D.W.; Worrall, D.S.; Ugi, S.; Hupfeld, C.J.; Olefsky, J.M. Insulin induces heterologous desensitization of g-protein-coupled receptor and insulin-like growth factor i signaling by downregulating beta-arrestin-1. *Mol. Cell Biol.* **2002**, *22*, 6272–6285. [CrossRef] [PubMed]
77. Chughtai, S. The nuclear translocation of insulin-like growth factor receptor and its significance in cancer cell survival. *Cell Biochem. Funct.* **2019**. [CrossRef]
78. Sarfstein, R.; Werner, H. Minireview: Nuclear insulin and insulin-like growth factor-1 receptors: A novel paradigm in signal transduction. *Endocrinology* **2013**, *154*, 1672–1679. [CrossRef]
79. Aleksic, T.; Chitnis, M.M.; Perestenko, O.V.; Gao, S.; Thomas, P.H.; Turner, G.D.; Protheroe, A.S.; Howarth, M.; Macaulay, V.M. Type 1 insulin-like growth factor receptor translocates to the nucleus of human tumor cells. *Cancer Res.* **2010**, *70*, 6412–6419. [CrossRef]
80. Romanelli, R.J.; LeBeau, A.P.; Fulmer, C.G.; Lazzarino, D.A.; Hochberg, A.; Wood, T.L. Insulin-like growth factor type-i receptor internalization and recycling mediate the sustained phosphorylation of akt. *J. Biol. Chem.* **2007**, *282*, 22513–22524. [CrossRef]
81. Sehat, B.; Andersson, S.; Girnita, L.; Larsson, O. Identification of c-cbl as a new ligase for insulin-like growth factor-i receptor with distinct roles from mdm2 in receptor ubiquitination and endocytosis. *Cancer Res.* **2008**, *68*, 5669–5677. [CrossRef]
82. Sehat, B.; Tofigh, A.; Lin, Y.; Trocme, E.; Liljedahl, U.; Lagergren, J.; Larsson, O. Sumoylation mediates the nuclear translocation and signaling of the igf-1 receptor. *Sci. Signal.* **2010**, *3*, ra10. [CrossRef]
83. Deng, H.; Lin, Y.; Badin, M.; Vasilcanu, D.; Stromberg, T.; Jernberg-Wiklund, H.; Sehat, B.; Larsson, O. Over-accumulation of nuclear igf-1 receptor in tumor cells requires elevated expression of the receptor and the sumo-conjugating enzyme ubc9. *Biochem. Biophys. Res. Commun.* **2011**, *404*, 667–671. [CrossRef]
84. Warsito, D.; Sjostrom, S.; Andersson, S.; Larsson, O.; Sehat, B. Nuclear igf1r is a transcriptional co-activator of lef1/tcf. *Embo. Rep.* **2012**, *13*, 244–250. [CrossRef] [PubMed]
85. Bodzin, A.S.; Wei, Z.; Hurtt, R.; Gu, T.; Doria, C. Gefitinib resistance in hcc mahlavu cells: Upregulation of cd133 expression, activation of igf-1r signaling pathway, and enhancement of igf-1r nuclear translocation. *J. Cell Physiol.* **2012**, *227*, 2947–2952. [CrossRef] [PubMed]
86. Aslam, M.I.; Hettmer, S.; Abraham, J.; Latocha, D.; Soundararajan, A.; Huang, E.T.; Goros, M.W.; Michalek, J.E.; Wang, S.; Mansoor, A. Dynamic and nuclear expression of pdgfralpha and igf-1r in alveolar rhabdomyosarcoma. *Mol. Cancer Res.* **2013**, *11*, 1303–1313. [CrossRef] [PubMed]
87. van Gaal, J.C.; Roeffen, M.H.; Flucke, U.E.; van der Laak, J.A.; van der Heijden, G.; de Bont, E.S.; Suurmeijer, A.J.; Versleijen-Jonkers, Y.M.; van der Graaf, W.T. Simultaneous targeting of insulin-like growth factor-1 receptor and anaplastic lymphoma kinase in embryonal and alveolar rhabdomyosarcoma: A rational choice. *Eur. J. Cancer* **2013**, *49*, 3462–3470. [CrossRef] [PubMed]
88. Lin, Y.; Liu, H.; Waraky, A.; Haglund, F.; Agarwal, P.; Jernberg-Wiklund, H.; Warsito, D.; Larsson, O. Sumo-modified insulin-like growth factor 1 receptor (igf-1r) increases cell cycle progression and cell proliferation. *J. Cell Physiol.* **2017**, *232*, 2722–2730. [CrossRef]

89. Asmane, I.; Watkin, E.; Alberti, L.; Duc, A.; Marec-Berard, P.; Ray-Coquard, I.; Cassier, P.; Decouvelaere, A.V.; Ranchere, D.; Kurtz, J.E. Insulin-like growth factor type 1 receptor (igf-1r) exclusive nuclear staining: A predictive biomarker for igf-1r monoclonal antibody (ab) therapy in sarcomas. *Eur. J. Cancer* **2012**, *48*, 3027–3035. [CrossRef]

90. Aleksic, T.; Browning, L.; Woodward, M.; Phillips, R.; Page, S.; Henderson, S.; Athanasou, N.; Ansorge, O.; Whitwell, D.; Pratap, S. Durable response of spinal chordoma to combined inhibition of igf-1r and egfr. *Front. Oncol.* **2016**, *6*, 98. [CrossRef]

91. Janku, F.; Huang, H.J.; Angelo, L.S.; Kurzrock, R. A kinase-independent biological activity for insulin growth factor-1 receptor (igf-1r): Implications for inhibition of the igf-1r signal. *Oncotarget* **2013**, *4*, 463–473. [CrossRef]

92. Sarfstein, R.; Pasmanik-Chor, M.; Yeheskel, A.; Edry, L.; Shomron, N.; Warman, N.; Wertheimer, E.; Maor, S.; Shochat, L.; Werner, H. Insulin-like growth factor-i receptor (igf-ir) translocates to nucleus and autoregulates igf-ir gene expression in breast cancer cells. *J. Biol. Chem.* **2012**, *287*, 2766–2776. [CrossRef]

MDPI

St. Alban-Anlage 66

4052 Basel

Switzerland

Tel. +41 61 683 77 34

Fax +41 61 302 89 18

www.mdpi.com

Cells Editorial Office

E-mail: cells@mdpi.com

www.mdpi.com/journal/cells